AF701625

...IFIC INFORMATION DEPT.

JAN 16 1986

BRISTOL-MYERS/EVANSVILLE

PROPERTY OF BRISTOL-MYERS SQUIBB CO.
RESEARCH LIBRARY
WALLINGFORD, CT

WALLINGFORD RESEARCH LIBRARY
BWFD014989

A Specialist Periodical Report

Biosynthesis

Volume 4

A Review of the Literature Published during 1974

Senior Reporter
J. D. Bu'Lock, *Department of Chemistry, University of Manchester*

Reporters
E. P. Abraham, *School of Pathology, University of Oxford*
P. A. Fawcett, *School of Pathology, University of Oxford*
J. R. Hanson, *School of Molecular Science, University of Sussex*
J. B. Harborne, *Department of Botany, University of Reading*
A. Kjaer, *Technical University of Denmark, Lyngby, Denmark*
E. Leete, *University of Minnesota, Minneapolis, U.S.A.*
T. Money, *University of British Columbia, Vancouver, Canada*
L. J. Mulheirn, *Milstead Laboratory of Chemical Enzymology, Sittingbourne*
P. Olesen Larsen, *Royal Veterinary and Agricultural University, Copenhagen, Denmark*
M. Tanabe, *Stanford Research Institute, California, U.S.A.*

PROPERTY OF BRISTOL-MYERS SQUIBB CO.
RESEARCH LIBRARY
WALLINGFORD, CT

© Copyright 1976

The Chemical Society
Burlington House, London, W1V 0BN

34714

ISBN: 0 85186 533 X
ISSN: 0301–0708
Library of Congress Catalog No 72–83455

Set in Monophoto Times and printed in Great Britain
by Page Bros (Norwich) Ltd, Norwich

REF. QP501 .B59 1976 v.4
12-10-85
U0755N219/X

Introduction

The eventual production of this volume can largely be credited to the momentum generated by the editor of the three previous 'Biosynthesis' Volumes, Ted Geissman, whose role is very readily acknowledged. In particular its overall scope and general pattern of subdivision are along previous lines. Because of the volume of research to be covered, we have separated out those aspects of sterol metabolism which come 'beyond cholesterol' as being relatively homogeneous in content and also in interest, so that the whole topic of isoprenoid biosynthesis now occupies three chapters in place of the previous two. Two topics that were last dealt with in the second Volume, *i.e.* up to 1972, are brought up to date in the chapters on polyketides and on non-protein amino-acids *etc.* Unfortunately, it has not proved possible to include an account of work in steroid hormone biosynthesis for 1974. This field is a particularly awkward one to handle. Partly because of its interdisciplinary status, and partly because it has reached a condition of such confusion and complexity as to require very careful critical analysis. Hopefully we shall attempt this in a later Volume.

However the field of biosynthesis is subdivided, there are topics which are liable to be left out, and others which are likely to be so split up that a coherent story is difficult to put together. Professor Geissman felt—and I concur—that such topics could usefully be covered by occasional special surveys which were not tied to the 'latest progress' formula, and this year we have included a special article of this type on the pencillins and cephalosporins, which is an old problem that has recently come into a new lease of life. Suggestions for further articles of this type will be particularly welcome.

Every contributor to this Volume has tacitly or expressly raised one or both of the following points. First, that wherever reasonably efficient precursor-incorporation can be contrived, applications of heavy-isotope n.m.r. methods will not merely facilitate biosynthetic studies, but will add considerably to their power and discrimination. Second, that as the study of overall precursor-relationships develops (dare one say, as it becomes easier?) attention is increasingly shifted towards either the more intimate details—step sequences, stereospecificity, enzyme mechanisms, *etc.*—or towards what I can only call the biological context. The former shift is I suppose towards what most of us would recognise as biochemistry, of a more or less refined sort according to taste; the latter shift is more difficult to categorize. It includes studies of the regulation of biosynthesis, out of which we can hope to learn what our 'natural products' mean to the long-suffering organisms that produce them for us; it would also seem to include studies of the role of the products of biosynthesis in ecosystem interactions, *i.e.* as chemical messengers, and perhaps also some of the more sensitive applications of chemical taxonomy. Does this shift presage the development of 'chemical biology' as something rather different to classical, or even post-classical, biochemistry? And if it does, how much of such a development should the Chemical Society seek to accommodate?

July 1975 J. D. BU'LOCK

Contents

1
Biosynthesis of Polyketides

BY T. MONEY

1 Introduction

The polyketides are a structurally diverse group of naturally occurring compounds produced by the acyl–polymalonate biosynthetic route. A description of the essential features of this route (Scheme 1) was given in a previous Report which covered the

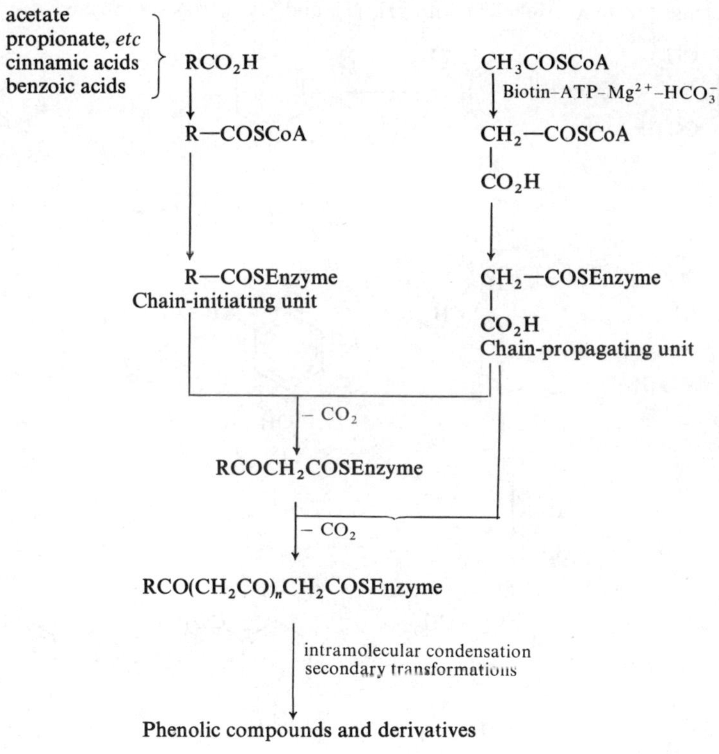

Scheme 1

literature to the end of 1972.[1] The various compounds considered in this Report have been classified into groups (tetraketide, pentaketide, *etc.*) according to the number of

[1] T. Money, in 'Biosynthesis', ed. T. A. Geissman (Specialist Periodical Reports), The Chemical Society, London, 1973, Vol. 2, p. 183.

C_2-units (acetate and malonate) involved in their biosynthesis,[2] and the literature has been covered for 1973 and 1974.

There has been a dramatic increase in the number of studies in which ^{13}C-labelled acetate has been used as a precursor,[3] including elegant and complementary variations based on the observation of ^{13}C–^{13}C couplings in products labelled alternatively from [1, 2-^{13}C] acetate or from a 1:1 mixture of [1-^{13}C]- and [2-^{13}C]-acetate; examples are noted in this Chapter and are fully described in Chapter 7.

A new text dealing with the biosynthesis of acetate-derived compounds contains an excellent chapter on polyketide biosynthesis.[4]

2 Aromatic Polyketides and Derivatives (Acetate as Chain-initiating Unit)

Tetraketides.—A full account of recent investigations on the biosynthesis of patulin (8) and co-metabolites[5] describes further experimental evidence in support of the metabolic grid shown below. Using replacement cultures of *Penicillium patulum* and appropriate precursors labelled with ^{2}H, ^{3}H, and ^{14}C it was established that the co-

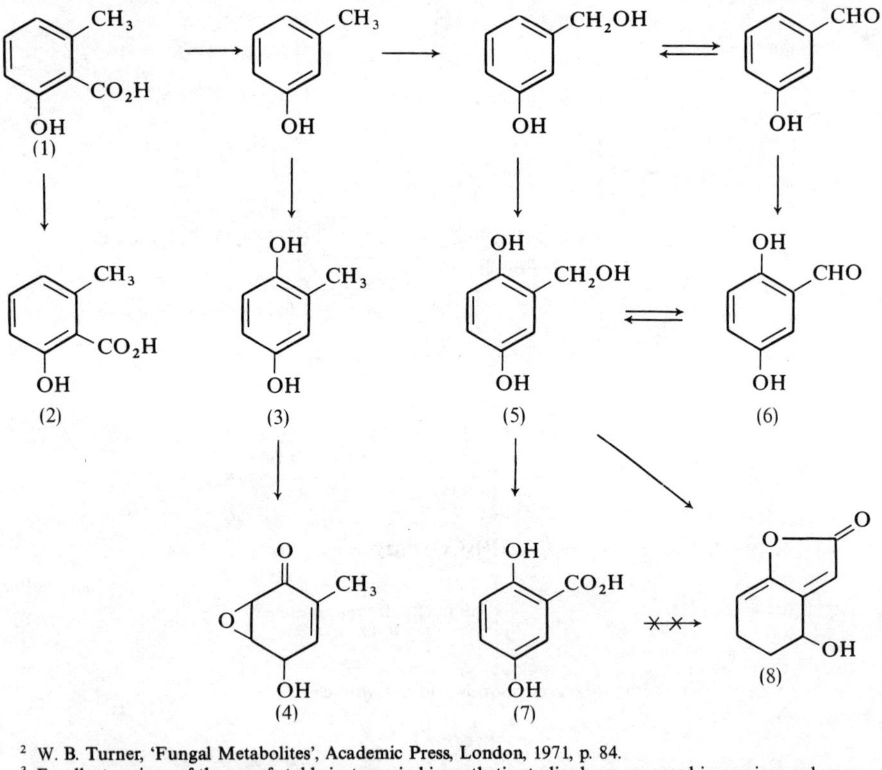

² W. B. Turner, 'Fungal Metabolites', Academic Press, London, 1971, p. 84.

³ Excellent reviews of the use of stable isotopes in biosynthetic studies have appeared in previous volumes: S. A. Brown, in 'Biosynthesis', ed. T. A. Geissman (Specialist Periodical Reports), The Chemical Society, London, 1972, Vol. 1, p. 6; M. Tanabe, *ibid.*, 1973, Vol. 2, p. 241; *ibid.*, 1975, Vol. 3, p. 247.

⁴ N. M. Packter, 'Biosynthesis of Acetate-derived Compounds', Wiley, London, 1972, p. 112.

⁵ A. I. Scott, L. Zamir, G. T. Phillips, and M. Yalpani, *Bioorg. Chem.*, 1973, 2, 124; for earlier data, see ref. 1, p. 186.

metabolites *m*-cresol, *m*-hydroxybenzyl alcohol, *m*-hydroxybenzaldehyde, and gentisaldehyde (6) [but not gentisic acid (7)] are intermediates in the biosynthetic sequence leading from 6-methylsalicylic acid (1) to patulin (8). Toluquinol (3) is not an intermediate in the sequence but can be converted into desoxyepoxydon (4) by the fungus. The mechanism of cleavage of the aromatic ring in late stages of patulin biosynthesis remains to be elucidated. (For ^{13}C studies on the ring cleavage in penicillic acid biosynthesis, see p. 214).

Further studies on the purification of 6-methylsalicylic acid (6-MSA) synthase and its susceptibility to inhibitors have been reported.[6] The results support the view that inhibition of enzymic activity by acetylenic thiol esters (*e.g.* 3-pentynoyl-NAC* and 2-hexynoyl-NAC) is similar to that previously demonstrated for unsaturated fatty acid synthase in *E. coli*,[7] namely specific inhibition of the dehydration step in the biosynthetic route (Scheme 2).

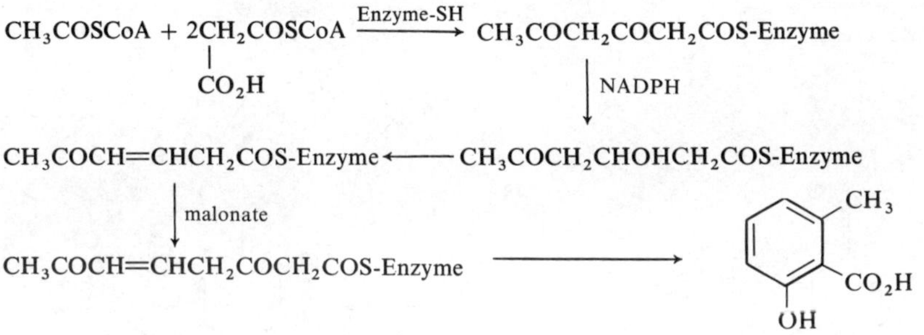

(1)

Scheme 2

Additional studies have shown that concentrations of acetylenic inhibitor which completely inhibit 6-MSA synthesis still allow NADPH oxidation to continue at a significant rate. This result supports the proposal that acetylenic inhibitors act after the reduction step and that the latter process occurs on a 6-carbon rather than an 8-carbon intermediate.[6,8]

In view of the important anti-cancer properties of mycophenolic acid (10; X = Me) considerable effort has been expended on the synthesis of analogues which could display enhanced or modified biological activity. A recent report[9] has shown that the enzymic system of *P. brevicompactum* can convert the halogenated phthalides (9a) and (9b) into the corresponding analogues of mycophenolic acid.

Incorporation studies using intact *Eupatorium rugosum* plants have shown that the aromatic ring in dehydrotremetone (12) is derived from acetate and the furan ring from mevalonate.[10] Further investigations using a cell-free homogenate of *E. rugosum* leaves demonstrated that isopentenyl pyrophosphate and tremetone (11) are efficient

* NAC, *N*-acetylcysteamine.

[6] A. I. Scott, L. C. Beadling, N. H. Georgopapadarou, and C. R. Subbarayan, *Bioorg. Chem.*, 1974, **3**, 238.

[7] R. R. Rando and K. Bloch, *J. Biol. Chem.*, 1968, **243**, 5627; G. M. Helmkamp and K. Bloch, *ibid.*, 1969, **244**, 6014; L. R. Kass and K. Bloch, *Proc. Nat. Acad. Sci. U.S.A.*, 1967, **58**, 1168; L. R. Kass, D. J. H. Brock, and K. Bloch, *J. Biol. Chem.*, 1967, **242**, 4418, 4432.

[8] *cf.* P. Dimroth, H. Walter, and F. Lynen, *European J. Biochem.*, 1970, **13**, 98.

[9] L. Canonica, B. Rindone, C. Scolastico, F. Aragozzini, and R. Craveri, *J.C.S. Chem. Comm.*, 1973, 222.

[10] T.-J. Lin, E. Ramstad, and P. Heinstein, *Phytochemistry*, 1974, **13**, 1809.

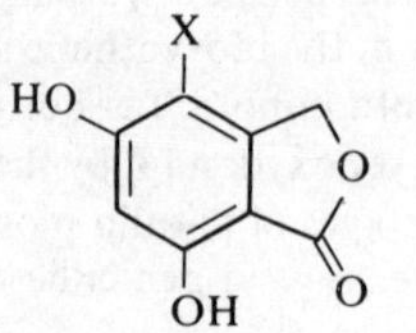

(9) a; X = Cl
b; X = Br

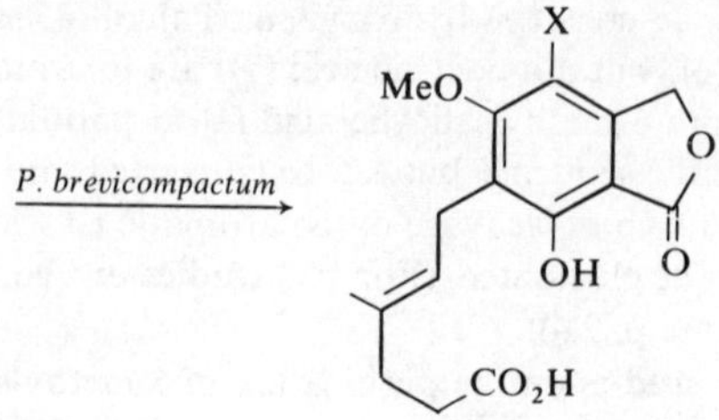

(10)

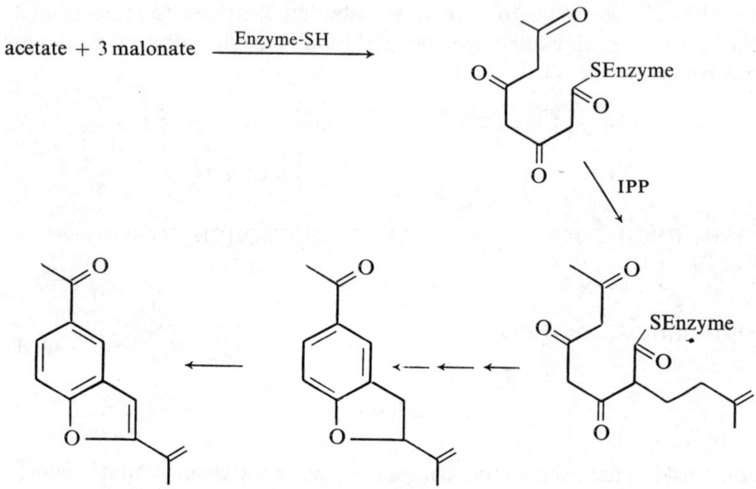

(12) (11)

precursors.[11] In contrast, dimethylallyl pyrophosphate and the acetophenones (13) and (14) were poorly utilized by the cell-free system and it has been suggested that isoprenylation of the enzyme-bound β-triketo-thiol ester intermediate occurs prior to aromatization.[11]

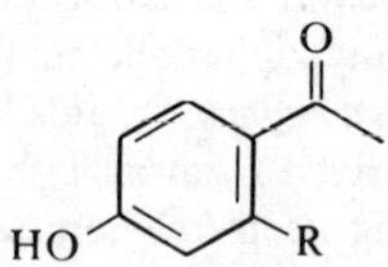

(13) R = H
(14) R = $CH_2CH{=}CMe_2$

Conflicting proposals concerning the biosynthesis of ring B in nidulin (15) and trisdechloronornidulin (16) (*syn.* yasmin, unguinol) have been resolved by studies[12] which demonstrate that the incorporation of acetate and methionine is consistent with biosynthetic Scheme 3.

[11] T.-J. Lin and P. Heinstein, *Phytochemistry*, 1974, **13**, 1817.
[12] J. Sierankiewicz and S. Gatenbeck, *Acta Chem. Scand.*, 1973, **27**, 2710.

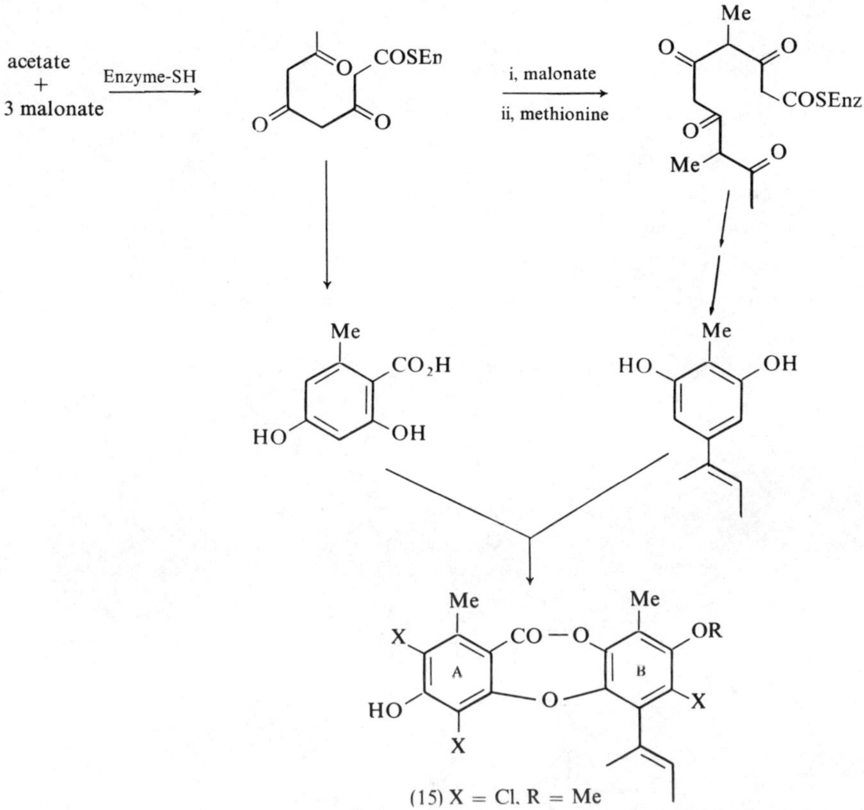

(15) X = Cl, R = Me
(16) X = R = H

Scheme 3

Pentaketides.—One of the first examples of the use of ^{13}C–^{13}C coupling in polyketide biosynthesis is described in a recent report.[13] [1,2-^{13}C]Acetate was used as precursor and the new technique was used to elucidate the biosynthetic route to mollisin (17).[13] Two possible pathways [paths *a* and *b* in Scheme 4] to this compound have previously been considerable, but the use of [1,2-^{13}C]acetate has shown that neither is correct and that the biosynthesis of mollisin follows a third route [path *c* in Scheme 4].

Administration of [1-^{13}C]acetate to *Phialophora lagerbergii* and the subsequent enhancement of signals due to C(1), C(3), C(4a), C(6), and C(8) in the ^{13}C n.m.r. spectrum of skytalone (18) has demonstrated that this compound is derived by the linear condensation of five acetate units;[14] *cf.* p. 228.

Hexaketides.—Incorporation studies using[1-^{13}C]-, [2-^{13}C]-, and [1,2-^{13}C]-acetate have shown that the biosynthesis of multicolic acid (21) and multicolosic acid (22) in *P. multicolor* involves the intermediate formation of 6-pentylresorcylic acid (19)

[13] H. Seto, L. W. Cary, and M. Tanabe, *J.C.S. Chem. Comm.*, 1973, 867.
[14] D. C. Aldridge, A. B. Davies, M. R. Jackson, and W. B. Turner, *J.C.S. Perkin I*, 1974, 1540.

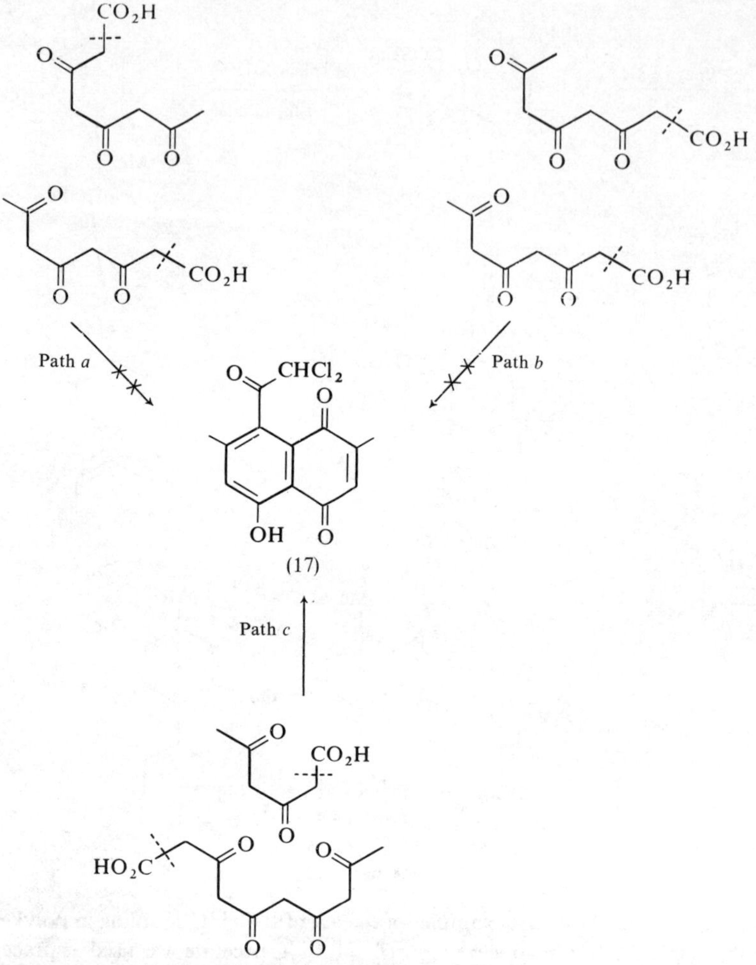

Scheme 4

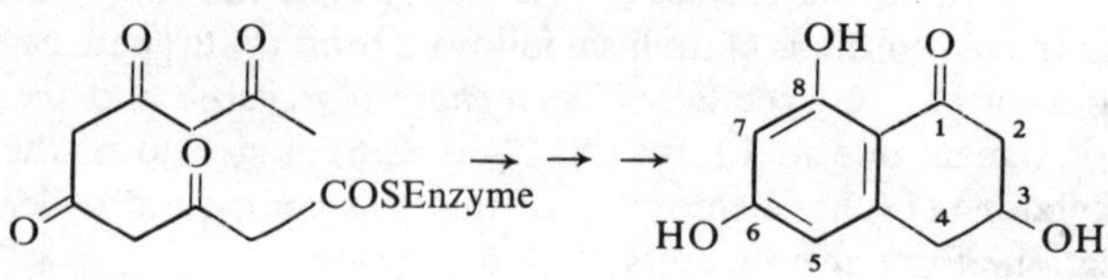

(18)

followed by cleavage of the C(4)—C(5) bond.[15] The ^{13}C–^{13}C couplings in the ^{13}C n.m.r. spectrum of multicolic acid derived from doubly labelled acetate showed that intact C_2 units were arranged as shown in formula (21), thus excluding the possibility

[15] J. A. Gudgeon, J. S. E. Holker, and T. J. Simpson, *J.C.S. Perkin I*, 1974, 636.

of cleavage at C(1)–C(2) in (19) and the intermediacy of a symmetrical intermediate such as 5-pentylresorcinol (20); a fuller account of the n.m.r. aspects is given on p. 216.

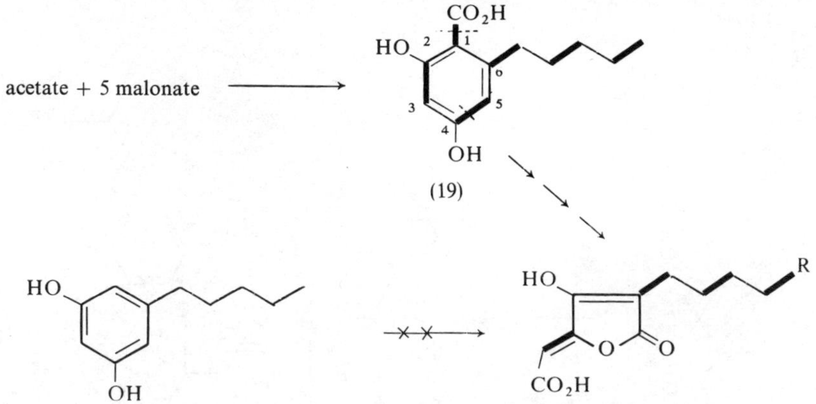

(19)

(20)

(21); R = CH_2OH
(22); R = CO_2H

Heptaketides.—The pattern of radioactivity in aloenin (23) after administration of [1-^{13}C]acetate, [2-^{14}C]malonate, and [*methyl*-^{14}C]methionine to *Aloe arborescens* has been determined and the results are consistent with the expected biosynthetic route.[16]

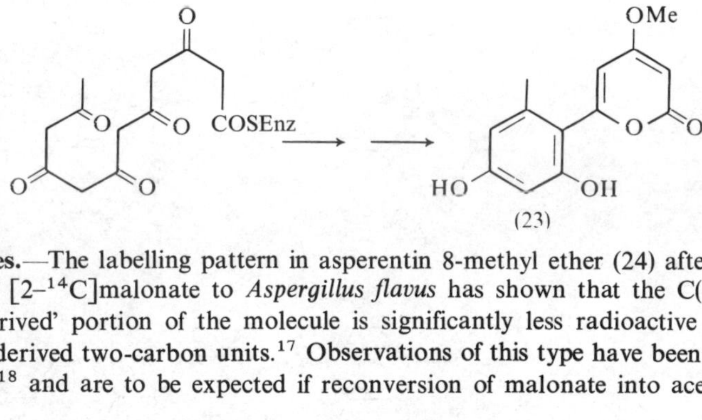

(23)

Octaketides.—The labelling pattern in asperentin 8-methyl ether (24) after administration of [2–^{14}C]malonate to *Aspergillus flavus* has shown that the C(6′)—C(7′) 'acetate-derived' portion of the molecule is significantly less radioactive than the malonate-derived two-carbon units.[17] Observations of this type have been recorded previously[18] and are to be expected if reconversion of malonate into acetate only

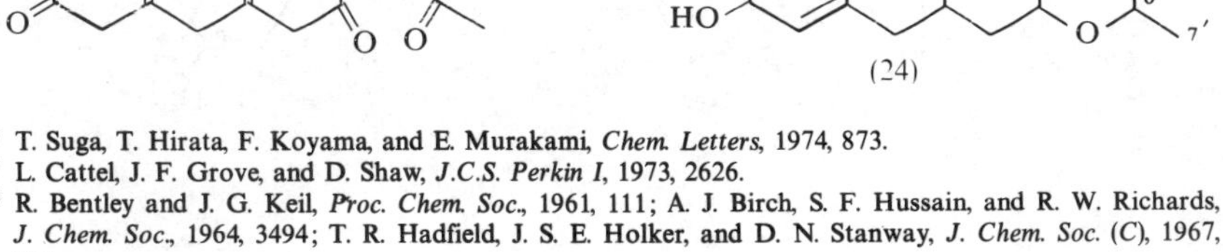

(24)

[16] T. Suga, T. Hirata, F. Koyama, and E. Murakami, *Chem. Letters*, 1974, 873.
[17] L. Cattel, J. F. Grove, and D. Shaw, *J.C.S. Perkin I*, 1973, 2626.
[18] R. Bentley and J. G. Keil, *Proc. Chem. Soc.*, 1961, 111; A. J. Birch, S. F. Hussain, and R. W. Richards, *J. Chem. Soc.*, 1964, 3494; T. R. Hadfield, J. S. E. Holker, and D. N. Stanway, *J. Chem. Soc. (C)*, 1967, 731.

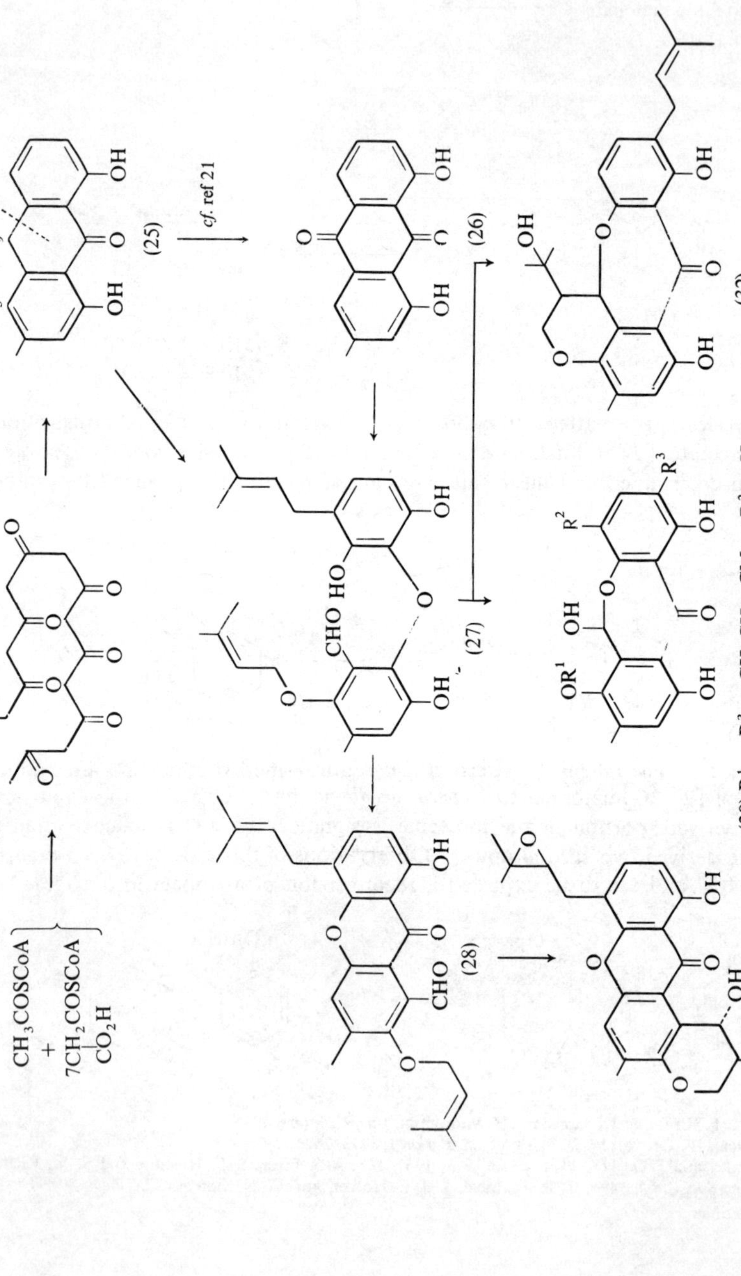

(30) $R^1 = R^3 = CH_2CH{=}CMe_2$; $R^2 = H$
(31) $R^1 = R^2 = CH_2CH{=}CMe_2$; $R^3 = H$

Scheme 5

occurs to a limited extent. The ^{13}C n.m.r. spectra of tajixanthone (29) (*Aspergillus variecolor*) derived from [1-^{13}C]- and [2-^{13}C]-acetate show enrichment patterns consistent with a biosynthetic pathway (Scheme 5) in which the xanthone nucleus is derived by ring cleavage of chrysophanol (26) or chrysophanol anthone (25).[19] A mevalonoid origin for the *C*- and *O*-prenyl groups was also established during these investigations.

Although ring cleavage of anthraquinone intermediates seems to be an established biosynthetic process[20] it has been suggested that in the biosynthesis of tajixanthone (29) the actual intermediate undergoing ring cleavage is the anthrone (25).[19,21] The product (27) of this reaction has also been proposed as an intermediate in the biosynthesis of arugosin A (30), B (31), and C (32), which co-occur with tajixanthone (29) in *A. variecolor*. For a full discussion, see p. 222.

Evidence for the direct involvement of emodinanthrone (33) in the biosynthesis of the range of anthraquinonoids [skyrin (36), iridoskyrin (37), (+)-rugulosin (38), and (−)-rubroskyrin (39)] of *P. brunneum* and *P. islandicum* has been reported.[22] The low incorporation of emodin (34) has also led to the suggestion that this compound is not itself on the main biosynthetic route. Normally, the absence of oxygen functionality at the expected positions in acetate-derived phenolic compounds is thought to arise by reduction and dehydration of the β-polyketo-thiol ester intermediate. However, the efficient incorporation of emodinanthrone (33) into islandicin (35), iridoskyrin (37), and rubroskyrin (39) seems to indicate that in this system direct reduction of the aromatic ring can also occur[22] (Scheme 6).

The biosynthesis of ochrephilone (40), a new metabolite of *P. multicolor*, has been elucidated by incorporation experiments using [1-^{13}C]-, [2-^{13}C]-, and [1,2-^{13}C]-acetate as precursor.[23] In the latter case ^{13}C–^{13}C coupling between C(1)—C(2), C(3)—C(4) *etc.* was observed, while a 1:1 mixture of [1-^{13}C]- and [2-^{13}C]-acetate resulted in coupling between C(2)–C(3), C(4)–C(5) *etc.* The results obtained (*cf.* p. 224) are consistent with the proposal[23] that ochrephilone (40) is formed by the condensation of two acetate-derived chains and the introduction of a C_1 unit (presumably from methionine) at C(4).

Decaketides.—Previous studies on the biosynthesis of the aflatoxins have shown that the entire carbon skeleton is derived from acetate.[24] Recent investigations provide support for the proposal[25] that averufin (41) and sterigmatocystin (42) are intermediates in the biosynthetic route (Scheme 7). Radioactive averufin (41) was prepared by adding [1-^{14}C]acetate to a mutant of *Aspergillus parasiticus* which was impaired in its ability to produce aflatoxins. When administered to wild-type mycelium the efficient conversion of averufin (41) into aflatoxins B_1 (44), G_1 (45), and B_2 (15,16-

[19] J. S. E. Holker, R. D. Lapper, and T. J. Simpson, *J.C.S. Perkin I*, 1974, 2135.

[20] (*a*) R. F. Curtis, C. H. Hassall, and D. R. Parry, *J.C.S. Perkin I*, 1972, 240; (*b*) B. Franck, F. Huper, D. Groger, and D. Erge, *Chem. Ber.*, 1968, **101**, 1954, 1970; (*c*) S. Gatenbeck, *Svensk kem. Tidskr.*, 1960, **72**, 188; (*d*) T. Thomas, 'Biogenesis of Antibiotic Substances', ed. Z. Vanek and Z. Hostalek, Academic Press, New York, 1965, p. 155.

[21] *cf.* T. Money, *Nature*, 1963, **199**, 592.

[22] U. Sankawa, Y. Ebizuka, and S. Shibata, *Tetrahedron Letters*, 1973, 2125.

[23] H. Seto and M. Tanabe, *Tetrahedron Letters*, 1974, 651.

[24] M. Biollaz, G. Buchi, and G. Milne, *J. Amer. Chem. Soc.*, 1970, **92**, 1035 and references cited therein.

[25] (*a*) M. O. Moss in 'Phytochemical Ecology', ed. J. B. Harbourne, Academic Press, New York, 1972, p. 125; (*b*) R. Thomas in ref. 20(*c*), p. 160.

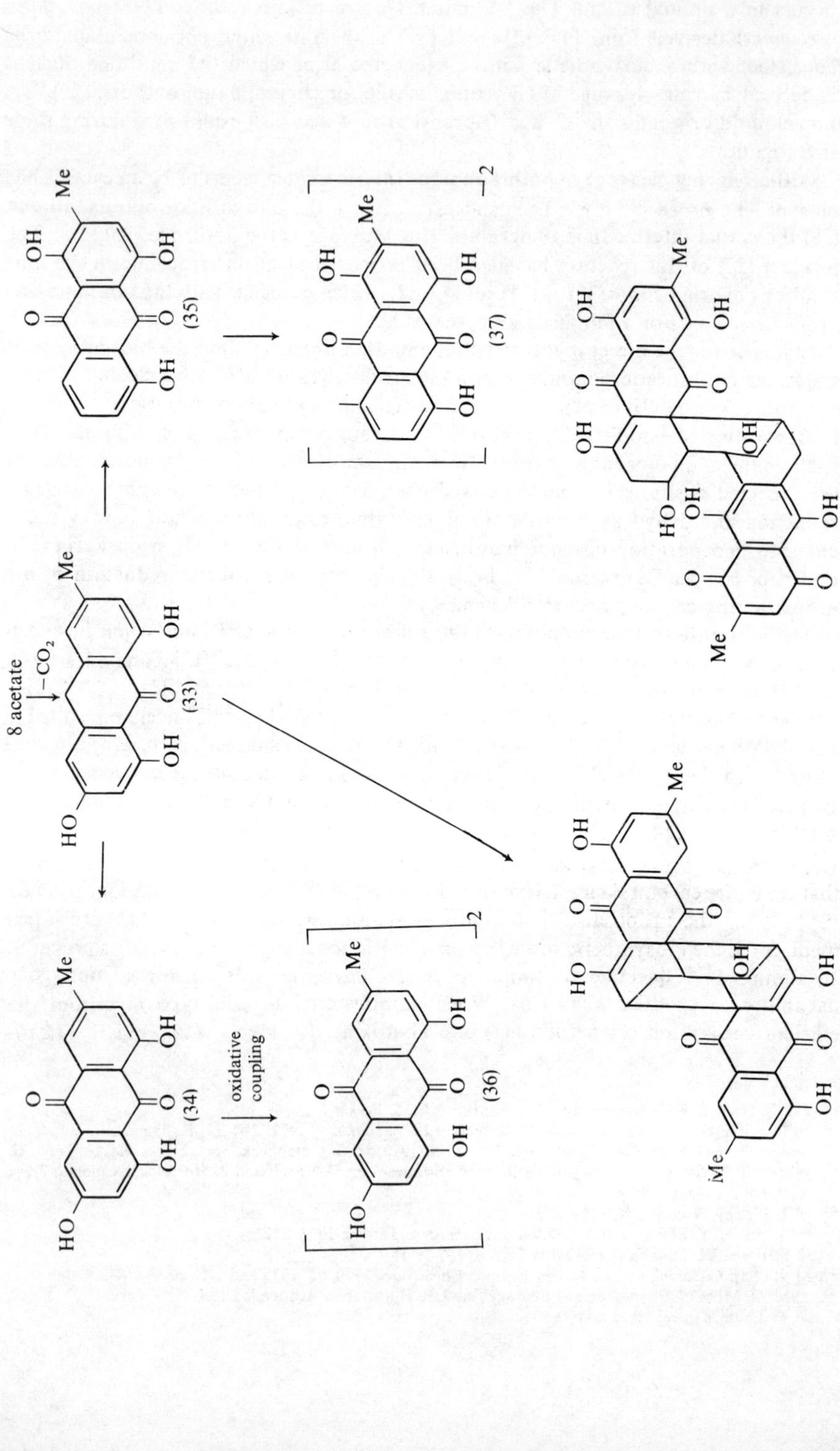

Scheme 6

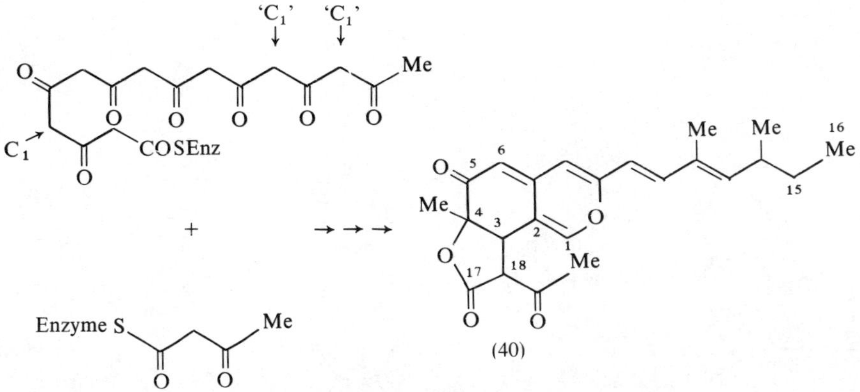

(40)

dihydro-B_1) was observed.[26] 6-Hydroxydihydrosterigmatocystin (43)* has previously been implicated in the biosynthesis of aflatoxin B_2 (15,16-dihydro-B_1) and G_2 (15,16-dihydro G_1)[27] and a recent report has shown that sterigmatocystin (42) can be efficiently converted into aflatoxin B_1 (44).[28] Related studies[29] using [1,2-^{13}C]acetate as precursor (see p. 214) have shown that sterigmatocystin (42) obtained from cultures of *A. versicolor* displays ^{13}C–^{13}C couplings consistent with the biosynthetic route shown in Scheme 7. A distinction between pathways (*a*) and (*b*) was made possible by the presence of ^{13}C–^{13}C-coupling between C(4)—C(5) and C(6)—C(7) (but see p. 247).

3 Aromatic Polyketides (Cinnamate as Chain-initiating Unit)

Polyketides whose biosynthesis involves the condensation of a C_6—C_3 unit (cinnamate and derivatives) with two or three acetate units are characteristic metabolites of higher plants and are rarely found in fungi. Recent tracer studies have, however, established both hispidin (47) (*Polyporus hispidus*)[30,31] and chlorflavonin (48) (*Aspergillus candidus*)[32] as true fungal metabolites. In the case of hispidin (47) the cumulative efforts of two research groups[30,31] have shown that phenylalanine, tyrosine, cinnamate, *p*-coumarate, caffeate, and acetate are specifically incorporated into hispidin (47) by cultures of *Polyporus hispidus*[30] and *Polyporus schweinitzii.*[31]

* Due to a different numbering system this compound is described as the 5-hydroxy-derivative in the original report.[27]

[26] M. T. Lin, D. P. H. Hsieh, R. C. Yao, and J. A. Donkersloot, *Biochemistry*, 1973, **12**, 5167; *cf.* M. T. Lin and D. P. H. Hsieh, *J. Amer. Chem. Soc.*, 1973, **95**, 1668.

[27] G. C. Elsworthy, J. S. E. Holker, J. M. McKeown, J. B. Robinson, and L. J. Mulheirn, *Chem. Comm.*, 1970, 1069.

[28] D. P. H. Hsieh, M. T. Lin, and R. C. Yao, *Biochem. Biophys. Res. Comm.*, 1973, **52**, 992; *cf.* J. S. E. Hokker and J. G. Underwood, *Chem. and Ind.*, 1964, 1865.

[29] H. Seto, L. W. Cary, and M. Tanabe, *Tetrahedron Letters*, 1974, 4491.

[30] P. W. Perrin and G. H. N. Towers, *Phytochemistry*, 1973, **12**, 589.

[31] G. M. Hatfield and L. R. Brady, *Lloydia*, 1973, **36**, 59.

[32] R. Marchelli and L. C. Vining, *J.C.S. Chem. Comm.*, 1973, 555.

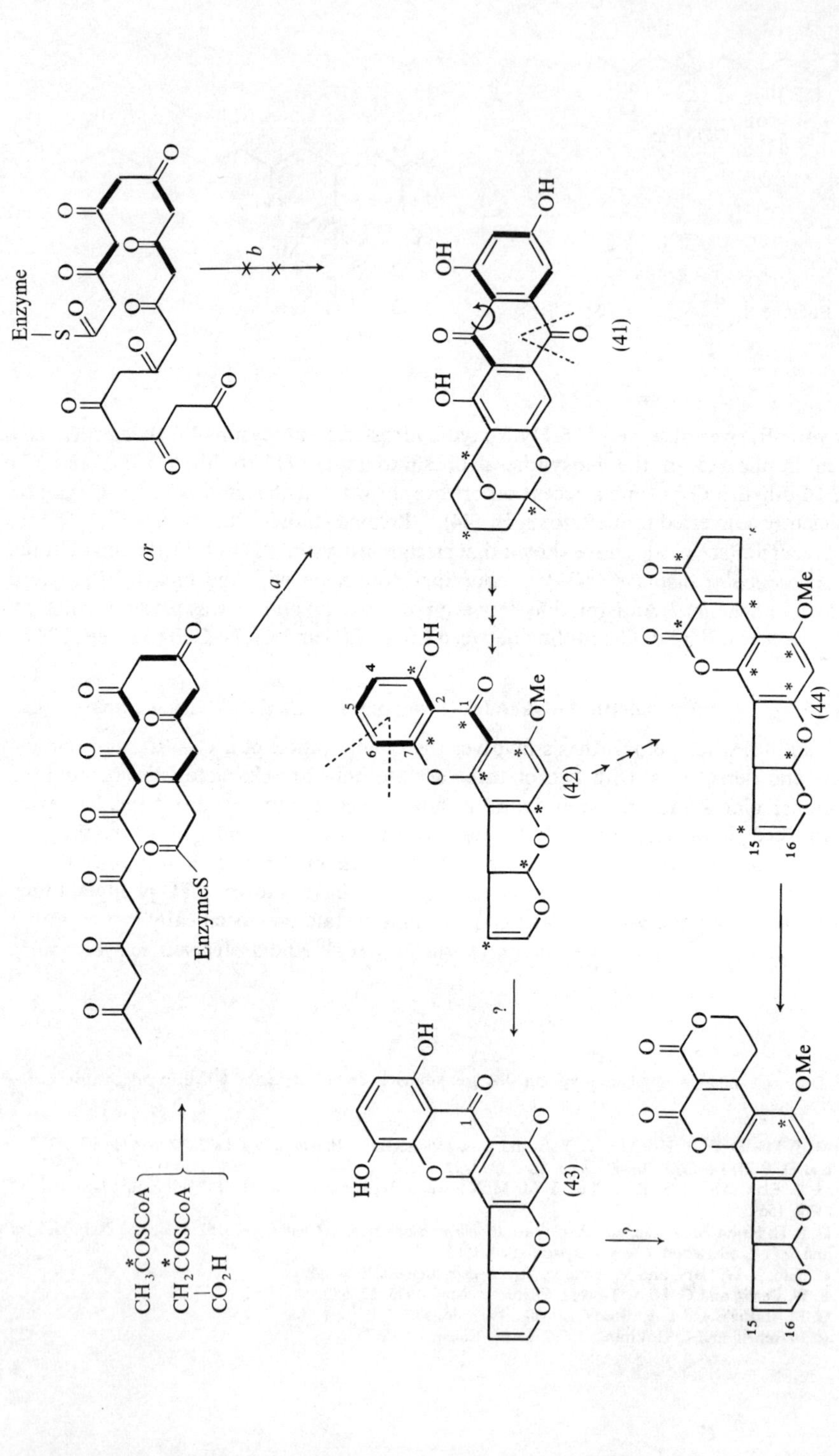

$\overset{*}{C}H_3COSCoA$
$\overset{*}{C}H_2COSCoA$
CO_2H
EnzymeS
Enzyme
S
or
a
b
(41)
(42)
(43)
(44)
(45)
OH
HO
OMe

Scheme 7

Evidence for the presence of bis-noryangonin (46) in the mycelial extract was also obtained and an enzyme preparation was able to convert this compound into hispidin.[30,33] One of the most interesting aspects of recent studies in this area is the discovery that light (380 and 440 nm) stimulates *p*-coumarate hydroxylase activity and styrylpyrone biosynthesis in cultures of *P. hispidus*.[34] The suggestion has been made[30] that hispidin could be derived by several alternative routes and these are summarized in Scheme 8. In a similar study[32] the specific incorporation of [3-^{14}C]-

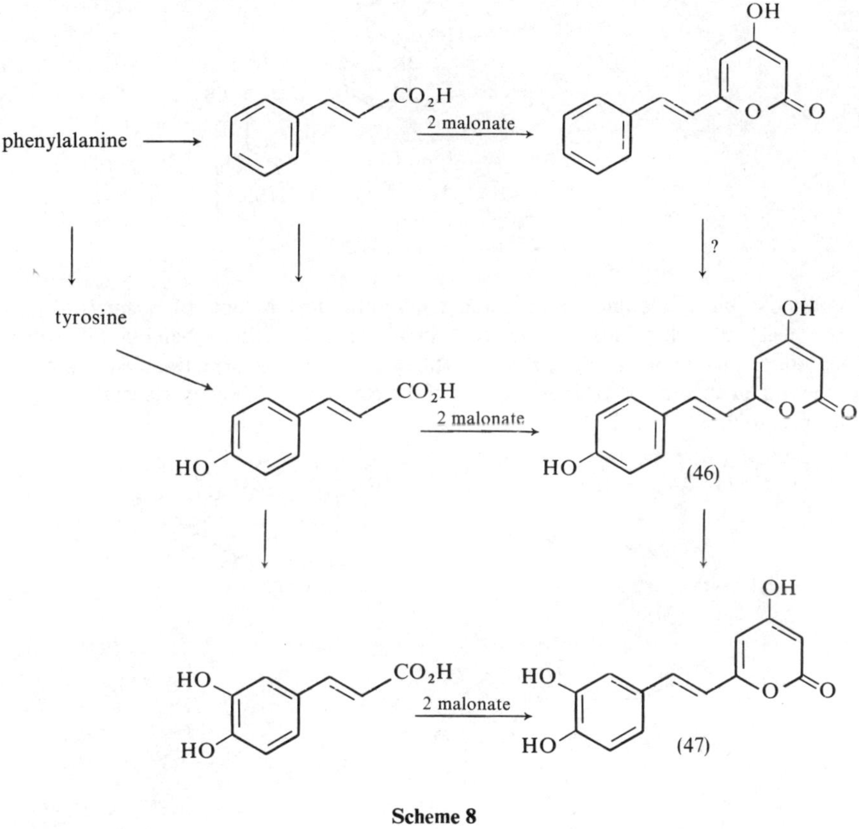

Scheme 8

phenylalanine into chlorflavonin (48) and the co-metabolite (49) has clearly demonstrated that these compounds are biosynthesized *de novo* by *Aspergillus candidus*; there are structural parallels for the terphenyl (49) but compound (48) is probably the only authenticated flavonoid from a fungus.

The labelling patterns found in eucomin (51) after feeding aqueous solutions of

[33] A. M. D. Nambuduri, C. P. Vance, and G. H. N. Towers, *Biochim. Biophys. Acta*, 1974, **343**, 148.

[34] C. P. Vance, E. B. Tregunna, A. M. D. Nambudiri, and G. H. N. Towers, *Biochim. Biophys. Acta*, 1974, **343**, 138; A. M. D. Nambudiri, C. P. Vance, and G. H. N. Towers, *Biochem. J.*, 1973, **134**, 891.

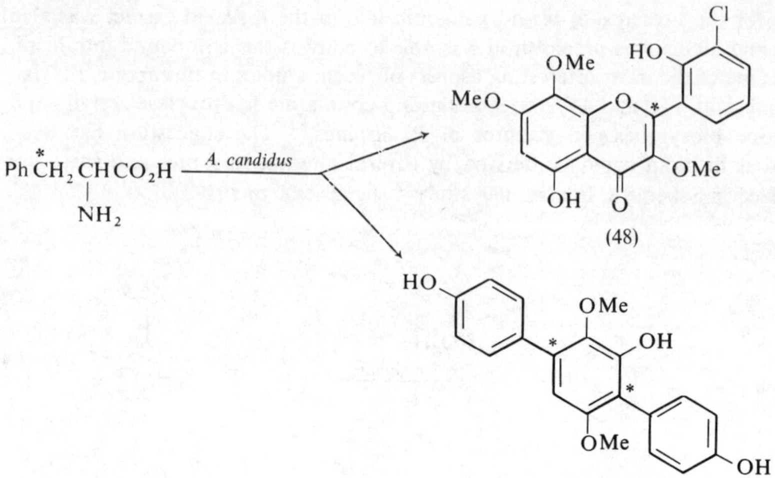

(48)

(49)

radioactive phenylalanine, acetate, and methionine to the roots of *Eucomis bicolor* are consistent with a biosynthetic route involving a 2-methoxychalcone (50) intermediate.[35] The involvement of the 2-methoxy-group in the formation of the heterocyclic ring in eucomin (51) is reminiscent of the reaction which converts isoflavonoids into rotenoids.

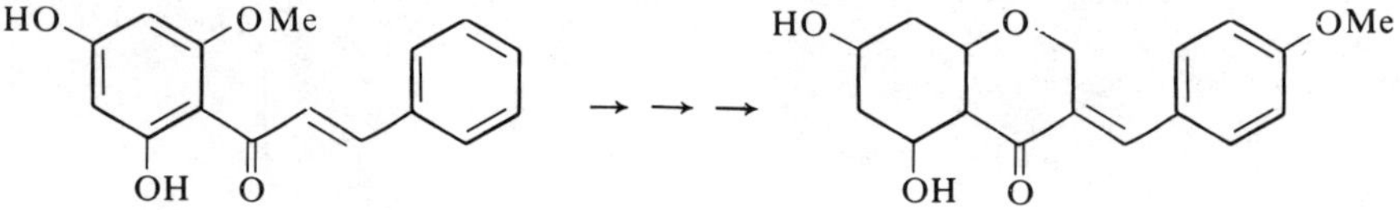

(50) (51)

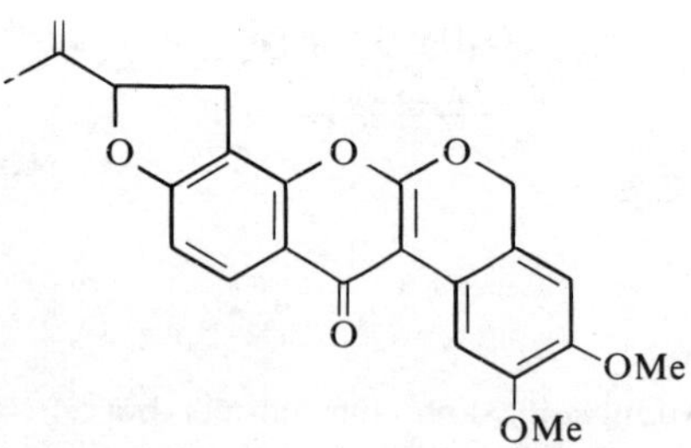

(52)

An authoritative account of recent major developments in isoflavone and rotenoid biosynthesis is provided in an excellent paper which specifically deals with the biosynthesis of amorphigenin (52) by *Amorpha fruticosa* seedlings.[36] Preliminary

[35] P. M. Dewick, *J.C.S. Chem. Comm.*, 1973, 438.
[36] L. Crombie, P. M. Dewick, and D. A. Whiting, *J.C.S. Perkin I*, 1973, 1285.

accounts of this work were published previously and an outline of the proposed biosynthetic scheme and the evidence supporting it has been given in a previous Report (see Vol. 2, p. 206).

Administration of [3-^{3}H, 3-^{14}C]cinnamic acid to the embryo fruit of *Aesculus carnea* results in the formation of (−)-epicatechin (55), procyanidin B-2 (58), and proanthyocyanidin A-2 (59) with radioactivity at the expected positions.[37] However, the level of radioactivity in each half of (58) and (59) was different, and this experi-

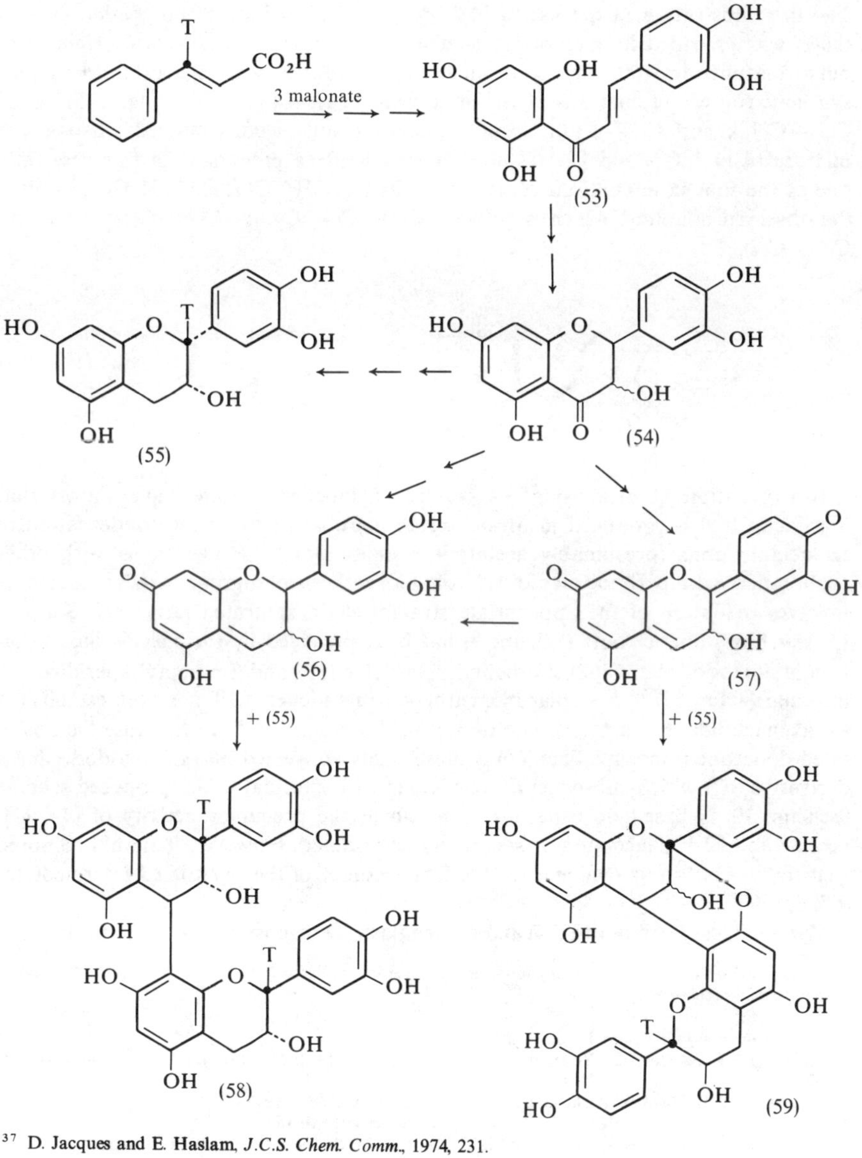

[37] D. Jacques and E. Haslam, *J.C.S. Chem. Comm.*, 1974, 231.

mental finding has prompted new mechanistic proposals to explain the biosynthesis of these dimeric compounds.

A review describing the importance of α-hydroxychalcones in flavonoid biosynthesis has been published;[38] flavonoid biosynthesis is dealt with by Harborne in Chapter 4.

4 Non-aromatic Polyketides*

The first illustration of the use of [1,2-^{13}C]-labelled acetate in polyketide biosynthesis was provided in a report[39a] in which the known penta-acetate origin[39b] of dihydrolatumcidin (60) was confirmed. In accordance with the established biosynthetic route, coupling was observed between C(2)—C(3), C(4)—C(4a), C(5)—C(6), C(7)—C(7a), and C(8)—C(9). Complementary results were obtained when a 1:1 mixture of [1-^{13}C]- and [2-^{13}C]-acetate was used as precursor. In this case only one of the four combinations of acetate units (. . . $\mathrm{CH_2^{13}CO}$ . . . $\mathrm{^{13}CH_2CO}$ . . .) gives the observed coupling, which is between C(3)—C(4), C(4a)—C(7a), C(5)—C(8), and C(6)—C(7).

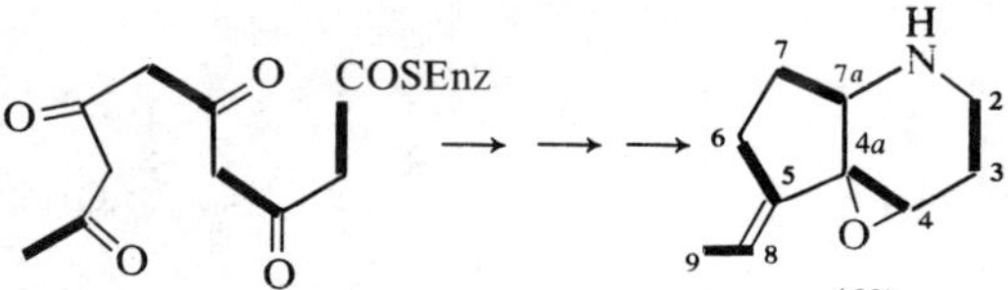

(60)

Incorporation studies using ^{14}C- and ^{13}C-labelled acetate have shown that nigrifactin (65) is produced in *Streptomyces nigrifaciens* by linear condensation of six acetate units (presumably acetate + 5 malonates).[40] However, as with other acetate-derived piperidine alkaloids,[41] the biosynthesis of nigrifactin (65) presumably involves oxidation of the appropriate straight-chain saturated fatty acid. Support for this biosynthetic route (Scheme 9) has been provided by the specific incorporation of 5-oxododecanal (62), 2-n-heptylpiperideine (63), and 2-n-heptylpiperidine (64) into nigrifactin (65).[40] A similar biosynthetic route (Scheme 10) has been established for avenaciolide (69), a metabolite of *Aspergillus avenaceus*.[42] In this case the postulated dodecanoyl intermediate (66) is presumably converted into a 3-oxododecanoyl derivative (67) which subsequently condenses with succinate. The proposed scheme (Scheme 10) is based on experiments in which the precursor activity of [1-^{13}C]-acetate and [2-^{13}C]acetate was separately determined. However, it should be noted that there is no direct evidence for the intermediacy of the postulated intermediates (67) and (68).

Cytochalasin B (phomin) (74) and cytochalasin D (zygosporin A) (75) are structur-

* It is possible that most of the compounds in this category are derived by oxidation of appropriate saturated fatty acid precursors.

38 D. G. Roux and D. Ferreira, *Phytochemistry*, 1974, **13**, 2039.
39 (*a*) H. Seto, T. Sato, and H. Yonehara, *J. Amer. Chem. Soc.*, 1973, **95**, 8461; (*b*) *J. Antibiotics*, 1973, **26**, 609.
40 T. Terashima, E. Idaka, Y. Kishi, and T. Goto, *J.C.S. Chem. Comm.*, 1973, 75.
41 E. Leete, *Accounts Chem. Res.*, 1971, **4**, 100, and references cited therein.
42 M. Tanabe, T. Hamasaki, Y. Suzuki, and L. F. Johnson, *J.C.S. Chem. Comm.*, 1973, 212.

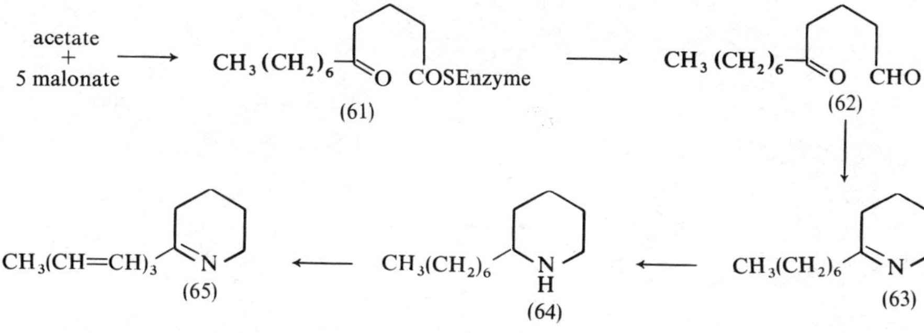

Scheme 9

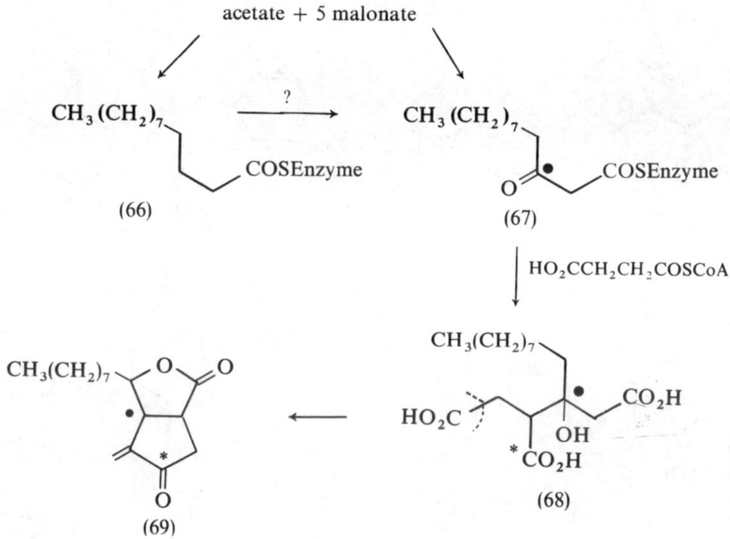

Scheme 10

ally related compounds isolated from *Phoma sp.* and *Zygosporium masonii* respectively. Incomplete results obtained from studies[43] using ^{14}C-labelled precursors led to the proposed biosynthetic scheme outlined in Scheme 11. Complementary results using ^{13}C-labelled acetate have provided strong support for these proposals and have added information on the biosynthetic origin of regions in the structures [*e.g.* C(1), C(9)] which were inaccessible by degradation.[44]

The biosynthesis of thermozymocidin (76) from nine acetate units and serine has been confirmed by recent incorporation experiments using [1-^{13}C]acetate.[45]

[43] C.-R. Lebet and C. Tamm, *Helv. Chim. Acta*, 1974, **57**, 1785 and references cited therein.

[44] W. Graf, J.-L. Robert, J. G. Vederas, C. Tamm, P. H. Solomon, I. Mivra, and K. Nakanishi, *Helv. Chim. Acta*, 1974, **57**, 1801.

[45] F. Aragozzini, M. G. Beretta, G. S. Ricca, G. Scolastico, and F. W. Wehrli, *J.C.S. Chem. Comm.*, 1973, 788.

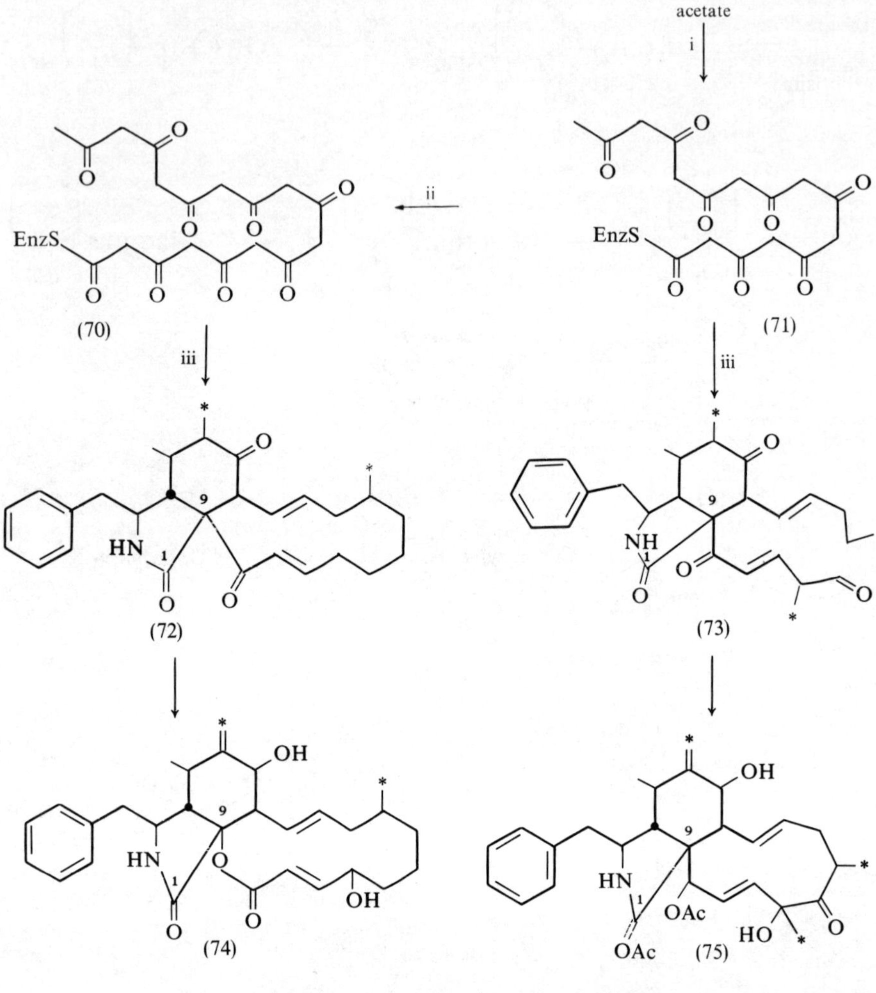

Reagents: 1, 7 malonate; ii, malonate: iii, $PhCH_2CH(NH_2)\dot{C}O_2H$, $\overset{*}{C}H_3SCH_2CH_2CH(NH_2)CO_2H$

Scheme 11

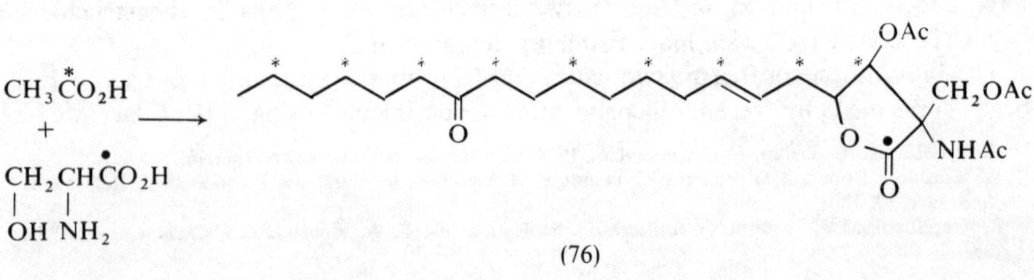

(76)

Analyses of the ^{13}C n.m.r. spectra of prodigiosin (79) after incorporation of ^{13}C-labelled acetate, alanine, proline, methionine, and serine have resulted in the proposed biosynthetic scheme[46] shown in Scheme 12. (For similar data on some prodigiosin analogues, see p. 211.)

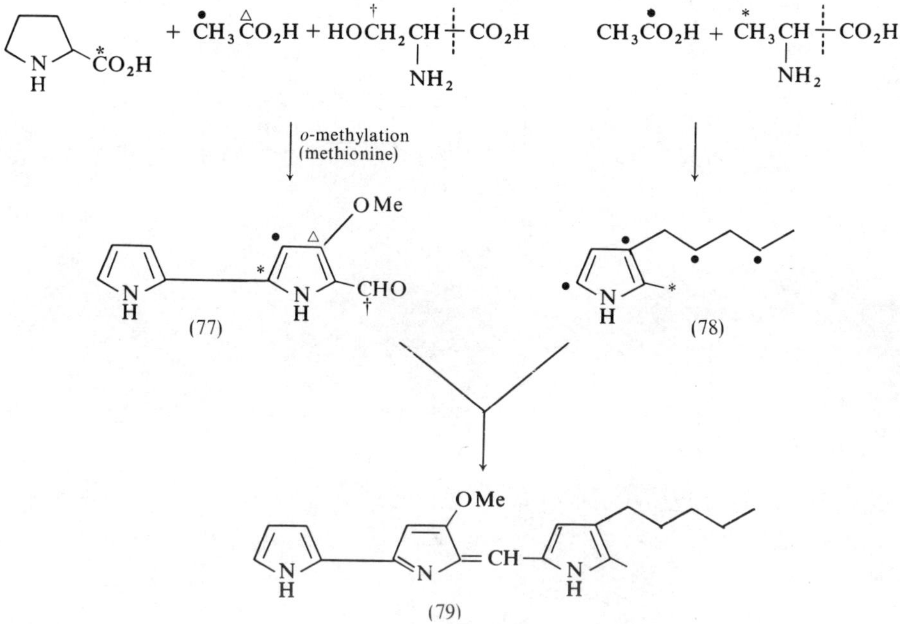

Scheme 12

A preliminary investigation into the biosynthesis of the insect metabolite pederin (80) has shown that much of the carbon framework is derived from acetate and that propionate (presumably as methyl malonate) also acts as a precursor.[47] Further information will be necessary, however, before an outline of the biosynthetic route can be provided. A full account of previously reported[1] studies on the biosynthesis of auerothin (81) has appeared.[48]

[46] H. H. Wasserman, R. J. Sykes, P. Peverada, C. K. Shaw, R. J. Cushley, and S. R. Lipsky, *J. Amer. Chem. Soc.*, 1973, **95**, 6874.

[47] C. Cardani, C. Fuganti, D. Ghiringhelli, P. Grasselli, M. Pavan, and M. D. Valcurone, *Tetrahedron Letters*, 1973, 2815.

[48] R. Cardillo, C. Fuganti, D. Ghiringhelli, D. Giangrasso, P. Graselli, and A. Santopietro-Amisano, *Tetrahedron*, 1974, **30**, 549.

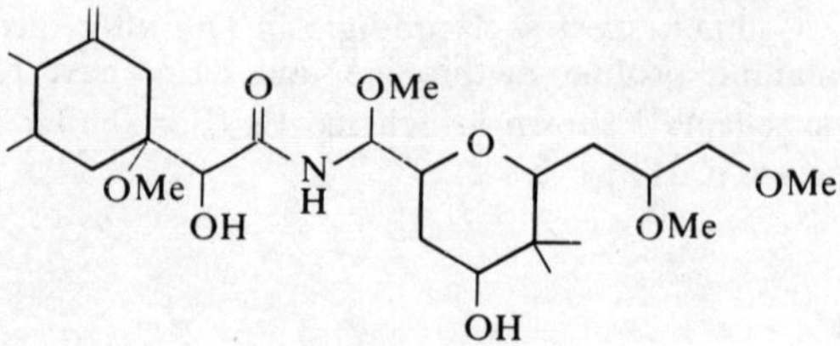

(80)

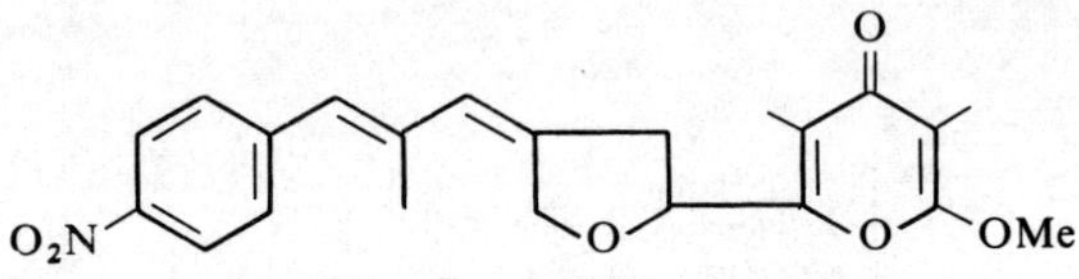

(81)

2
The Biosynthesis of C_5—C_{25} Terpenoid Compounds

BY J. R. HANSON

1 Introduction

This Report for 1974 follows the pattern of the previous reports.[1] A number of reviews covering various general aspects of terpenoid biosynthesis have appeared.[2–4] During the year, significant use has been made of carbon-13 methods in terpenoid biosynthesis in ways which complement the use of radio-active isotopes; in particular, ^{13}C–^{13}C coupling constants have been used to demonstrate the integrity of units through a biosynthesis (also see Chapter 7).

2 Mevalonic Acid

Mevaldate reductase, which is present in liver, catalyses the reduction of 3(*R*)- and 3(*S*)-enzyme-mevaldic acid to mevalonate. On the other hand, 3-hydroxy-3-methyl-glutaryl-CoA (HMG-CoA) reductase mediates the reduction of the real intermediate, the hemithioacetal of 3-hydroxy-3-methylglutaryl-coenzyme A, to mevalonate. Whilst both enzymes utilize the *pro*-4(*R*)-hydrogen atom of NADPH as the source of hydrogen, mevaldate reductase affords a *pro*-5(*R*)-labelled mevalonate whilst HMG-CoA reductase affords *pro*-5(*S*)-labelled material. Mevaldate reducate has been partly purified[5] and it has properties similar to those of other animal tissue aldehyde reductases, suggesting that it might be reclassified.

The substrate stereochemistry of 3-hydroxy-3-methylglutaryl synthase has been determined[6] by studying the condensation of chiral acetate, (*R*)-[$^2H_1$, $^3H_1$]acetyl-coenzyme A and (*S*)-[$^2H_1$, $^3H_1$]acetyl-coenzyme A, with acetoacetyl-coenzyme A. Analysis of the resultant mevalonate showed that the reaction was stereospecific and that it was associated with a hydrogen isotope effect. If the effect was normal, then the condensation had proceeded with inversion of configuration at the methyl group.

3-Hydroxy-3-methylglutaryl-coenzyme A reductase is a rate-controlling enzyme in cholesterol biosynthesis, and consequently there have been a number of studies of its properties[7,8] and of the features which inhibit its function.[9,10] These are fully

[1] J. R. Hanson, in 'Biosynthesis', ed. T. A. Geissman (Specialist Periodical Reports), The Chemical Society, London, 1972, Vol. 1, p. 41; 1973, Vol. 2, p. 1; 1975, Vol. 3, p. 1.
[2] J. W. Cornforth, *Tetrahedron*, 1974, **30**, 1515.
[3] C. D. Poulter, *J. Agric. Food Chem.*, 1974, **22**, 167.
[4] 'Natural Product Chemistry', ed. K. Nakanishi, Academic Press, New York, 1974, Vol. 1.
[5] A. S. Beedle, H. H. Rees, and T. W. Goodwin, *Biochem. J.*, 1974, **139**, 205.
[6] J. W. Cornforth, G. T. Phillips, B. Messner, and H. Eggerer, *European J. Biochem.*, 1974, **42**, 591.
[7] M. J. R. Higgins, D. Brady, and H. Rudney, *Arch. Biochem. Biophys.*, 1974, **163**, 271.
[8] M. S. Brown, S. E. Dana, and M. D. Soperstein, *J. Biol. Chem.*, 1974, **249**, 6585.
[9] C. Madhosingh, B. B. Migicovsky, and I. M. Wood, *F.E.B.S. Letters*, 1974, **46**, 20.
[10] R. A. Heller and R. G. Gould, *J. Biol. Chem.*, 1974, **249**, 5254.

covered in Chapter 3, p. 31, of these Reports. When sweet potato is infected with *Ceratocystis fimbriata*, terpenes such as ipomeamarone are vigorously synthesized in the adjacent tissue. HMG-CoA reductase is found in this tissue.[11] This enzyme system has also been found in *Neurospora crassa*.[12] The pH optima of the mevalonate kinases of chloroplast preparations from the French bean are at pH 7.5 whilst leaf and cotyledon preparations show optima at lower pH.[13] Mevalonate kinase has also been isolated[14] from *Pinus pinaster* seedlings. A cell-free extract which converts mevalonate into isopentenyl pyrophosphate has been obtained from the leaves of *Nepeta cataria*.[15] These enzymes are found in the 30 000 *g* supernatant. A report has appeared[16] of the recycling of carbon atoms from [2-^{14}C]mevalonate into fatty acids, possibly *via* 3-hydroxy-3-methylglutaryl-CoA and free acetoacetate.

A cell-free homogenate of *Eupatorium rugosum* leaves has been prepared and utilized to study the biosynthesis of dehydrotremetone (2) *via* an isopentenylacetopheneone (1)[17] (see p. 3).

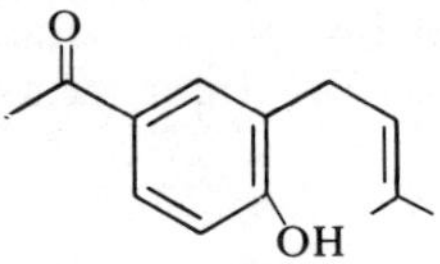

(1)

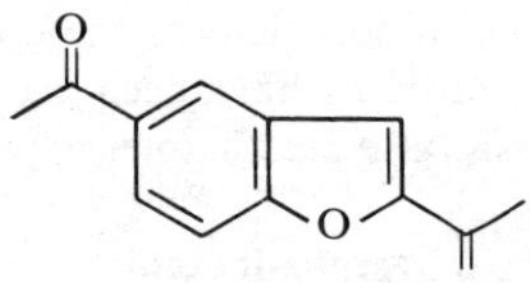

(2)

3 Monoterpenoids

The *cis–trans* isomerization of geraniol and geranyl phosphate has been studied[18] using enzyme systems derived from carrot and peppermint. This system required a flavin coenzyme and light. Partial reduction of the flavin prior to incubation allowed the isomerization reaction to proceed in the absence of light. The biosynthesis of geraniol from mevalonate in *Pelargonium roseum* leaves has been studied.[19] As with other examples, the isoprene unit derived directly from isopentenyl pyrophosphate contains a substantial proportion of the label. A time-course study showed that citronellol was formed after geraniol. For many years there have been suggestions that the amino-acid leucine might act as a terpene precursor, and a low incorporation of L-leucine into linalool has now been observed.[20] In contrast to the mevalonate results, the portion derived from dimethylallyl pyrophosphate was selectively labelled, suggesting that there may indeed be several routes to the DMAPP pool. The formation of linalool and 4-terpineol from [2-^{14}C]mevalonic acid has been observed in *Pinus pinaster* seedlings.[21]

[11] H. Suzuki, K. Oba, and J. Uritani, *Agric. and Biol. Chem. (Japan)*, 1974, **38**, 2053.
[12] R. I. Imblum and V. W. Rodwell, *J. Lipid Res.*, 1974, **15**, 211.
[13] H. M. Hill and L. J. Rogers, *Phytochemistry*, 1974, **13**, 763.
[14] D. Suarez, E. Garcia-Peregrin, and F. Mayor, *Phytochemistry*, 1974, **13**, 1059.
[15] M. R. Downing and E. D. Mitchell, *Phytochemistry*, 1974, **13**, 1419.
[16] J. Edmond and G. Popjak, *J. Biol. Chem.*, 1974, **249**, 66.
[17] T. J. Lin, E. Ramstad, and P. Heinstein, *Phytochemistry*, 1974, **13**, 1809, 1817.
[18] W. E. Shine and W. D. Loomis, *Phytochemistry*, 1974, **13**, 2095.
[19] T. Suga and T. Shishibori, *Bull. Chem. Soc. Japan*, 1973, **46**, 3545.
[20] T. Suga, T. Hirata, T. Shishibori, and K. Tange, *Chem. Letters*, 1974, 189.
[21] A. Machado, E. Garcia-Peregrin, and F. Mayor, *Plant Science Letters*, 1974, **2**, 83.

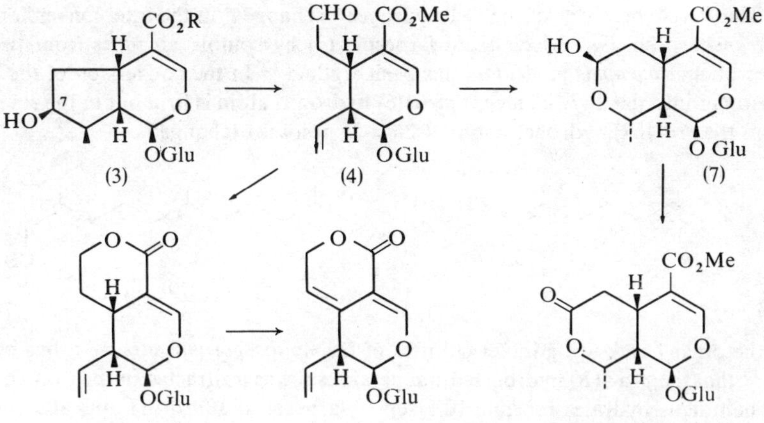

(3) (4) (7) (5) (6) (8)

Full papers have appeared on the biosynthesis of the seco-iridoid glucosides.[22] Tracer experiments revealed that loganin (3;R = Me) was incorporated into gentiopicroside (6), probably *via* secologanin (4) and sweroside (5), and into morroniside (7). Secologanin (4) also acted as a precursor of (7). The biosynthetic pathway to the oleuropein glucosides was examined further using 7-deoxyloganic acid, loganin (3;R = Me), secologanin (4), and kingiside (8). The conversion of loganic acid (3;R = H), formed biosynthetically from mevalonate, into loganin (3;R = Me) and secologanin (4) in *Vinca rosea* has been described.[23]

4 Sesquiterpenoids

Prenyl transferases from a number of sources have been examined in evaluating the regulatory role that they may play in higher isoprenoid biosynthesis. The purification and characterization of two forms of geranyl transferase from *Ricinus communis* have been described.[24] Both systems catalyse the specific formation of *trans*:*trans*-farnesyl pyrophosphate from isopentenyl pyrophosphate and geranyl pyrophosphate. The substrate specificity of prenyl transferase is relatively low with respect to the structure of the allylic pyrophosphate. However, the structural requirement for the condensing partner is more rigorous, and among a range of homologues of isopentenyl pyrophosphate, only 3-ethylbut-3-enyl pyrophosphate and 4-methylpent-4-enyl pyrophosphate react.[25] The reaction of the latter with geranyl pyrophosphate resulted in the exclusive formation of a non-allylic *cis*-homofarnesyl pyrophosphate (see p. 37).

Biogenetic proposals for the origin of many cyclic sesquiterpenoids require the isomerization of the C-2 double bond of all-*trans*-farnesyl pyrophosphate at some

[22] H. Inouye, S. Ueda, K. Inoue, and Y. Takeda, *Chem. and Pharm. Bull* (*Japan*), 1974, **22**, 676.
[23] R. Guarnaccia, L. Botta, and C. J. Coscia, *J. Amer. Chem. Soc.*, 1974, **96**, 7079.
[24] T. R. Green and C. A. West, *Biochemistry*, 1974, **23**, 4720.
[25] K. Ogura, A. Saito, and S. Seto, *J. Amer. Chem. Soc.*, 1974, **96**, 4037.

stage. The stereochemistry of the C-1 hydrogen exchange[26] in the interconversion of *trans*:*trans*-farnesol (9) and *cis*:*trans*-farnesol (10) by soluble enzymes from tissue cultures of *Andrographis paniculata* has been studied.[27] In the conversion of the all-*trans*-isomer into the 2-*cis*-isomer, a *pro*-1(*S*) hydrogen atom is lost, but in the reverse reaction the *pro*-1(*R*) hydrogen atom of 2-*cis*-farnesol is exchanged.

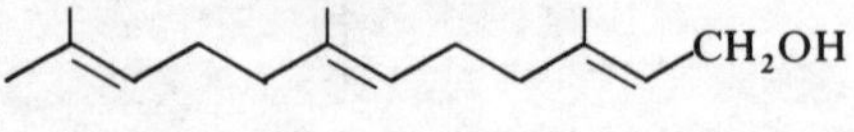

(9)

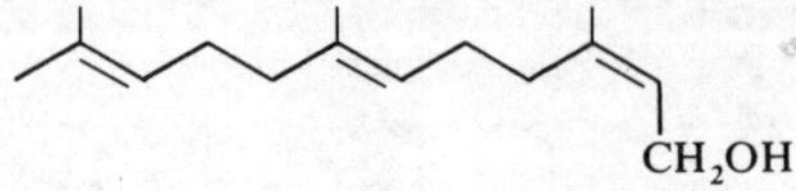

(10)

On the other hand, using intact cultures of *Helminthosporium sativum*, it has been shown[28] that the *pro*-1(*R*) hydrogen atom of farnesol was abstracted in the isomerization. The fungus hydrates racemic 10,11-epoxyfarnesol to afford the optically *minus* metabolites from either of the enantiomers of the substrate.[29] The fungus will also isomerize the C-2 double bond.[30] The full paper has appeared describing the biosynthesis of the triprenyl-phenol siccanin (12) from *trans*-γ-monocyclofarnesyl pyrophosphate (11) and orsellinic acid in a cell-free system from *H. siccans.*[31]

The value of using [$^{13}C_2$]acetate, and observing the coupling patterns in a metabolite, has been illustrated[32] with the triprenyl-phenol ascochlorin (13). Only five out of the six ^{13}C–^{13}C couplings were observed in the terpenoid part of the molecule, confirming that a ^{13}C—^{13}C bond was cleaved in the biosynthesis. The C-14—C-1 coupling

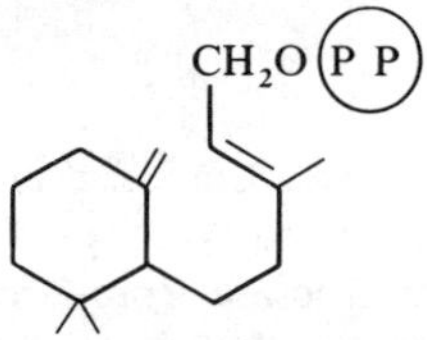

(11)

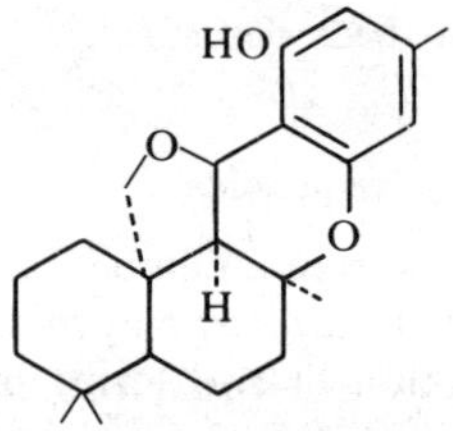

(12)

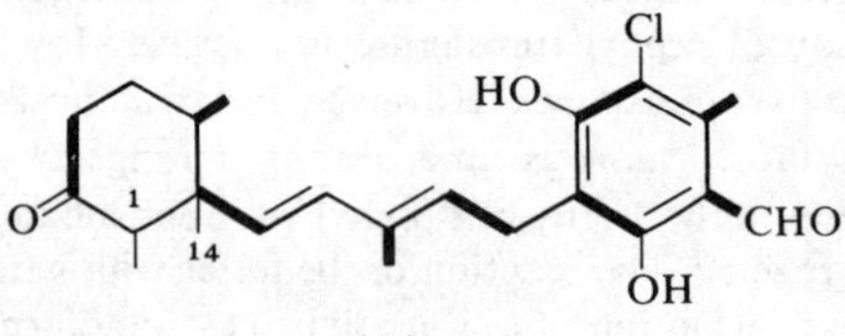

(13)

[26] K. H. Overton and F. M. Roberts, *Phytochemistry*, 1974, **13**, 2741.
[27] K. H. Overton and F. M. Roberts, *J.C.S. Chem. Comm.*, 1974, 385.
[28] K. Imai and S. Marumo, *Tetrahedron Letters*, 1974, 4401.
[29] Y. Suzuki, K. Imai, and S. Marumo, *J. Amer. Chem. Soc.*, 1974, **96**, 3703.
[30] L. W. van Broekhoven, M. W. van Maarschalkwerweerd, R. J. J. Ch. Lousheg, and C. A. Salemink, *Tetrahedron Letters*, 1974, 2909.
[31] K. T. Suzuki and S. Nozoe, *Bio-organic Chemistry*, 1974, **3**, 72.
[32] M. Tanabe and K. T. Suzuki, *J.C.S. Chem. Comm.*, 1974, 445.

was absent in the labelled metabolite, and both the C-14 and C-1 signals appeared as enriched singlets. This established that fission of the C-14—C-1 bond had occurred, as well as a stereospecific migration of the C-14 methyl group to C-6; see p. 236.

The co-occurrence of a number of sesquiterpene hydrocarbons in *Abies balsamea* has been discussed[33] in terms of the cyclization of farnesyl pyrophosphate.

The stereochemistry of the mevalonate labelling pattern of ring C of the 12,13-epoxytrichothecenes calonectrin (16), trichodermol (17), and trichothecolone (18) has been determined.[34] The results are in accord with the folding of farnesyl pyrophosphate (14) to afford the trichothecene (15), and they imply that hydroxylation at C-3 and C-4 has proceeded with retention of configuration. All three *pro*-5(*R*)-mevalonoid hydrogen atoms of all-*trans*-farnesyl pyrophosphate (14) are retained in the formation of the trichothecenes. The hydrogen atom which is introduced at C-1 of farnesyl pyrophosphate on isomerization of the double bond is lost again during the subsequent steps. When trichothecolone (18) was biosynthesized[35] from [2-^{13}C]-mevalonic acid, C-4, C-8, and C-14 were enriched, indicating that farnesyl pyrophosphate was folded in the manner (14).

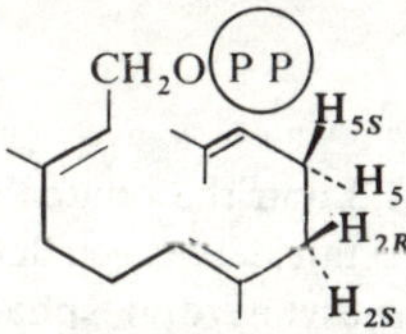

(14)

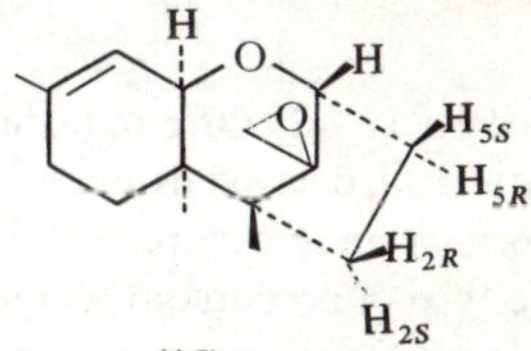

(15)

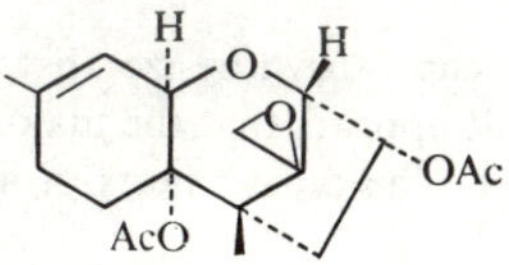

(16)

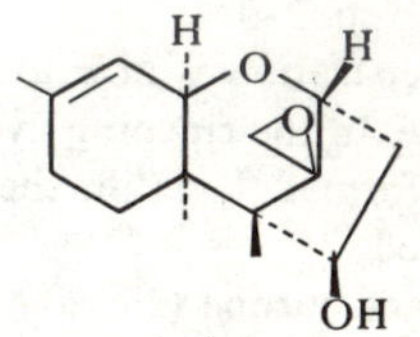

(17)

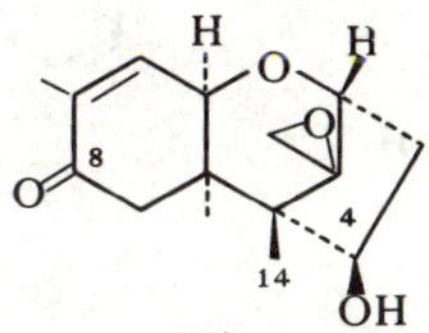

(18)

[33] C. J. Lee, K. Snajberk, and E. Zavarin, *Phytochemistry*, 1974, **13**, 179.
[34] R. Evans, J. R. Hanson, and T. Marten, *J.C.S. Perkin I*, 1974, 857.
[35] J. R. Hanson, T. Marten, and M. Siverns, *J.C.S. Perkin I*, 1974, 1033.

SCIENTIFIC INFORMATION DEPT
BRISTOL-MYERS/EVANSVILLE

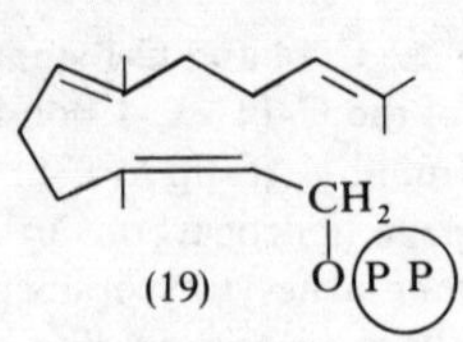

(19)

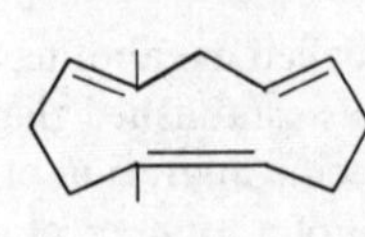

(20)

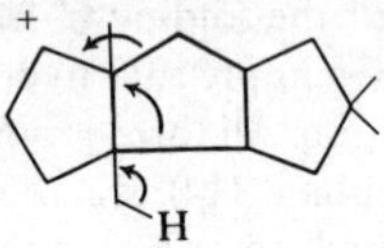

(21)

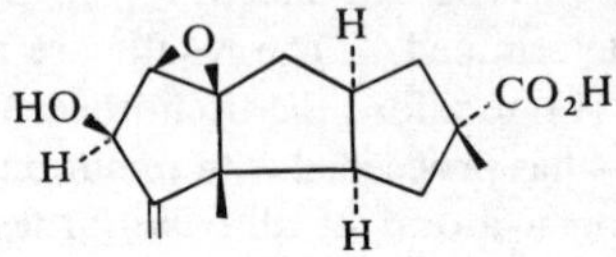

(22)

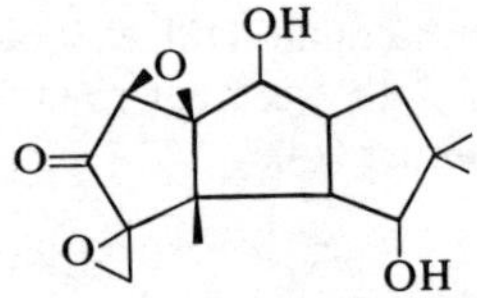

(23)

The carbon-13 labelling patterns of hirsutic acid (22) and the related C-5 ketone complicatic acid, derived from [1-^{13}C]- and [2-^{13}C]-acetate, have been determined.[36] The patterns were in accord with the cyclization of farnesyl pyrophosphate (19) *via* a humulene type of precursor (20) and the carbonium ion (21). In the case of the related coriolins [*e.g.* (23)], an examination of the coupling patterns obtained when [1,2-$^{13}C_2$]-acetate was used as the substrate has shown[37] that this pathway also occurs in this system. The ^{13}C studies of all these sesquiterpenes [(18), (22), (23), *etc.*] are reported fully in Chapter 7, p. 240.

A cell-free preparation is now available that can biosynthesize abscisic acid (24) from mevalonate. In the ripening avocado fruit it appears that the major amount of abscisic acid is formed[38] within the chloroplast. The biosynthesis of abscisic acid has been reviewed.[39]

Dehydroipomeamarone (25) is[40] an intermediate in the biosynthesis of ipomeamarone in the sweet potato that has been infected by *Ceratocystis fimbriata.*

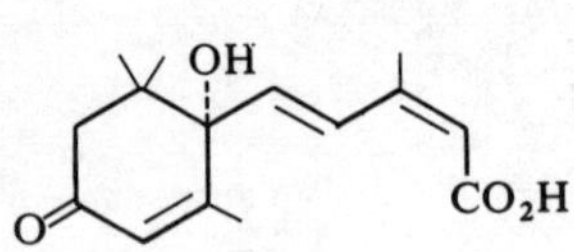

(24)

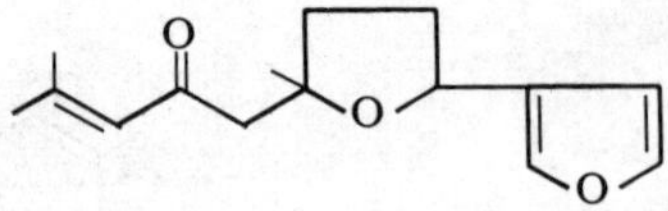

(25)

[36] T. C. Feline, G. Mellows, R. B. Jones, and L. Phillips, *J.C.S. Chem. Comm.*, 1974, 63.
[37] M. Tanabe, K. T. Suzuki, and W. C. Jankowski, *Tetrahedron Letters*, 1974, 2271.
[38] B. V. Milborrow, *Phytochemistry*, 1974, **13**, 131.
[39] B. V. Milborrow in 'Chemistry and Biochemistry of Plant Hormones', ed. V. C. Runeckles, E. Sondheimer, and D. C. Walton, Academic Press, New York, 1974.
[40] I. Oguni and I. Uritani, *Plant Physiol.*, 1974, **53**, 649.

5 Diterpenoids

During germination of *Curcurbita pepo*, the activity of the prenyl transferase that is ultimately responsible for gibberellin biosynthesis, geranylgeranyl pyrophosphate synthetase, increases markedly.[41] The biosynthesis of enmein and oridonin from 15-oxygenated kaurenoids has been examined.[42] *ent*-Kaur-16-en-15-one and the corresponding 15β-alcohol are incorporated but not *ent*-kaur-15-ene or the 15α-alcohol.

One of the major areas of current biosynthetic interest is that of the gibberellin plant hormones. Some recent aspects of their biosynthesis have been reviewed.[43] A detailed analysis has been presented[44] of those metabolites of *Gibberella fujikuroi* which are derived from mevalonic acid. Mutant studies with *G. fujikuroi* have provided some valuable insights into gibberellin biosynthesis. The position of the metabolic block for gibberellin biosynthesis in a particularly useful mutant, B1-41a, has been shown[45] to lie between *ent*-kaur-16-en-19-al (26) and *ent*-kaur-16-en-19-oic acid (27). The full paper has appeared[46] describing the partial synthesis of gibberellin A_{14} aldehyde (28), which is an intermediate in gibberellin biosynthesis. The conversion of

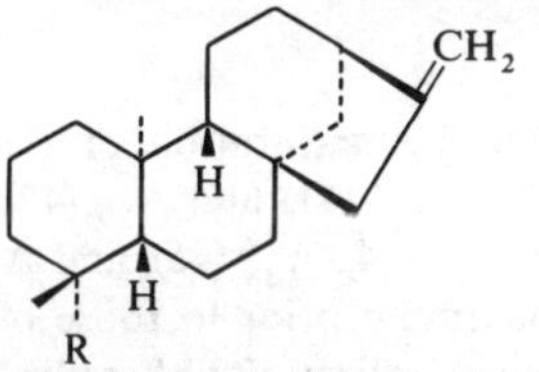

(26) R = CHO
(27) R = CO_2H

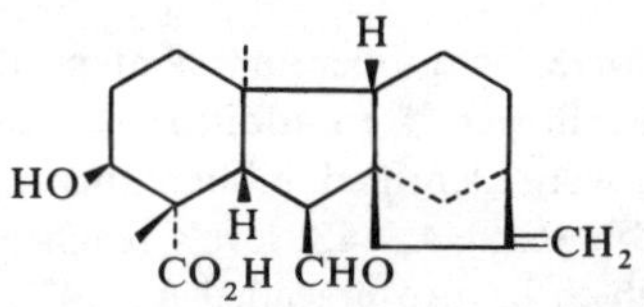

(28)

gibberellin A_{14} into other gibberellins in seedlings of dwarf pea, *Pisum sativum*, has been examined in a time-course study.[47] In this system gibberellin A_{14} (29) was converted into gibberellins A_{18} (30), A_{38} (31), A_{23} (32), A_1 (33), and A_8 (34), suggesting that in this case bridgehead hydroxylation preceded the loss of the angular C-20 group, which is the converse of the sequence found in the fungus. A cell-free system from *Curcurbita maxima* has been found[48] which, in the absence of Mn^{2+}, will convert gibberellin A_{12} aldehyde (35) and gibberellin A_{12} (36) into gibberellins A_{15} (37), A_{24} (38), A_{36} (39), and A_{37} (40). In the presence of Mn^{2+}, gibberellin biosynthesis from mevalonate [*via ent*-kaur-16-ene, *ent*-kaur-16-en-19-ol, *ent*-kaur-16-en-19-oic acid, and *ent*-7α-hydroxykaur-16-en-19-oic acid] is inhibited at the gibberellin A_{12} aldehyde level.

41 T. Shinka, K. Ogura, and S. Seto, *Phytochemistry*, 1974, **13**, 2103.
42 T. Fujita, S. Takao, Y. Nagao, and E. Fujita, *J.C.S. Chem. Comm.*, 1974, 666.
43 J. MacMillan in ref. 39.
44 J. MacMillan and C. M. Wels, *Phytochemistry*, 1974, **13**, 1413.
45 J. R. Bearder, J. MacMillan, C. M. Wels, M. B. Chaffey, and B. O. Phinney, *Phytochemistry*, 1974, **13**, 911.
46 P. Hedden, J. MacMillan, and B. O. Phinney, *J.C.S. Perkin I*, 1974, 587.
47 R. C. Durley, I. D. Railton, and R. P. Pharis, *Phytochemistry*, 1974, **13**, 547.
48 J. E. Graebe, P. Hedden, P. Gaskin, and J. MacMillan, *Phytochemistry*, 1974, **13**, 1433.

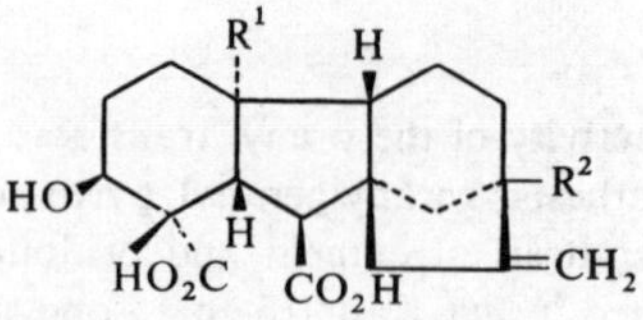

(29) R^1= Me, R^2 = H
(30) R^1= Me, R^2 = OH
(32) R^1= CO_2H, R^2 = OH

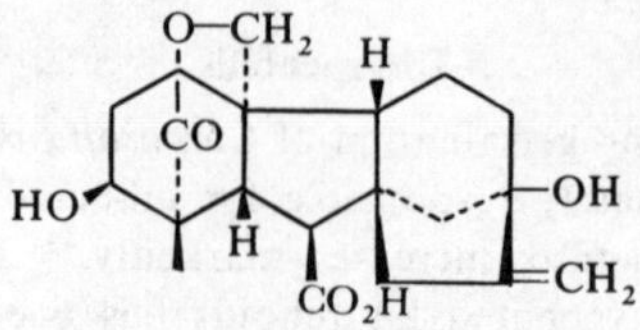

(31)

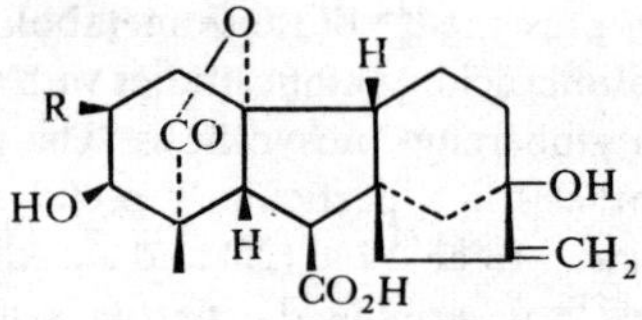

(33) R = H
(34) R = OH

The changes in gibberellin content through the maturation of *Pisum sativum* have been followed.[49] In addition to gibberellins A_{20} (41) and A_{29} (42), which had been previously identified, gibberellins A_9 (43), A_{17} (44), A_{38} (45), and A_{44} (46) were found. Gibberellin A_9 (43) levels reached a maximum prior to those of gibberellin A_{20} (41), which in turn preceded A_{29} (42). The metabolism of gibberellin A_9 in dwarf

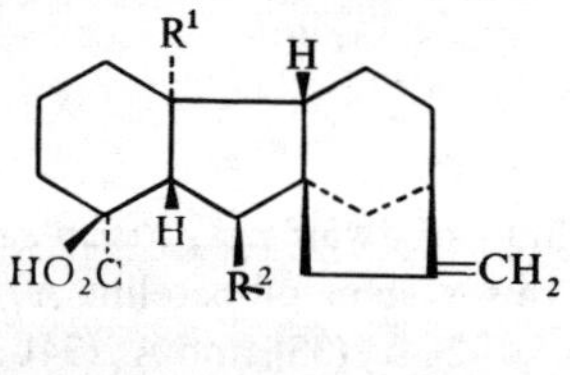

(35) R^1= Me, R^2 = CHO
(36) R^1= Me, R^2 = CO_2H
(38) R^1= CHO, R^2 = CO_2H

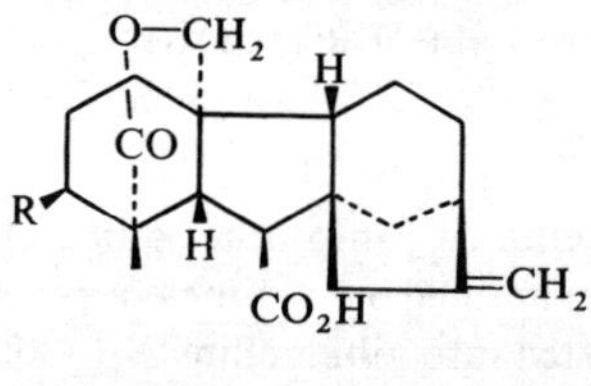

(37) R = H
(40) R = OH

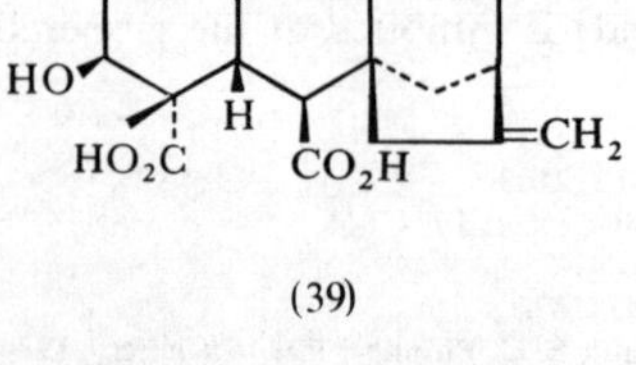

(39)

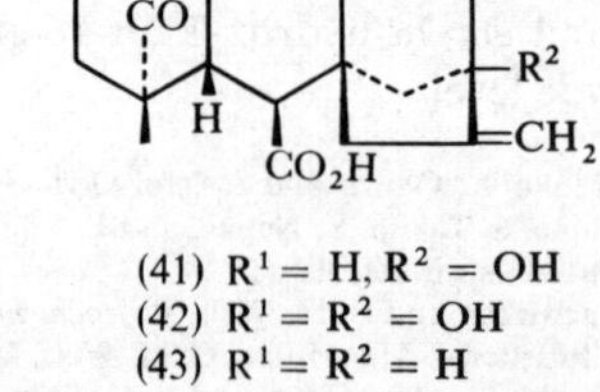

(41) R^1 = H, R^2 = OH
(42) R^1 = R^2 = OH
(43) R^1 = R^2 = H

[49] V. M. Frydman, P. Gaskin, and J. MacMillan, *Planta*, 1974, **118**, 123.

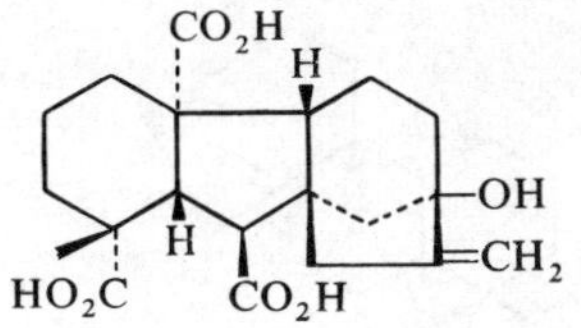

(44)

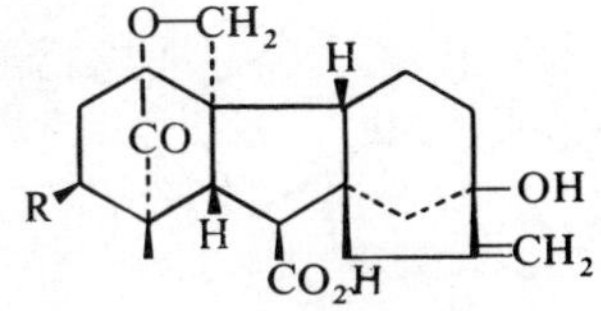

(45) R = OH
(46) R = H

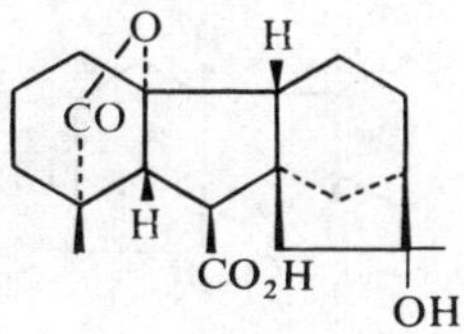

(47)

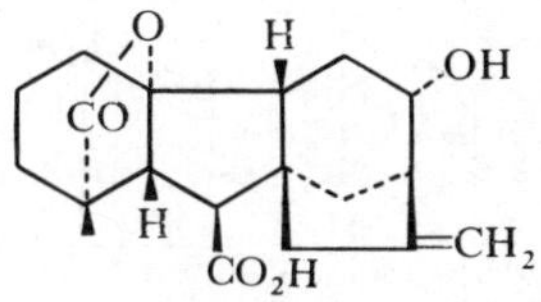

(48)

peas grown in the dark and in light has been studied.[50] The gibberellin A_9 was converted into gibberellin A_{10} (47), dihydrogibberellin A_{31} (48), and gibberellins A_{20} and A_{29}. The conversion of gibberellin A_{20} into gibberellin A_{29} appeared to be inhibited by light. This step has been examined in germinating seeds.[51]

The translocation and metabolism of gibberellic acid have been studied.[52] The decomposition products gibberellenic acid and isogibberellin A_3, together with 3-*O*-β-D-glucosylgibberellic acid, were identified. The conversion of gibberellin A_1 (33) into gibberellin A_8 (34) has been examined in a cell-free system.[53]

Other kaurenoid diterpenes are found in a number of higher plants. The microbiological transformation by *G. fujikuroi* of some of the more readily available of these has been studied, on the one hand with the aim of preparing novel gibberellins or intermediates which could be converted into them, and on the other hand with the aim of blocking gibberellin biosynthesis. The conversion of *ent*-kaur-16-enes into gibberellic acid was blocked by esterification of C-19 as succinate ester and by epimerization at C-3. Thus *ent*-3β-hydroxykaur-16-en-19-yl succinate (49) gave a good conversion into the 7β-hydroxy derivative in *G. fujikuroi*.[54] An interesting feature from the point of view of gibberellin biosynthesis is that blocking of the C-19 oxygen function with an ester grouping inhibited ring-contraction. The succinate ester of *ent*-kaura-2,16-dien-19-ol (50) is more labile, and a series of gibberellins, including 13-deoxygibberellin A_5 (51) and 13-deoxygibberellin A_6 (52), were produced. At the C-20 level, the C-2 olefins corresponding to gibberellins A_{12}, A_{24}, and A_{25} [compounds (53)—(55), respectively] were produced,[55] together with the 2,3-epoxide of gibberellin A_{12}

[50] I. D. Railton and D. M. Reid, *Plant Science Letters*, 1974, **2**, 157; 1974, **3**, 207; 1974, **3**, 303; I. D. Railton, R. C. Durley, and R. P. Pharis, *Plant Physiol.*, 1974, **54**, 6.
[51] I. D. Railton, N. Murofushi, R. C. Durley, and R. P. Pharis, *Phytochemistry*, 1974, **13**, 793.
[52] Y. Asakawa, K. Tamari, K. Inoue, and J. Kaji, *Agric. and Biol. Chem. (Japan)*, 1974, **38**, 713; Y. Asakawa, K. Tamari, A. Shoji and J. Kaji, *ibid.*, p. 719.
[53] L. Rappaport and R. J. Patterson, *Planta*, 1974, **119**, 183.
[54] P. R. Jefferies, J. R. Knox, and T. Ratajczak, *Phytochemistry*, 1974, **13**, 1423.
[55] H. J. Bakker, I. F. Cook, P. R. Jefferies, and J. R. Knox, *Tetrahedron*, 1974, **30**, 3631.

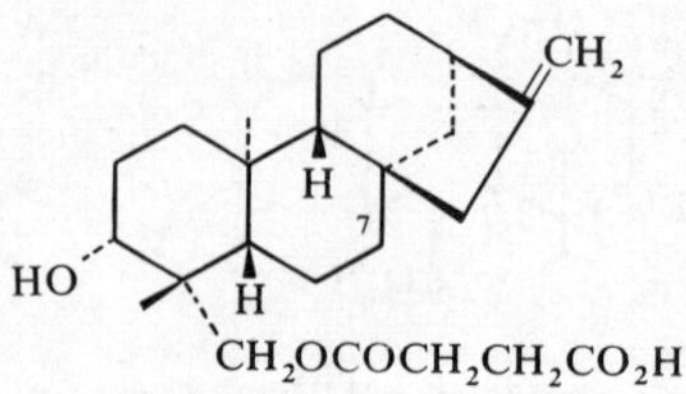

(49)

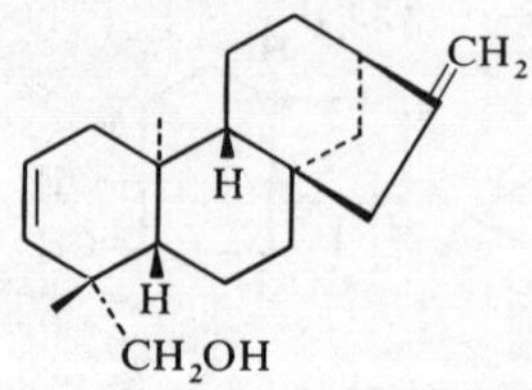

(50)

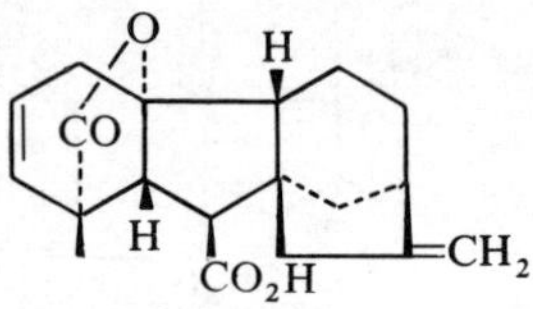

(51)

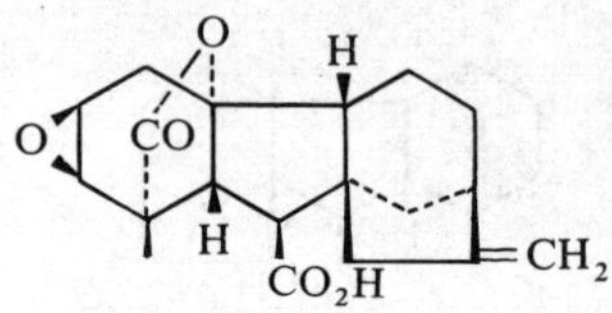

(52)

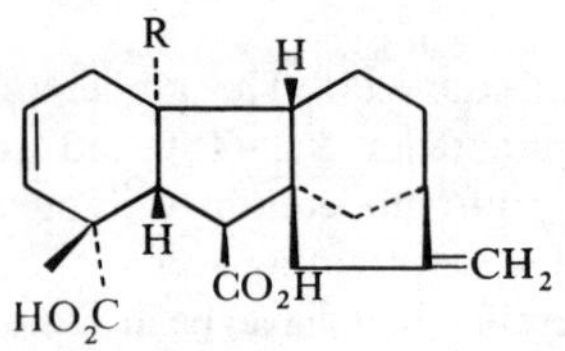

(53) R = CH_3
(54) R = CHO
(55) R = CO_2H

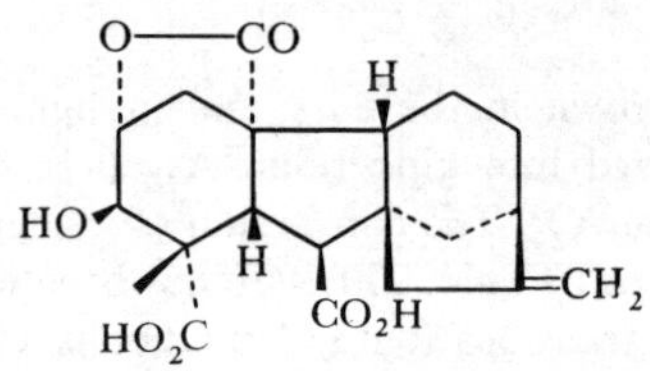

(56)

and a lactone (56) arising from the epoxide. Hydroxylation of the 2′-ethoxycarbonylethyl ester of *ent*-kaur-16-en-19-oic acid has been utilized[56] for an efficient microbiological synthesis of *ent*-7α-hydroxykaur-16-en-19-oic acid. The full paper describing the preparation of fluorogibberellic acid and fluorogibberellin A_9, by the microbiological transformation of fluorogibberellin A_{12} aldehyde with *G. fujikuroi*, has appeared.[57] The transformation of *ent*-kaur-16-enes by other micro-organisms with the object of preparing gibberellin intermediates has also been described.[55, 58] Thus 7α-hydroxykaurenolide was hydroxylated[58] at the 11α- and 13-positions by *Rhizopus arrhizus*.

6 Sesterterpenoids

A useful review has appeared,[59] including a section on the biosynthesis of these compounds.

[56] K. D. Croft, E. L. Ghisalberti, P. R. Jefferies, J. R. Knox, T. J. Mahoney, and P. N. Sheppard, *Tetrahedron*, 1974, **30**, 3663.
[57] J. H. Bateson and B. E. Cross, *J.C.S. Perkin I*, 1974, 1131.
[58] J. R. Hanson, G. Savona, and M. Siverns, *J.C.S. Perkin I*, 1974, 2001.
[59] G. A. Cordell, *Phytochemistry*, 1974, **13**, 2343.

3
Biosynthesis of Triterpenoids, Steroids, and Carotenoids

BY L. J. MULHEIRN

1 Introduction

The 1974 literature on biosynthesis of higher terpenoids is surveyed in this Report. The scope of the review is basically similar to that of the corresponding chapters[1—3] in Volumes 1—3. The reader is referred to earlier volumes in this series for background information on many of the topics which are updated here. One again, a major theme of work reported in 1974 has concerned the operation and regulation of individual enzymic processes of the main pathways. The sterospecificity of enzyme reactions, and its implications, have been discussed and graphically illustrated by reference to the stereochemistry of reactions involved in conversion of acetate into cholesterol.[4] The current state of knowledge on the enzymes which mediate these reactions has also been reviewed.[5]

2 Mevalonic Acid

Considerable attention has been given to regulatory effects in cholesterol synthesis, particularly in various mammalian tissues, in which the control steps lie between acetyl-CoA and mevalonic acid. The rate of cholesterogenesis in rat liver slices has been measured using acetate and octanoate as sources of C_2 units.[6] Acetate is incorporated into a cytosolic pool of acetyl-CoA, which is subsequently used for sterol synthesis but is not in equilibrium with the mitochondrial pool of acetyl-CoA involved in ketogenesis. The maximal rate of acetate incorporation into cholestrol is significantly lower than that of octanoate, suggesting that activation of acetate may be rate-limiting in this situation. Acetate incorporation into sterols is also inhibited by some synthetic progestational and oestrogenic compounds, though their site of action was not determined.[7] This observation may explain the inefficiency of acetate as substrate for sterol biosynthesis in placental tissue.

[1] H. H. Rees and T. W. Goodwin, in 'Biosynthesis', ed. T. A. Geissman (Specialist Periodical Reports), The Chemical Society, London, 1972, Vol. 1, p. 59.
[2] H. H. Rees and T. W. Goodwin, in 'Biosynthesis', ed. T. A. Geissman (Specialist Periodical Reports), The Chemical Society, London, 1973, Vol. 2, p. 16.
[3] H. H. Rees and T. W. Goodwin, in 'Biosynthesis', ed. T. A. Geissman (Specialist Periodical Reports), The Chemical Society, London, 1974, Vol. 3, p. 14.
[4] J. W. Cornforth, *Tetrahedron*, 1974, **30**, 1515.
[5] J. L. Gaylor in 'Biochemistry of Lipids', ed. T. W. Goodwin, MTP International Review of Science, Biochemistry Series, Butterworths, London, 1974, Vol. 4, p. 1.
[6] J. M. Dietschy and J. D. McGarry, *J. Biol. Chem.*, 1974, **249**, 52.
[7] V. V. Patwardhan and A. Lanthier, *Canad. J. Biochem.*, 1974, **52**, 28.

Alk-3-ynoyl-CoA esters have been shown to act as potent inhibitors of pig heart cytoplasmic thiolase.[8] But-3-ynoyl-CoA is more active than higher homologues. Since acetoacetyl-CoA, which is required for cholesterol biosynthesis, could be synthesized from acetyl-CoA by hepatic cytoplasmic β-ketoacyl-CoA thiolase (E.C.2.3.1.9), the acetylenic esters have been tested as inhibitors of cholesterol biosynthesis.[9] Complete inhibition of acetate incorporation into sterols was achieved, while mevalonate conversion into cholesterol and the synthesis of fatty acids from acetate were not reduced. The results are consistent with inhibition of thiolase, and they also complement recent evidence against a second pathway of cholesterol biosynthesis involving condensation of acetyl-CoA and malonyl-CoA.[3]

The condensation of (*R*)-[$^2H_1$,$^3H_1$]acetyl-CoA (1) and its (*S*)-enantiomer with acetoacetyl-CoA (2) on 3-hydroxy-3-methylglutaryl-CoA (HMG-CoA) synthase has been shown to be sterospecific, involving an intramolecular hydrogen isotope effect.[10] Analysis of the stereochemical course of the condensation was achieved by conversion of the product into mevalonate and incorporation into cholesterol. The orientation of tritium atoms at C-1, -7, and -15 of cholesterol (5), derived from C-2 of mevalonate and hence from the original chiral acetate molecule, was assayed by microbial conversion of the sterol into androsta-1,4-diene-3,17-dione (6). Assuming a normal isotope effect, the results demonstrate that the condensation proceeds with inversion of

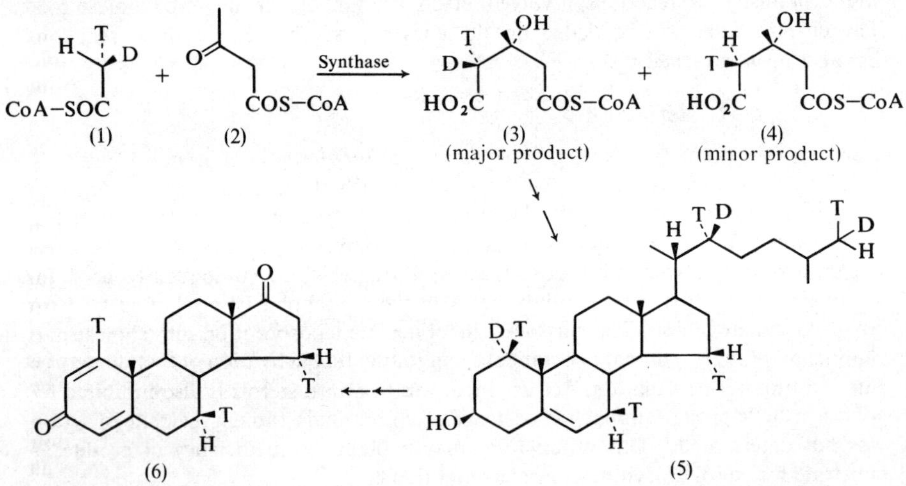

Scheme 1

configuration at the methyl group of acetate. Scheme 1 shows analysis of the major radiolabelled product from (*S*)-acetyl-CoA (1). The condensation on HMG-CoA synthase from yeast has been interpreted as involving acylation of the enzyme by acetyl-CoA (probably *via* a thioester linkage) followed by condensation with aceto-

[8] P. C. Holland, M. G. Clark, and D. P. Bloxham, *Biochemistry*, 1973, **12**, 3309.
[9] D. P. Bloxham, *Biochem. Soc. Trans.*, 1974, **2**, 32.
[10] J. W. Cornforth, G. T. Phillips, B. Messner, and H. Eggerer, *European J. Biochem.*, 1974, **42**, 591.

acetyl-CoA.[11] The covalently bound HMG-CoA derivative is then rapidly hydrolysed. The antibiotic cerulenin is reported to be a specific inhibitor of this enzyme in yeast.[12]

Further factors involved in control of the circadian rhythm of HMG-CoA reductase in rat liver have been defined. The characteristic five- to eight-fold increase in activity around midnight involves rapid enzyme synthesis, which ceases during the subsequent decline in reductase activity. Epinephrine, administered during periods of basal or declining mevalonate formation, produces a rapid increase in reductase activity.[13] Stimulation of enzyme synthesis is inferred from the countervailing effect of actinomycin D. Administration of cholesterol to rats at the time of peak activity produces a significantly lower basal level 4 to 8 hours later, suggesting possible stimulation of enzyme breakdown or inactivation by the sterol. Decreased microsomal reductase activity also coincides with increased cholesterol ester content, though the significance of this correlation is still unclear.

Suppression of HMG-CoA reductase in cultured human fibroblasts by cholesterol and a large group of related compounds has been studied.[14] Eleven compounds proved more potent than cholesterol, the only common structural requirement for activity being the presence of a free 3-hydroxy-group. Ketone functions at C-6, -7, or -25 enhance activity. In fibroblasts which are genetically resistant to repression of this enzyme by cholesterol, activity of keto-sterols was similar to that in normal cells. Suppression of cholesterol synthesis by 7-ketocholesterol resulted in marked inhibition of cell growth, which was prevented by addition of mevalonate or cholesterol, but not acetate, suggesting that HMG-CoA reduction is rate-limiting for both

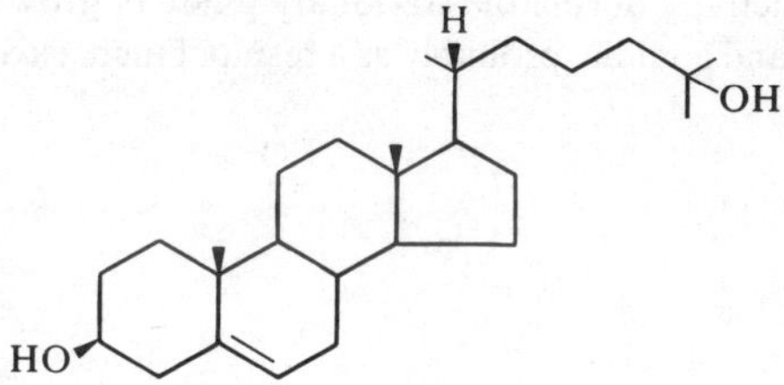

(7)

cholesterol formation and cell growth. Similar results are reported on suppression of the enzyme in mouse foetal liver cells and L-cell cultures by side-chain-hydroxylated steroids.[15] Inhibition decreases with the position of hydroxylation in the sequence 25 > 20α > kryptogenin > 22α > 22β; 25-hydroxycholesterol (7) is 20 times as active as cholesterol. Pregn-5-ene-3,20-diol and its 20-methyl and -ethyl homologues are ineffective.

The interaction of hormones which regulate HMG-CoA reductase in rat liver has been examined. While insulin stimulates the diurnal variation of enzyme activity,

[11] B. Middleton and P. K. Tubbs, *Biochem. J.*, 1974, **137**, 15.
[12] T. Ohno, T. Kesado, J. Awaya, and S. Omura, *Biochem. Biophys. Res. Comm.*, 1974, **57**, 1119.
[13] P. A. Edwards and R. G. Gould, *J. Biol. Chem.*, 1974, **249**, 2891.
[14] M. S. Brown and J. L. Goldstein, *J. Biol. Chem.*, 1974, **249**, 7306.
[15] A. W. Kandutsch and H. W. Chen, *J. Biol. Chem.*, 1974, **249**, 6057.

glucagon and cyclic AMP (adenosine 3′,5′-cyclic monophosphate) are strongly inhibitory.[3] L-Tri-iodothyronine stimulates reductase activity in hypophysectomized rats to 3—4 times the normal maximal level.[16] The markedly lower response in normal and hypothyroid rats suggests that at least one other hormone subject to pituitary regulation may be involved in control of the enzyme synthesis or activity. This view of regulation by hormone interplay is supported by the observation that both insulin and L-tri-iodothyronine are required to stimulate reductase activity in hypophysectomized diabetic rats.[16] Also, the rapid stimulation by insulin and the delayed response produced by L-tri-iodothyronine are both effectively blocked by either glucagon or hydrocortisone, indicating that both inhibitors act at sites controlling enzyme synthesis. Adrenalectomy is reported to have no effect on cholesterol biosynthesis in rat liver.[17]

The similarity of kinetic, electrophoretic, and stability properties of solubilized HMG-CoA reductase from normal rat liver and hepatoma[18] is consistent with the view that the absence of feedback inhibition by cholesterol in the hepatoma results not from changes in the enzyme but from failure of the sterol to control synthesis of the enzyme. The solubilized enzyme from liver exhibits reversible cold-inactivation.[19] This is the first time that this property has been found in an enzyme of microsomal origin, although it is not found in the microsome-bound enzyme. Kinetic data suggest that the active form of the solubilized enzyme resembles the bound enzyme, while protein-concentration effects indicate that enzyme–subunit interactions have a marked influence on activity.

Microsomal HMG-CoA reductase from *Neurospora crassa* requires NADPH, and shows maximal activity during the stationary phase of growth and initial formation of aerial hyphae and conidia, probably as a result of increased enzyme synthesis.[20]

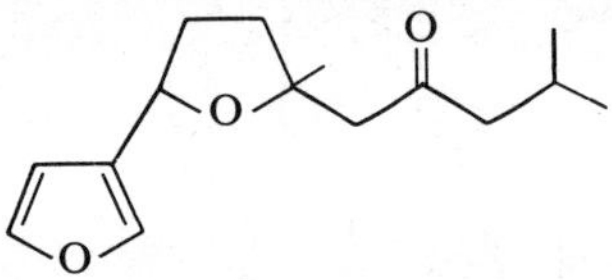

(8)

In contrast to *Phycomyces*, no feedback inhibition by ergosterol or β-carotene was detected. The activity of the enzyme in the root tissue of sweet potato is low.[21] Fungal infection promotes a rapid increase in activity in adjacent non-infected tissue, coinciding with production of the terpenoid furan ipomeamarone (8).

The kidney has been pinpointed as the major site of *in vivo* metabolism of circulating mevalonate, producing squalene and lanosterol rather than cholesterol.[22] Examination

[16] R. E. Dugan, G. C. Ness, M. R. Lakshmanan, C. M. Nepokroeff, and J. W. Porter, *Arch. Biochem. Biophys.*, 1974, **161**, 499.
[17] F. O. Nervi and J. M. Dietschy, *Biochim. Biophys. Acta*, 1974, **369**, 351.
[18] M. S. Brown, S. E. Dana, and M. D. Siperstein, *J. Biol. Chem.*, 1974, **249**, 6585.
[19] R. A. Heller and R. G. Gould, *J. Biol. Chem.*, 1974, **249**, 5254.
[20] R. L. Imblum and V. W. Rodwell, *J. Lipid Res.*, 1974, **15**, 211.
[21] H. Suzuki, K. Oba, and I. Uritani, *Agric. and Biol. Chem. (Japan)*, 1974, **38**, 2053.
[22] K. H. Hellström, M. D. Siperstein, L. A. Bricker, and L. J. Luby, *J. Clin. Invest.*, 1973, **52**, 1303.

of specific tissues in rat and dog reveals that most of the activity resides in the glomeruli of the renal cortex.[23] The fate of mevalonate in various tissues has been examined by subcutaneous injection of [2-^{14}C]mevalonate in developing rats.[24] Fatty acids of brain, spinal cord, and skin are rapidly labelled, while less extensive incorporation is observed in kidney and lung tissue, and no labelling of fatty acids is detected in the liver. The observation that the product of mevalonate breakdown is diluted in a large precursor pool before incorporation into fatty acids suggests that the ketone bodies acetoacetate and β-hydroxybutyrate are labelled from mevalonate

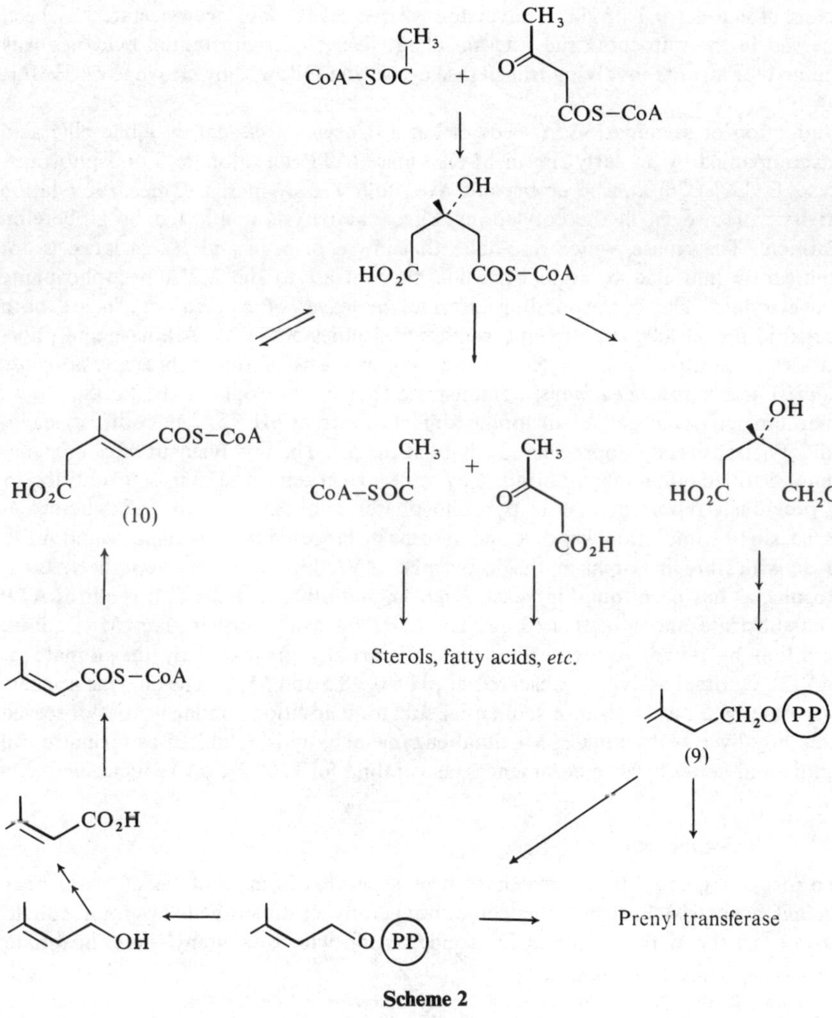

Scheme 2

[23] P. Raskin and M. D. Siperstein, *J. Lipid Res.*, 1974, **15**, 20.
[24] J. Edmond and G. Popják, *J. Biol. Chem.*, **249**, 66.

in ectodermal tissues. Absence of the pathway in liver is thought to result from lack of the 3-oxoacid:succinyl-CoA transferase [E.C.2.8.3.5] required for conversion of acetoacetate into its CoA derivative. A shunt mechanism is proposed for mevalonate conversion into ketone bodies (Scheme 2) involving hydrolysis of dimethylallyl pyrophosphate (9) (which could accumulate as a result of prenyl transferase inhibition) followed by oxidation and carboxylation to give *trans*-3-methylglutaconyl-CoA (10). Support for the overall scheme comes from studies of the fate of labelled β-hydroxybutyrate, acetoacetate, acetate, and mevalonate in developing rats.[25] Mevalonate is a good substrate for sterol biosynthesis only in liver and kidney, while β-hydroxybutyrate is the preferred substrate for both sterol and fatty acid production in the three organs of ectodermal origin. Conversion of free HMG into acetoacetate has been observed in the mitochondrial fraction of rat liver.[26] Circumstantial evidence was obtained for a route involving transferase activation followed by cleavage on HMG-CoA lyase.

Induction of germination in seeds of hazel (*Corylus avellana*) by gibberellic acid is accompanied by an early rise in MVA kinase (ATP:mevalonate 5-phosphotransferase, E.C.2.7.1.36) level in embryonic axes, followed by increased decarboxylation activity.[27] However, in the cotyledons, kinase activity is unaffected by gibberellin treatment. The kinase, which resembles that in castor bean and *Hevea* latex, is not significantly inhibited by iodoacetamide, in contrast to the MVA pyrophosphate decarboxylase. The corresponding enzymes in leaves of *Nepeta cataria* are both located in the 30 000*g* supernatant fraction.[28] Studies of the MVA kinase and phosphatases in green leaves, cotyledons, and chloroplasts of french beans (*Phaseolus vulgaris*) and maize (*Zea mays*) demonstrate that in chloroplasts the kinase shows maximum activity at pH 7.5, dropping almost to zero at pH 5.5.[29] In contrast, leaves and cotyledons retain appreciable activity at pH 5.5. The involvement of isoenzymes cannot be ruled out. Kinase inhibition by farnesol, geraniol, and sterols, in addition to the previously reported effect of pyrophosphates, is observed at pH 7.5, whereas at pH 5.5, slight stimulation was detected in some instances. Acid phosphatase and ATPase activities are maximal at the lower pH. MVA kinase in *Neurospora crassa* is cytosolic, as has been found in yeast, *Euglena*, and other systems.[20] It requires ATP as co-substrate and is activated preferentially by Mg^{2+} rather than Mn^{2+} ions. Inhibition by *p*-hydroxymercuribenzoate is partially prevented by mevalonate or MgATP. Optimal activity is observed at pH 8.0—8.5 and 55 °C. The enzyme appears to operate by a mechanism of sequential substrate addition similar to that observed in the hog-liver MVA kinase. Maximal enzyme activity is exhibited as formation of conidia and aerial hyphae commences, as is found for HMG-CoA reductase.

3 Squalene

Two forms of geranyl transferase have been separated from seedlings of castor bean (*Ricinus communis*). Both enzymes can utilize geranyl or dimethylallyl pyrophosphate, but not farnesyl pyrophosphate, for condensation with isopentenyl pyrophosphate.

[25] J. Edmond, *J. Biol. Chem.*, 1974, **249**, 72.
[26] R. Deana, R. Meneghello, L. Manzi, and C. Gregolin, *Biochem. J.*, 1974, **138**, 481.
[27] P. R. Shewry, N. J. Pinfield, and A. K. Stobart, *Phytochemistry*, 1974, **13**, 341.
[28] M. R. Downing and E. D. Mitchell, *Phytochemistry*, 1974, **13**, 1419.
[29] H. M. Hill and L. J. Rogers, *Phytochemistry*, 1974, **13**, 763.

at pH 6.8 and a requirement for Mg^{2+} ions. Variations of apparent *Km* with protein concentration may reflect a modulation of activity by reversible protein interactions.[30] The activity of farnesyl pyrophosphate synthetase in seeds of *Cucurbita pepo* decreases sharply after the early stages of germination. In contrast, geranylgeranyl pyrophosphate synthetase activity increases during this period.[31]

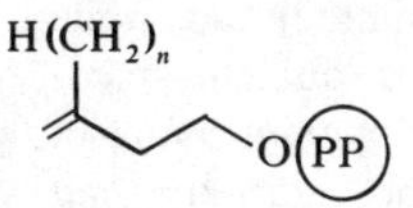

(11) a; $n = 1$
b; $n = 2$

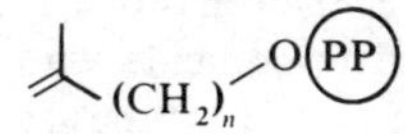

(12) a; $n = 1$
b; $n = 2$
c; $n = 3$

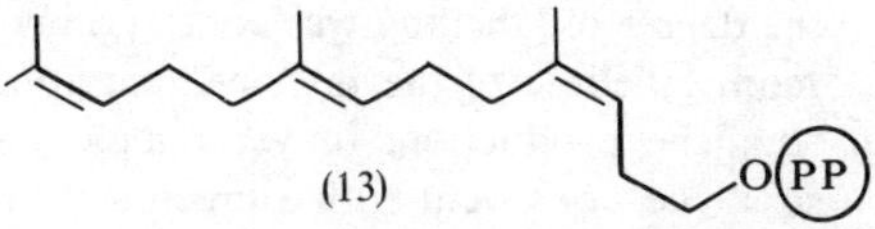

(13)

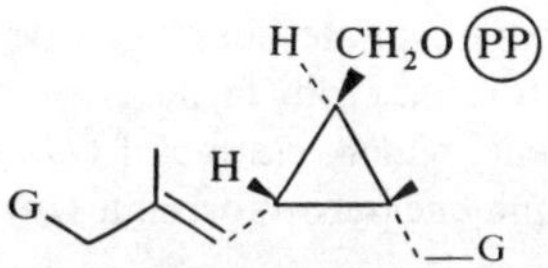

(14)

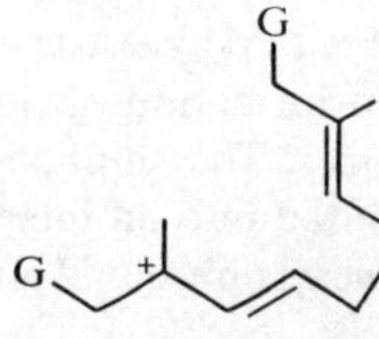

(15)

Examination of the substrate specificity of prenyltransferase (farnesyl pyrophosphate synthetase, E.C.2.5.1.1) has been extended using homologues of isopentenyl pyrophosphate (11a). Previous results demonstrated that the active site of the enzyme imposes stringent limitations on the size of the alkyl group, since only 3-ethylbut-3-enyl pyrophosphate (11b) was active as a substrate, in the absence of isopentenyl pyrophosphate, giving homologues of *trans,trans*-farnesol.[32] Chain elongation in the series (12a—c) demonstrated that only (12b) is capable of replacing the natural substrate of transferase from pig liver.[33] However, the product of condensation with geranyl pyrophosphate was shown to be exclusively the 2-*cis*,6-*trans*-homofarnesyl pyrophosphate (13). These observations suggest that the site occupied by the methyl group of (11a) is capable of accepting a bulkier group without changing the substrate conformation, while the longer chain of (12b) can only be accommodated by twisting the molecule sufficiently to permit the pyrophosphate moiety to occupy its normal position. This hypothesis suggests that removal of the 3-*pro-R* hydrogen atom of (12b) should be observed, in contrast to loss of the 2-*pro-R* proton of isopentenyl pyrophosphate, though this remains to be tested. The failure of the homologue (12b) to condense with dimethylallyl pyrophosphate lends support to the theory that two separate sites are involved in formation of geranyl and farnesyl pyrophosphates in pig liver.

The position of presqualene pyrophosphate (14) in the biosynthetic sequence from farnesol to squalene has been clarified.[34] Earlier studies had involved isolation of this

[30] T. R. Green and C. A. West, *Biochemistry*, 1974, **13**, 4720.
[31] T. Shinka, K. Ogura, and S. Seto, *Phytochemistry*, 1974, **13**, 2103.
[32] T. Koyama, K. Ogura, and S. Seto, *Chem. Letters*, 1973, 401.
[33] K. Ogura, A. Saito, and S. Seto, *J. Amer. Chem. Soc.*, 1974, **96**, 4037.
[34] F. Muscio, J. P. Carlson, L. Kuehl, and H. C. Rilling, *J. Biol. Chem.*, 1974, **249**, 3746.

compound from enzyme preparations lacking reduced pyridine nucleotides. This led to the suggestion that the metabolite might not be a normal product of farnesyl pyrophosphate coupling but rather the result of a reversible side-reaction, activated in the absence of cofactors and possibly involving stabilization of the true intermediate carbonium ion (15).[3] Evidence for the direct involvement of (14) in squalene biosynthesis is now provided by its isolation from intact rat liver and from a yeast microsome system containing NADPH.[34] Also, pulse labelling of rat liver with mevalonate gave rapid conversion into (14) (maximal at 0.5—1 minute) and, while the degree of labelling was considerably lower than that observed in squalene, maximum labelling of the intermediate appeared to coincide with the maximal rate of squalene production. In yeast microsomes, (14) is synthesized more rapidly than squalene, consistent with a precursor–product relationship.

The reported copper-catalysed cyclopropanation of olefinic compounds by sulphur ylides[35] has led to the suggestion that both cyclopropanation of olefins by *S*-adenosylmethionine and formation of presqualene pyrophosphate could involve analogous enzymic processes. The sulphonium salt (16) formed from farnesyl pyrophosphate might be converted into an intermediate carbene, which, *via* metal-catalysed attack on a second farnesyl unit, could give rise to presqualene pyrophosphate (14) (Scheme 3).

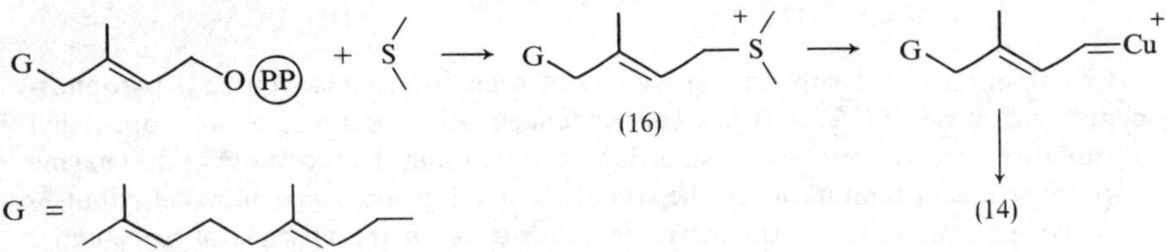

Scheme 3

The mechanism by which presqualene pyrophosphate is converted into squalene has been examined using model carbonium ion reactions.[36] The results indicate[36,37] that, in the postulated sequence (17)—(20), rearrangement of the cyclobutyl carbonium ion (18) to squalene (20) is energetically favoured, but that regiospecific ring expansion of the initial carbonium ion (17) is less favoured and may proceed only when alternative routes are blocked in an enzymic environment. The chemical properties of the cations and the stereochemical criteria imposed by earlier studies of squalene formation have been accommodated in the sequence outlined in Scheme 4.[36] It is proposed that, on the enzyme, the pyrophosphate anion acts as a template for the rearrangements, maintaining proximity of the intermediate carbonium ions and prohibiting routes involving energetically unfavourable charge separation. The orientation of the NADPH molecule could also help to control the rearrangement since the correct orientation of substrate and cofactor for hydrogen transfer, with overall inversion of configuration, would only be achieved in ion (19). It is calculated that the sequence

[35] T. Cohen, G. Herman, T. M. Chapman, and D. Kuhn, *J. Amer. Chem. Soc.*, 1974, **96**, 5627.
[36] C. D. Poulter, O. J. Muscio, and R. J. Goodfellow, *Biochemistry*, 1974, **13**, 1530.
[37] C. D. Poulter, *J. Agric. Food. Chem.*, 1974, **22**, 167.

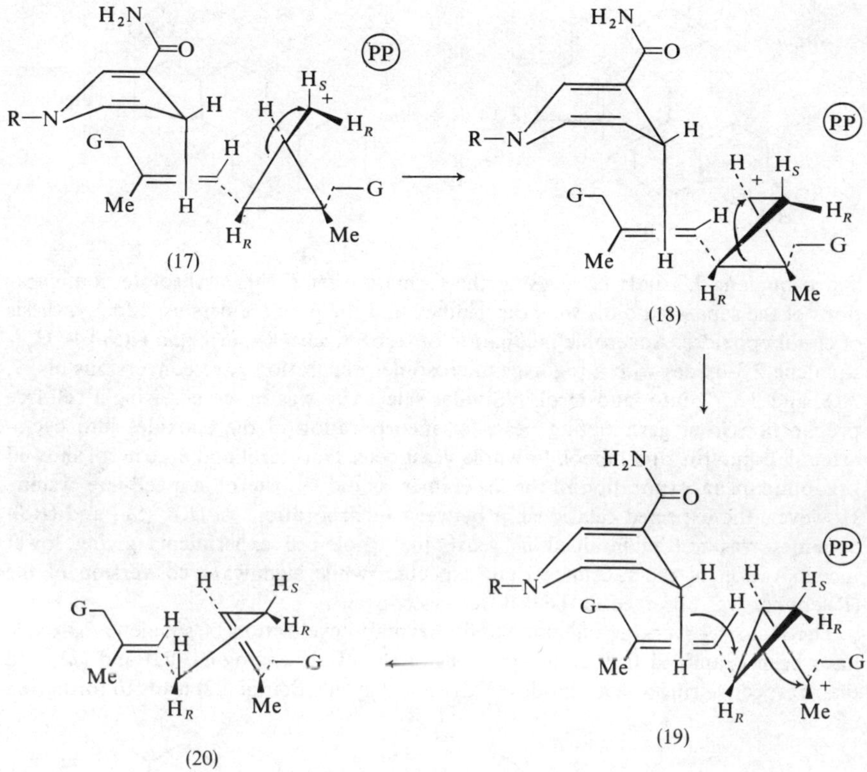

[(PP) = pyrophosphate]

Scheme 4

would involve average movements in any direction within the enzyme cavity of only 1.5 Å, which is approximately the same as the overall size difference between squalene and its precursor.

4 Formation and Cyclization of Squalene 2,3-Oxide

Enzyme preparations of brain exhibit low conversion of mevalonate into cholesterol and a build-up of squalene. It has now been shown[38] that epoxidase activity in brain is comparable to that in liver, and is not a control point for cholesterol synthesis. The *in vitro* results do not reflect the true picture in intact brain.

While the role of squalene 2,3-oxide as the precursor of most 3-oxygenated triterpenes is well established,[1] no definitive evidence had been provided as to the stereochemistry of the enzymically active isomer. The assumption that the natural product is the (3*S*)-isomer has now been confirmed by resolution of the enantiomeric di-

[38] D. W. Vidrine and H. J. Nicholas, *Biochem. Biophys. Res. Comm.*, 1974, **59**, 340.

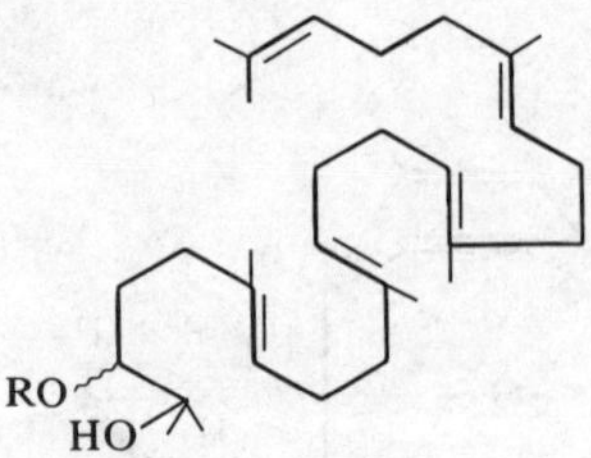

(21) a; R = H
b; R =

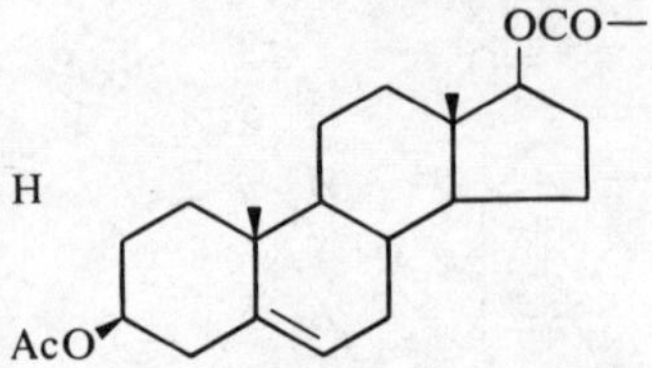

hydrosqualene-2,3-diols (21a) using the formate ester (21b).[39] Absolute configurations of the separated diols were determined and the pure epimers used for synthesis of chiral epoxides. Anaerobic incubation of respectively (*R*)-, (*S*)-, and (*RS*)-[4-$^3H_2$]-squalene 2,3-oxides with a pig liver microsome preparation gave conversions of 1.9, 76.8, and 39.7% into lanosterol.[40] Similar selectivity was observed, using a cell-free preparation from germinating peas, for incorporation of the epoxides into cycloartenol, β-amyrin, and lupeol. In whole yeast cells, lanosterol and ergosterol showed predominant incorporation of the (*S*)-epimer, as did lanosterol in a cell-free system. However, the expected relationship between incorporations of (*R*)-, (*S*)-, and (*RS*)-epoxides was not maintained in yeast, the whole-cell experiments giving lower incorporation of the racemate than expected, while significant conversion of the (*R*)-epimer was observed in the cell-free system.

The factors governing chemical and enzymatic cyclization of squalene 2,3-oxide have been examined further by synthesis of the cyclic analogues (22) and (23) and their respective ring A–B antipodes.[41] Chemical cyclization of (22) leads to formation

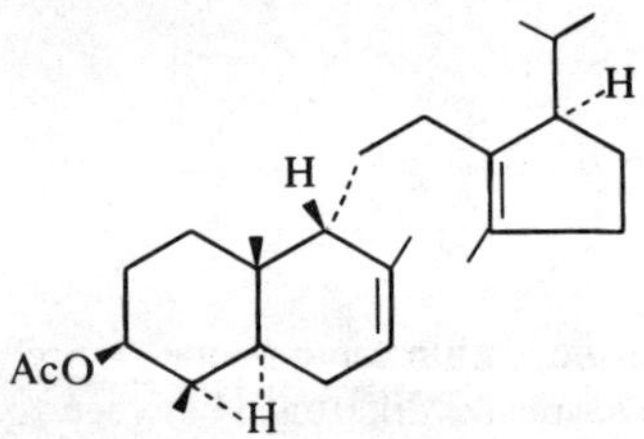

(22)

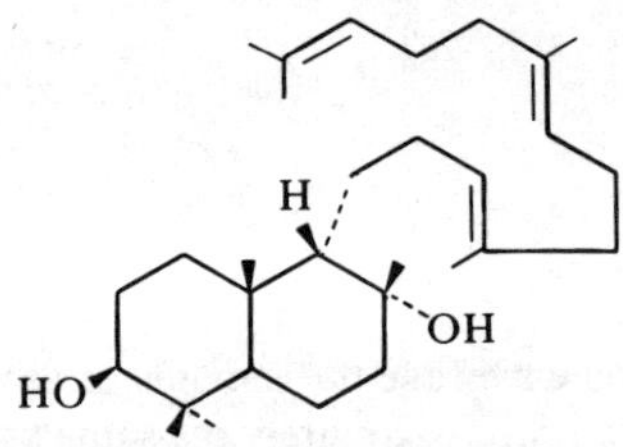

(23)

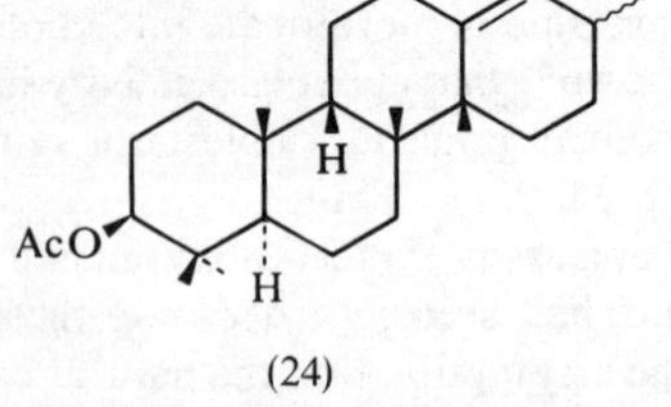

(24)

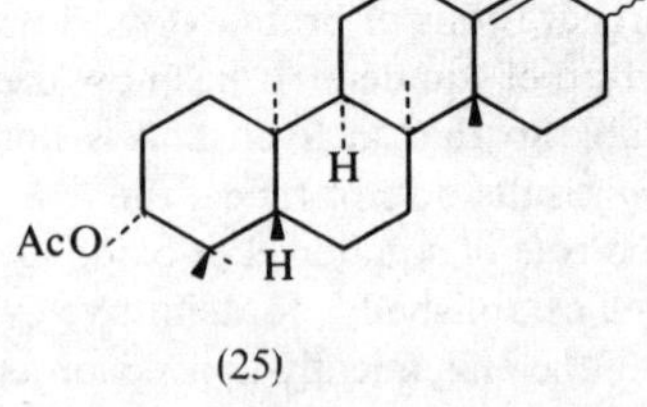

(25)

[39] R. B. Boar and K. Damps, *Tetrahedron Letters*, 1974, 3731.
[40] D. H. R. Barton, T. R. Jarman, K. G. Watson, D. A. Widdowson, R. B. Boar, and K. Damps, *J.C.S. Chem. Comm.*, 1974, 861.
[41] E. E. van Tamelen, A. Grieder, and R. G. Lees, *J. Amer. Chem. Soc.*, 1974, **96**, 2253.

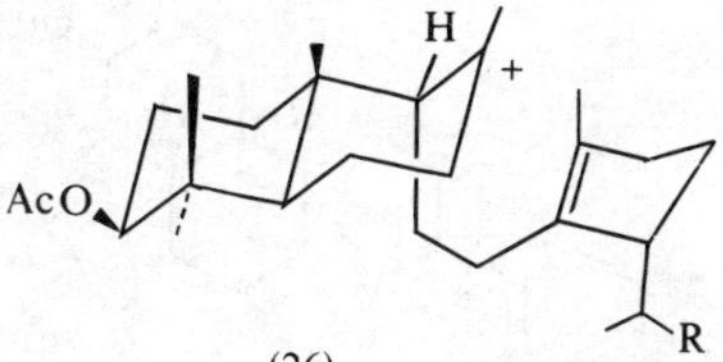

(26)

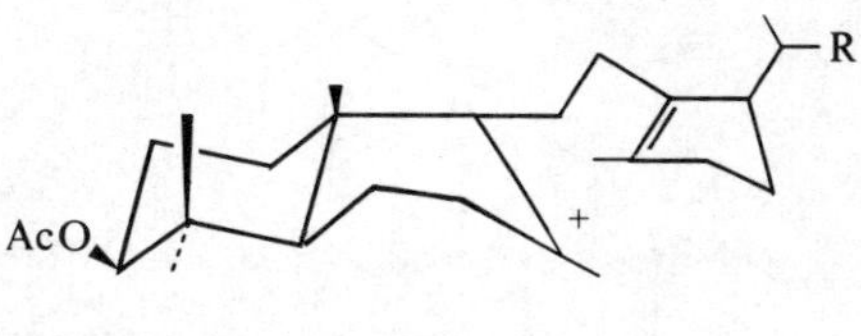

(27)

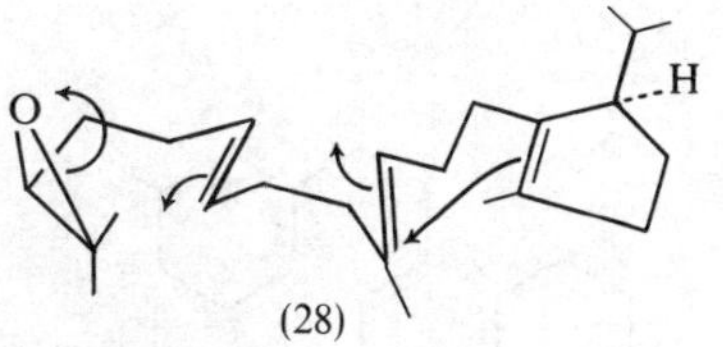

(28)

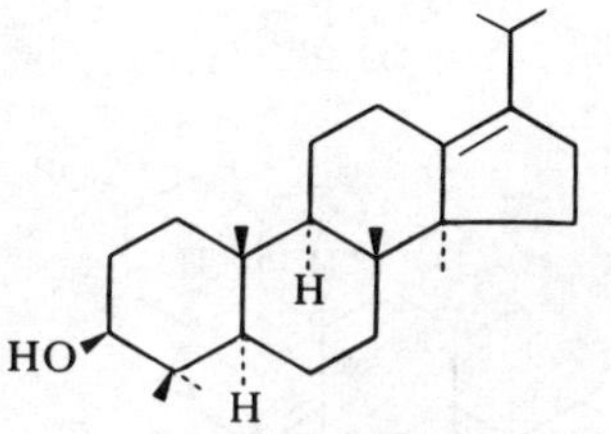

(29)

of (24) and (25) in yields of up to 75%.[42] The stereochemistry at the B,C,D-ring junctions in the products suggests that cyclization proceeds *via* a C-8 carbonium ion (26) having the stable chair conformation. The failure of the reaction to yield any products having the protosterol, lanosterol, or parkeol ring systems suggests that the boat conformer of the intermediate carbonium ion (27) is energetically disfavoured. Consequently, the earlier reported non-enzymic cyclization of the epoxide (28) to the iso-euphenol analogue (29) must either proceed with totally synchronous formation of rings A, B, and C or involve the 'frozen' unstable conformer (27; OH for OAc), which does not equilibrate with (26). Assuming that enzymic formation of lanosterol utilizes the chemical propensities of the substrate, these results support the idea of a concerted cyclization process for formation of rings A, B, and C. The lack of conversion of diol (23) into lanosterol in several enzyme systems is considered consistent with the non-involvement of a ground-state bicyclic intermediate in the process.[42]

The degree of stereoelectronic control involved in annelation reactions leading to triterpenes is underlined by the elegant biogenetic-like synthesis of *dl*-serratenediol (30a).[43] The bicyclic tetraene (31) gave the pentacyclic intermediate (30b) and its isomer (32) in 20% yield; also the acyclic tetraene (33) yields *dl*-D-homo-steroids (34a) and (34b) in 30% yield.[44]

Full details have been reported on experiments demonstrating the non-involvement of the C-22 hydrogens of the protosterol fusidic acid (35) in the termination of the squalene 2,3-oxide→protosterol cyclization process.[45]

5 Sterol Formation in Vertebrates

Cholesterol Biosynthesis.—Aspects of the enzymology of sterol biosynthesis have been reviewed in detail.[5] The exact sequence of double-bond migrations and de-

42 E. E. van Tamelen, R. G. Lees, and A. Grieder, *J. Amer. Chem. Soc.*, 1974, **96**, 2255.
43 G. D. Prestwich and J. N. Labovitz, *J. Amer. Chem. Soc.*, 1974, **96**, 7103.
44 W. S. Johnson, K. Wiedhaup, S. F. Brady, and G. L. Olson, *J. Amer. Chem. Soc.*, 1974, **96**, 3979.
45 R. C. Ebersole, W. O. Godtfredsen, S. Vangedal, and E. Caspi, *J. Amer. Chem. Soc.*, 1974, **96**, 6499.

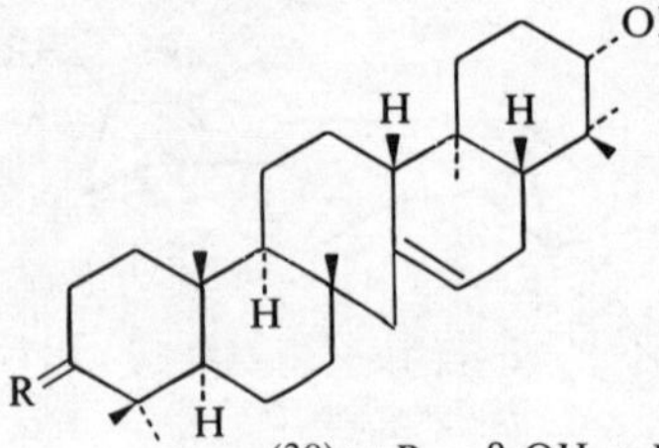

(30) a; R = β-OH, α-H
b; R = Me_2C

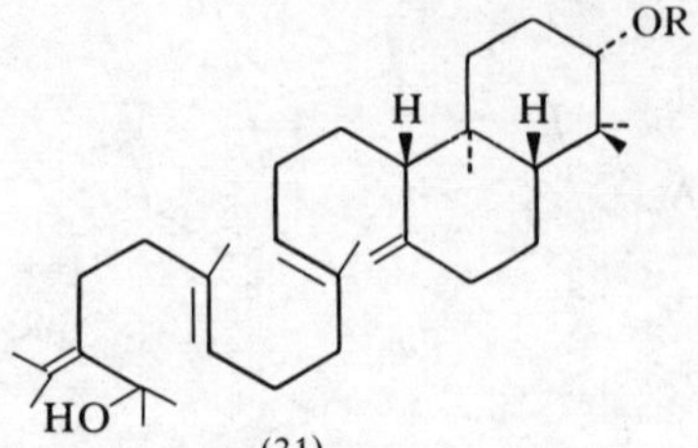

(31)

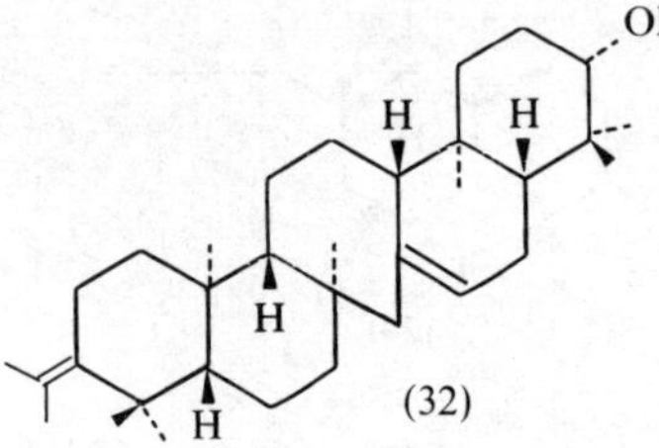

(32)

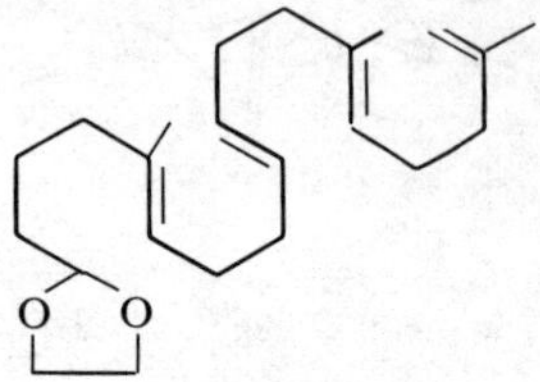

(33)

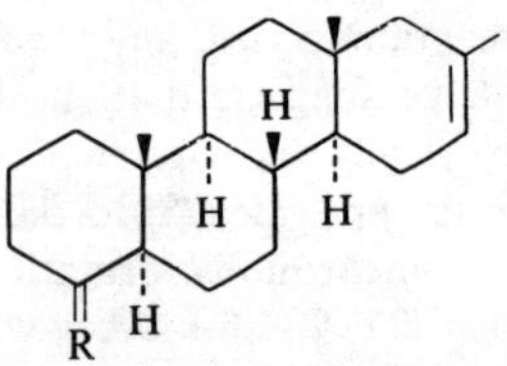

(34) a; R = α-H, β-OCH_2CH_2OH
b; R = α-OCH_2CH_2OH, β-H

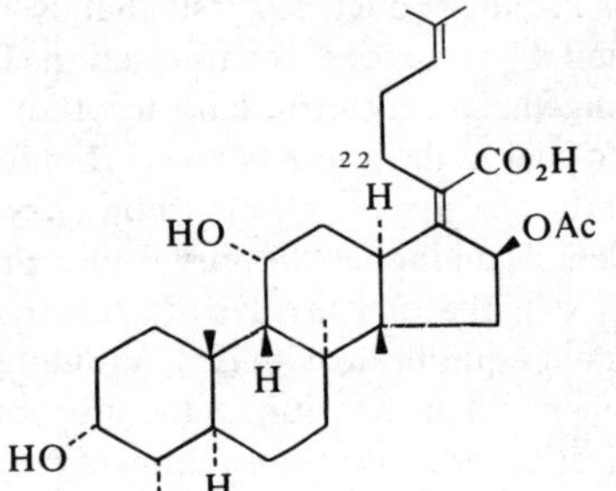

(35)

methylations involved in the conversion of lanosterol into cholesterol by rat liver microsomes has been a subject of prolonged study.[1] In particular, much debate has centred on the possible involvement of $\Delta^{8(14)}$-sterols. While 4,4-dimethylcholest-8(14)-enol (36) is converted into cholesterol in cell-free systems, strong evidence against its involvement in the normal biosynthetic pathway was obtained by radioactive incorporation and trapping experiments. Cholest-8(14)-enol (37) is also convertible into cholesterol and into cholesta-8,14-dienol *in vitro*.[1] New evidence[46] suggests that the normal 14-demethylation product, 4,4-dimethylcholesta-8,14-dienol (38), undergoes reduction of Δ^{14} prior to C-4-demethylation (Scheme 5), the demethylated product, cholest-8-enol (39), then being converted into cholesterol. The extent of labelling of (37) and (40) in this experiment was very low. 4,4-Dimethylcholest-8(14)-enol (36) is efficiently demethylated, though the process is strongly inhibited by added diene (38), suggesting competition for the enzymes involved in demethylation and a low substrate specificity. The presence of unlabelled cholesta-8,14-dienol (40) in a similar

[46] G. F. Gibbons, *Biochem. J.*, 1974, **144**, 59.

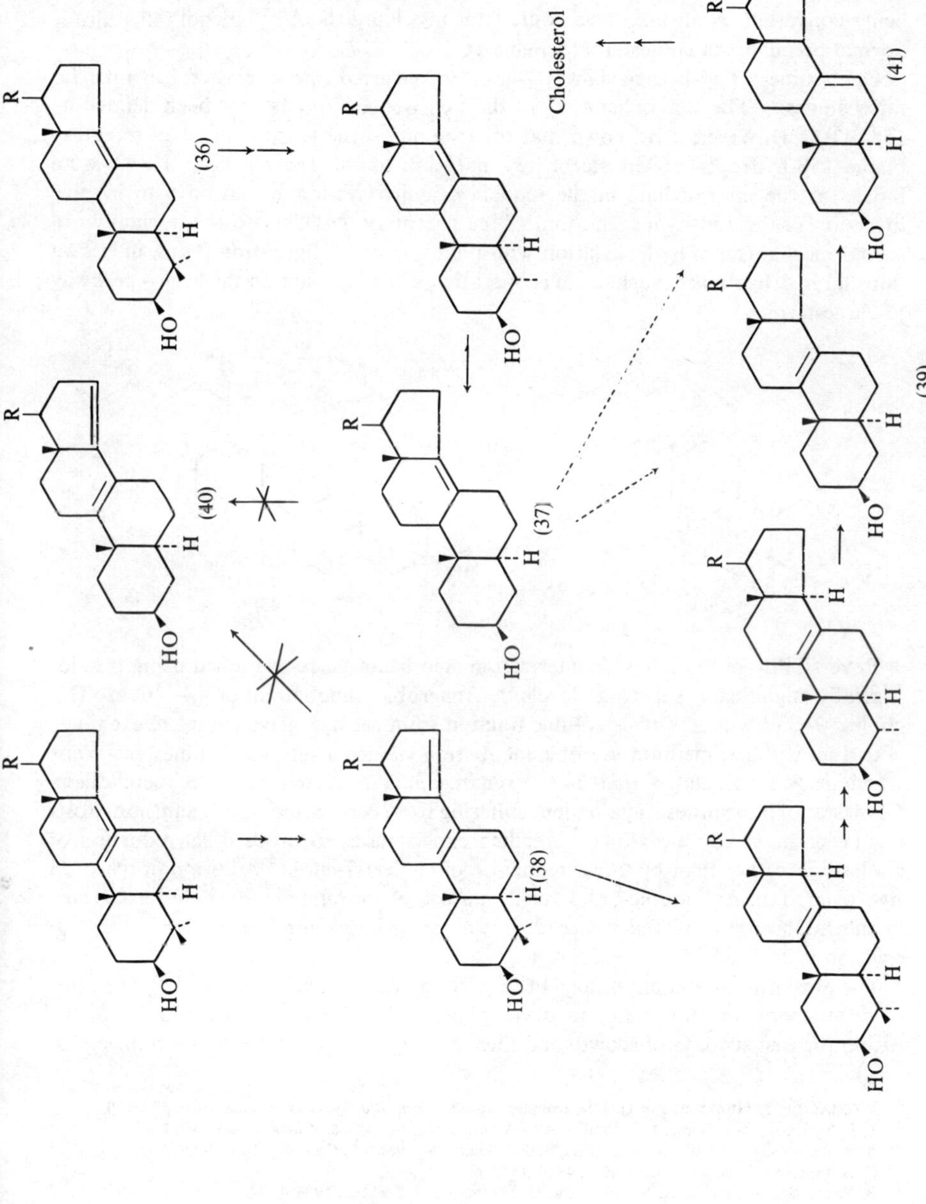

Scheme 5

incubation decreased the amount of desmethylsterol synthesized from (36), but failed to change the ratio of radioactivity in the demethylated sterols (37) and (41) as would have been expected if the sequence $\Delta^{8(14)} \rightarrow \Delta^{8,14} \rightarrow \Delta^{7}$ were operating. Thus, in addition to the normal route of sterol synthesis in rat liver, the system is capable of C-4-demethylation of exogenous 4,4-dimethylcholest-8(14)-enol, the product (37) being converted by an unknown route [not involving the $\Delta^{8,14}$-dienol (40)] into a normal precursor of cholesterol (Scheme 5).

One epimer of 15-hydroxylanost-7-enol is converted into cholesterol in rat liver microsomes.[47] The stereochemistry of the hydroxy-group has now been defined as 15β (42).[48] However, it is known that the true biosynthetic route involves retention of the 15β-hydrogen of lanosterol (*via* inversion to the 15α-position). If (42) were indeed a true intermediate in the sequence, hydroxylation would have to involve inversion of the 15β-hydrogen atom of the precursor, contrary to the normally observed mechanism of hydroxylation with retention of configuration. Thus, unless an unusual hydroxylation mechanism is operative, diol (42) is not on the *in vivo* pathway to cholesterol.

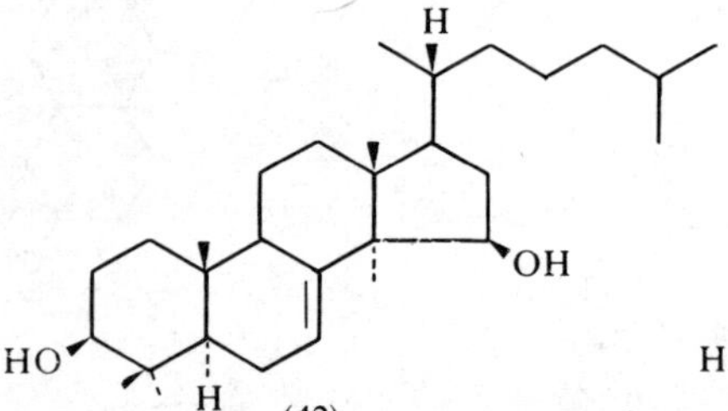

(42)

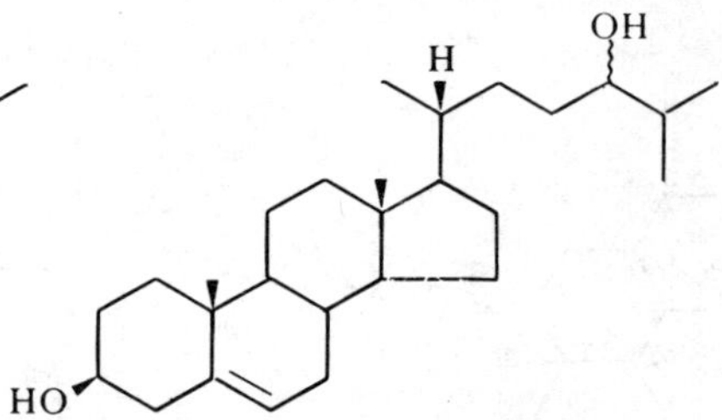

(43)

Reversibility of the Δ^{8}–Δ^{7} isomerization step has been re-examined using physiologically significant substrate levels.[49] Anaerobic incubation of [4-^{14}C,9α-^{3}H]-cholest-7-enol with a 10000*g* soluble fraction from rat liver gave complete exchange of tritium with the medium. Aerobic incubation yielded a sample of cholesterol from which it was calculated that 34% exchange had occurred prior to metabolism. Cholestanol biosynthesis in a patient suffering from cerebrotendinous xanthomatosis has been shown, by mevalonate labelling experiments, to proceed *via* reduction of cholesterol rather than by direct reduction of cholest-7-enol.[50] Addition of the C-25 hydrogen to the Δ^{24} double-bond during lanosterol conversion into cholesterol *in vivo* in the rat occurs from the *si*-face,[51] as was previously found using a rat liver S_{10} fraction.

The biosynthesis and physiology of steroids in brain has been reviewed.[52] The ratio of desmosterol to cholesterol in developing rat brain reaches a maximum of 0.5 within the first six days of growth and then steadily declines as myelination proceeds.

[47] J. A. Martin, S. Huntoon, and G. J. Schroepfer jun., *Biochem. Biophys. Res. Comm.*, 1970, **39**, 1170.
[48] T. E. Spike, A. H.-J. Wang, I. C. Paul, and G. J. Schroepfer jun., *J.C.S. Chem. Comm.*, 1974, 477.
[49] A. Scala, M. Galli-Kienle, M. Anastasia, and G. Galli, *European J. Biochem.*, 1974, **48**, 263.
[50] G. S. Tint and G. Salen, *J. Lipid Res.*, 1974, 15, 256.
[51] B. Yagen, J. S. O'Grodnick, E. Caspi, and C. Tamm, *J.C.S. Perkin I*, 1974, 1994.
[52] D. M. Bowen, A. N. Davison, and R. B. Ramsey, in ref. 5, p. 152.

The demonstration that this ratio is similar in both neuronal and glial cells has led to the suggestion that desmosterol is an important moderator of the properties of developing myelin and neuronal membranes.[53] Cerebrosterol (43) (cholest-5-ene-3,24-diol) has been identified as a minor sterol of both developing and mature rat brains.[54] The metabolite appears to accumulate during maturation and, while detectable in all cell fractions, is found in highest concentration in myelin and nerve-end fractions. Brain homogenates and microsomal fractions convert cholesterol into cerebrosterol in 0.15—0.18% yield.

The effects of detergents on the enzymes of sterol biosynthesis in rat brain preparations are to cause a marked shift in distribution of radioactivity incorporated from [2-^{14}C]mevalonate into squalene, sterols, and steryl esters. This results from changes in turnover rates of squalene and methyl-sterols.[55] A major neutral compound similar to geranylgeraniol is also produced. In contrast, phospholipase A enhances incorporation of mevalonate into sterols without production of the diterpene-like compound.

Unusually high levels of desmosterol and its sulphate (40–90% respectively of total sterol fractions) have been detected in hamster spermatozoa.[56] Desmosterol has previously been detected only in trace amounts in animal cells, with the exception of the immature brain, as noted above. The presence of an inhibitor of the Δ^{24}-reductase, or a low concentration of the enzyme, may be responsible for this accumulation in spermatozoa, the physiological significance of which is unknown.

Oral administration of ethanol to rats results in the accumulation of cholesteryl esters and triglycerides in the liver, while the free cholesterol level is normal. *In vitro* experiments indicate that the esterification rate is unchanged but that ester hydrolysis is significantly reduced.[57] Conversion of mevalonate into squalene and sterols is low in livers of suckling rats but increases rapidly after weaning. This rise is prevented by a cholesterol-enriched diet.[58] Similar low incorporation of mevalonate is obtained in livers of fasted adult rats, while the increased incorporation normally observed on re-feeding the rats is again countered by a cholesterol-supplemented diet. It is thought that, after weaning or starvation, liver growth is accompanied by enhanced sterol synthesis. Cholesterol effectively inhibits one or more enzymes between mevalonate and squalene and probably also at a site beyond squalene. The possibility of regulation by variations in the amount of sterol carrier protein (SCP) was also examined. The S_{105} fraction from livers of weaned rats did not increase conversion of squalene in microsomes from the livers of suckling rats, while microsomes from weaned rats showed significant conversion of squalene into sterols on addition of the S_{105} fraction from either suckling or weaned rats. Thus, control is not mediated by SCP levels.

Sterol Carrier Proteins.—The complex and sometimes apparently conflicting results on the role of the non-catalytic SCPs in conversion of squalene into cholesterol have been reviewed in detail.[59] At least three proteins now appear to be involved in the

[53] R. G. Dennick, M. Grant, and P. D. G. Dean, *F.E.B.S. Letters*, 1974, **45**, 126.
[54] Y. Y. Lin and L. L. Smith, *Biochim. Biophys. Acta*, 1974, **348**, 189.
[55] R. B. Ramsey, A. Atallah, M. Fredericks, and H. J. Nicholas, *Biochem. Biophys. Res. Comm.*, 1974, **61**, 170.
[56] G. Bleau and W. J. A. Vanden Heuvel, *Steroids*, 1974, **24**, 549.
[57] N. Takeuchi, M. Ito, and Y. Yamamura, *Lipids*, 1974, **9**, 353.
[58] R. C. Johnson and S. N. Shah, *Lipids*, 1974, **9**, 962.
[59] T. J. Scallen, M. V. Srikantaiah, B. Seetharam, E. Hansbury, and K. L. Gavey, *Fed. Proc.*, 1974, **33**, 1733.

overall process and, as purification procedures improve, marked specificities of proteins for various steps or sections of the pathway are emerging. Scallen and co-workers have isolated a protein, designated SCP_1, which has been purified 160-fold.[2] Its molecular weight is 28 000, although dissociation into five or six subunits is observed in 6M-guanidine hydrochloride. The order of addition of reactants which gives best results appears to be initial formation of a squalene–SCP_1 complex, in the presence of phosphatidylserine and FAD, followed by binding to the microsomes. A further addition of SCP_1 is required for maximal conversion. Using washed microsomes, efficient conversion of squalene into sterols (containing 20% cholesterol) was observed, while with triply washed microsomes the major product appeared to be a triterpene (possibly lanosterol), the sterol content being considerably reduced. This suggests that other factors associated with the microsomes may also take part in the overall process.

The carrier protein SCP_2, purified approximately 25-fold, is active in conversion of cholesta-5,7-dienol into cholesterol but shows little or no promotion of squalene metabolism.[59] This protein shows some similarity to the carrier protein reported by Ritter and Dempsey.[1,3] However, neither preparation has yet been shown to comprise a single protein. Indeed, the preferential reduction in conversion of cholest-7-enol *vis-à-vis* cholesta-5,7-dienol on heating the Ritter–Dempsey material indicates the presence of more than one active species in this preparation. A third, partially purified, protein from liver S_{105} fraction has also been reported.[60] It has been obtained free of SCP_1 and SCP_2 activities, is heat-labile, and, in contrast to the other proteins, stimulates conversion of 4,4-dimethylcholest-8-enol into C_{27} steroids. Whether this protein is the heat-labile component of the Ritter–Dempsey material which is responsible for its activity with precursors of cholest-7-enol remains to be examined. The S_{105} fraction of rat liver, prepared in 20mM-phosphate buffer, stimulates conversion of squalene into lanosterol and of both lanosterol and desmosterol into cholesterol. This fraction is also active with microsomes from heart, kidney and brain. Stimulation of squalene conversion is reduced on heat treatment.[61] The supernatants from heart, kidney, and brain prepared in 100mM buffer are active with microsomes from all sources but, when prepared in 20mM buffer, no stimulation of squalene conversion in liver is detected. Thus the heat-labile factor or factors in liver preparations may be absent from the extrahepatic supernatants when they are prepared in dilute buffer. The study of the conversion of 4,4-dimethylcholesta-8,14-dienol and 4,4-dimethylcholest-8(14)-enol in rat liver, discussed earlier, also included an investigation of the role of the S_{104} fraction in metabolism of substrates.[46] While conversion rates of 4,4-dimethylsterols were enhanced by factors of 2—4, no major change in the sequence of events or product distribution was observed. The S_{105} fraction from human placenta is capable of binding squalene but is not an efficient activator of placental or rat liver microsomal squalene epoxidase, compared to rat-liver S_{105}.[62]

6 Triterpenoids and Steroids in Higher Plants, Algae, and Fungi

Triterpenoid and Steroid Biosynthesis.—*Higher Plants.* The structures, distribution,

[60] B. Seetharam, E. Hansbury, M. V. Srikantaiah, and T. J. Scallen, *Fed. Proc.*, 1974, 33, 1573.
[61] R. C. Johnson and S. N. Shah, *Arch. Biochem. Biophys.*, 1974, **164**, 502.
[62] M. Astruc, C. Tabacik, B. Descomps, and A. Crastes de Paulet, *F.E.B.S. Letters*, 1974, 47, 66.

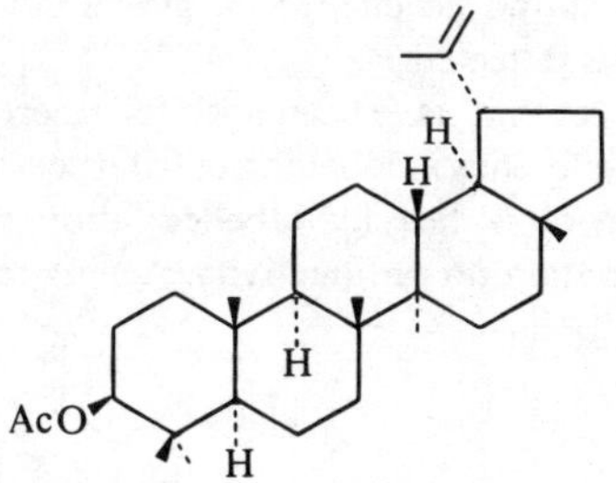

(44)

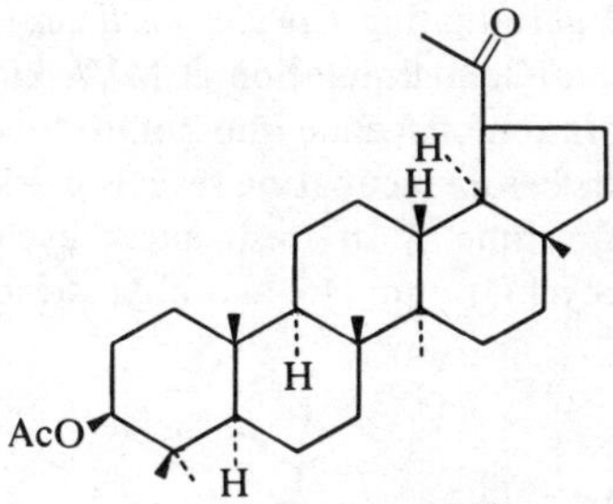

(45)

and biosynthesis of triterpenoid saponins and their genins have been reviewed.[63] Incorporation of stereospecifically tritiated mevalonates into α-amyrin, β-amyrin, ψ-taraxasterol, lupeol, and other triterpenes of *Calendula officinalis* flowers[64] is broadly in accord with the biosynthetic hypothesis proposed by Ruzicka and Eschenmoser. On the basis of isotope ratios in lupeyl acetate (44) and its oxidation product (45), equal distribution of the mevalonate C-2 carbon between C-29 and C-30 of lupeol is inferred, suggesting random proton-elimination from the putative C-20 carbonium ion or its equivalent stabilized precursor.

The current state of knowledge of phytosterol alkylation mechanisms has been reviewed and the phylogenetic distribution of C-24-alkylated sterols has been updated.[65] Sterol distribution has also been surveyed with regard to their probable role, at least in part, as structural components of membranes.[66] The evolutionary significance of the lanosterol–cycloartenol and cholesterol–phytosterol bifurcations has also been discussed.

Several studies bearing on the regulation of phytosterol biosynthesis have been reported. Naphthaleneacetic acid stimulates mevalonate incorporation into sterol significantly in the zone of elongation of mung-bean hypocotyl, though triterpene labelling is unaffected.[67] This result supports the suggestion that one major regulatory point in phytosterol production occurs after cycloartenol synthesis. Further, the growth retardation of tobacco seedlings by Amo 1618 [2′-isopropyl-4′-(trimethylammonium chloride)-5′-methylphenylpiperidine carboxylate], CCC[β-chloroethyltrimethylammonium chloride], and Phosphon D [tributyl-2,4-dichlorobenzylphosphonium chloride] is parallelled by decreased incorporation of mevalonate into sterols.[68] All three retardants markedly reduce incorporation into 4-desmethylsterol. Amo 1618 and CCC have little effect on squalene and triterpene levels but cause increased incorporation into squalene 2,3-epoxide. In contrast, Phosphon D, while reducing phytosterol labelling from mevalonate, promotes labelling, not of squalene 2,3-epoxide, but of 4,4-dimethyl- and 4-methyl-sterols. The retarding effects of these three compounds can be overcome by gibberellic acid or by added sterol. The sterol

[63] S. K. Agarwal and R. P. Rastogi, *Phytochemistry*, 1974, **13**, 2623.
[64] J. Sliwowski and Z. Kasprzyk, *Phytochemistry*, 1974, **13**,1441.
[65] L. J. Goad, J. R. Lenton, F. F. Knapp, and T. W. Goodwin, *Lipids*, 1974, **9**, 582.
[66] W. R. Ness, *Lipids*, 1974, **9**, 596.
[67] J. M. C. Geuns and J. C. Vendrig, *Phytochemistry*, 1974, **13**, 919.
[68] T. J. Douglas and L. G. Paleg, *Plant Physiol.*, 1974, **54**, 238.

content of germinating *Corylus avellana* seeds shows no change on gibberellic acid treatment, though stimulation of MVA kinase is detected.[69]

Incorporation of acetate into potato tuber slices that have been aged for a period of 0—24 hours before incubation reveals marked difference in labelling of triterpenes and sterols with time.[70] In fresh slices cycloartenol is heavily labelled, though 24-methylenecycloartanol (46) and phytosterols contain no radioactivity. Ageing results

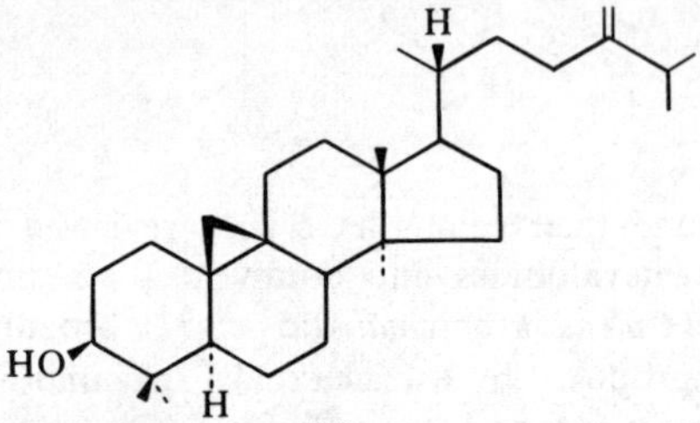

(46)

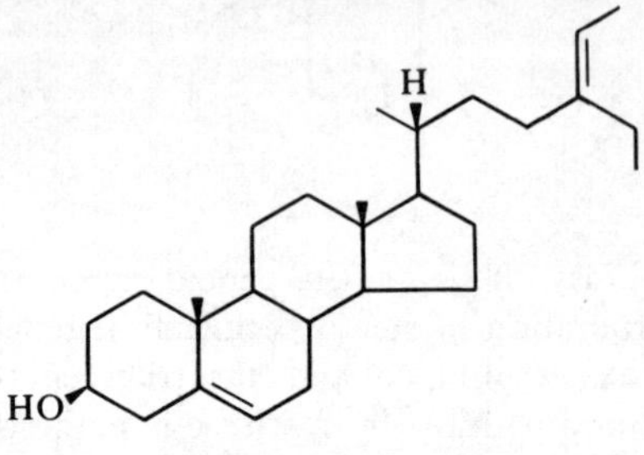

(47)

in significant radiolabelling of metabolites of cycloartenol, notably (46) and isofucosterol (47). Free phytosterol content also increases fivefold over the 24 hour period. It is thought that, during the process of ageing, protein synthesis is stimulated and membrane biogenesis recommences, with concomitant sterol synthesis. The appearance of C-24 methyltransferase activity only after the onset of ageing and the observation that the specific activity of cholesterol is considerably higher than that of either sitosterol or stigmasterol are particularly interesting results. The exact substrate for the enzyme which opens the 9β,19β-cyclopropane ring of plant triterpenes has been

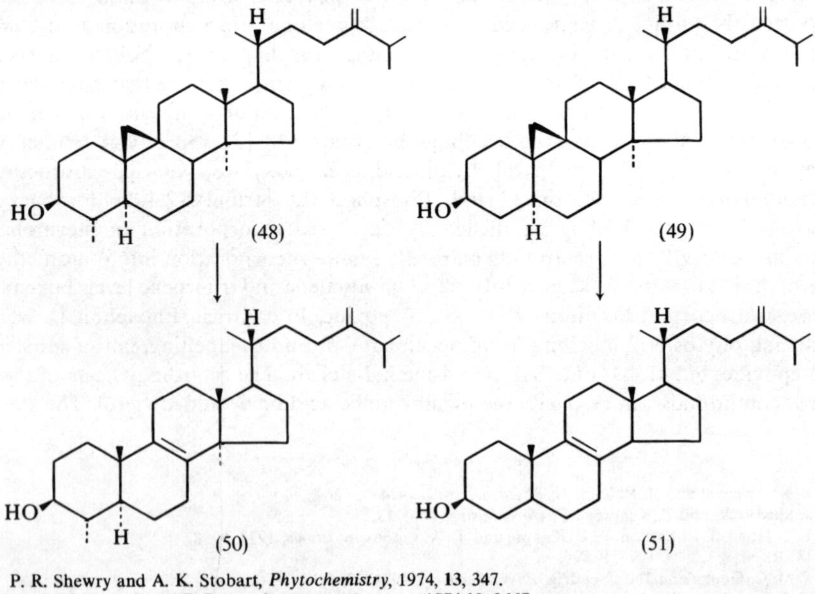

(48) (49)

(50) (51)

[69] P. R. Shewry and A. K. Stobart, *Phytochemistry*, 1974, **13**, 347.
[70] M. A. Hartmann and P. Benveniste, *Phytochemistry*, 1974, **13**, 2667.

examined using a cell-free preparation from tissue cultures of bramble (*Rubus fruticosus*).[71] Cycloeucalenol (48) and 24-methylenepollinasterol (49) were converted into obtusifoliol (50) and 4-desmethylobtusifoliol (51) respectively by the microsomal fraction, without addition of ATP or NADH. No significant stimulation by the supernatant was observed. 24-Methylenecycloartanol (46) was not transformed, suggesting that a 4β-methyl group cannot be accommodated by the enzyme. A tentative model of the enzyme active site and ring-opening mechanism has been proposed (Scheme 6).[71] Microsomes from germinating peas and tobacco tissue culture are also reported

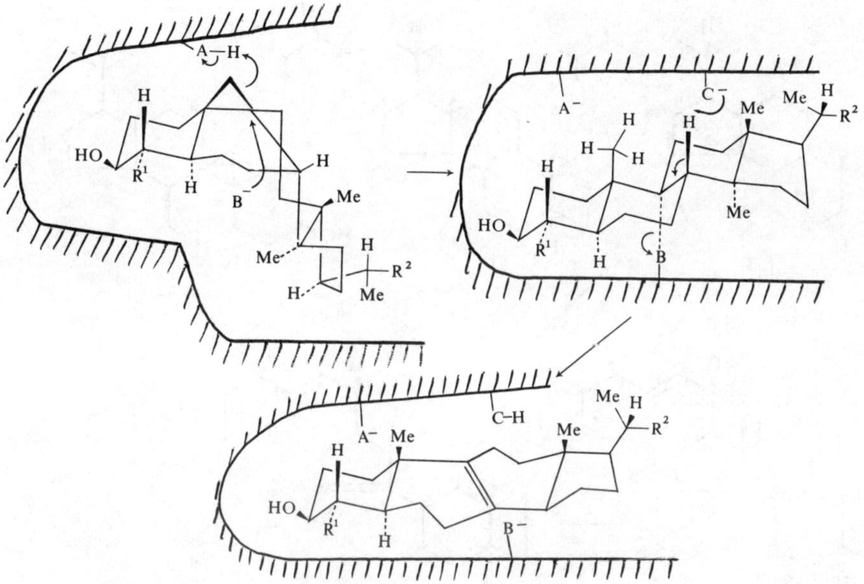

Scheme 6

to transform cycloeucalenol into obtusifoliol, while rabbit liver microsomes and yeast enzymes are inactive.

Stereospecifically tritiated mevalonates [2*R*, 2*S*, 4*R*, and 5*RS*] have been incorporated into the phytosterols of *Calendula officinalis* flowers.[72] Although the labelled positions were not located by degradation of the sterols, the results suggest, on the assumption of a labelling pattern similar to that reported in other systems, that formation of the Δ^7 double-bond of all the sterols involves loss of the 7β-hydrogen atom, as has been observed in rat liver, *Camellia sinensis*, and *Ochromonas* species (though the 7α-hydrogen is removed in yeast). The Δ^{22} double-bond formation in stigmasterol proceeds with removal of the 22-*pro-S* and 23-*pro-S* hydrogens (assuming a *cis*-elimination, by analogy with all known Δ^{22} dehydrogenations). The stereochemistry

[71] R. Heintz and P. Benveniste, *J. Biol. Chem.*, 1974, **249**, 4267.
[72] J. Sliwowski and Z. Kasprzyk, *Phytochemistry*, 1974, **13**, 1451.

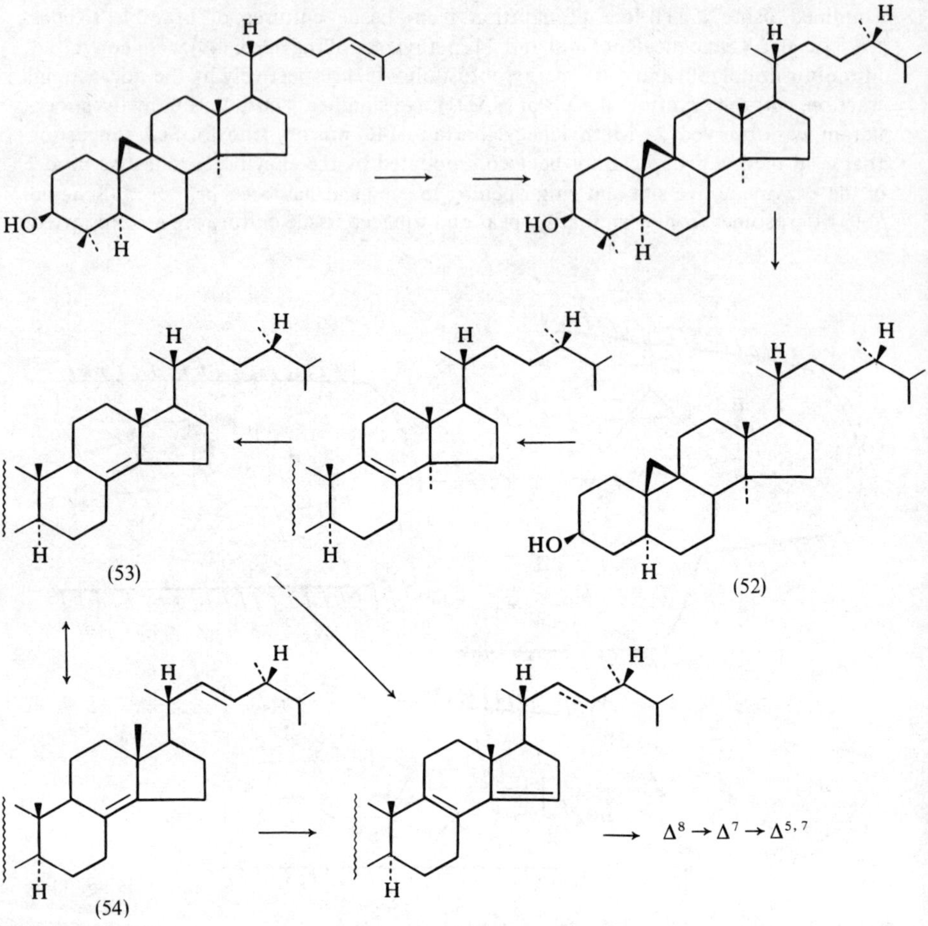

Scheme 7

of this elimination appears to be the same as that observed in ergosterol biosynthesis in fungi, while similar dehydrogenations in sterols of algae and *Camellia sinensis* involve loss of the *pro-R* hydrogens. Side-chain alkylation proceeds with elimination of the C-24 hydrogen, as has been observed in several other higher plants. The Δ^5 double-bond is introduced, as usual, with loss of the 6α-proton. In contrast, the results indicate that C-14-demethylation proceeds in a novel manner without involvement of the normal Δ^{14} intermediate, since neither hydrogen from C-15 appears to be lost in formation of the sterols.

Algae and Diatoms. When *Chlorella sorokiniana* is grown in the presence of 1 p.p.m. triparanol succinate, the sterol content is reduced by 42%. Ergosterol is replaced by

5α-ergost-8-enol as the major sterol, and several new sterols containing Δ^{8}, $\Delta^{8(14)}$, and $\Delta^{8,14}$ double-bonds are produced.[73] Compounds (52)—(54) and ergosta-8,14,22-trienol are reported for the first time as biosynthetic products. The observed sterol distribution in the presence and absence of the inhibitor is accommodated in a postulated biosynthetic sequence (Scheme 7). However, in view of the doubtful role of $\Delta^{8(14)}$-sterols in the *de novo* biosynthetic pathway in animals, the key position of (53) in Scheme 7 must be considered tentative in the absence of direct experimental support from tracer studies.

The major inhibition points of sterol biosynthesis by triparanol and AY9944 in three species of *Chlorella* have been compared.[74] Incorporation of [2,4-$^{3}H_{4}$]14α-methyl-5α-ergost-8-enol and [2,4-$^{3}H_{4}$]5α-ergosta-8,14-dienol into ergost-5-enol by *C. ellipsoidea* confirms the presence, in this species, of a 14-demethylase system which is not dependent on the presence of a 4α-methyl group, and provides evidence for the operation of a Δ^{14}-reduction pathway associated with the demethylation process.[75]

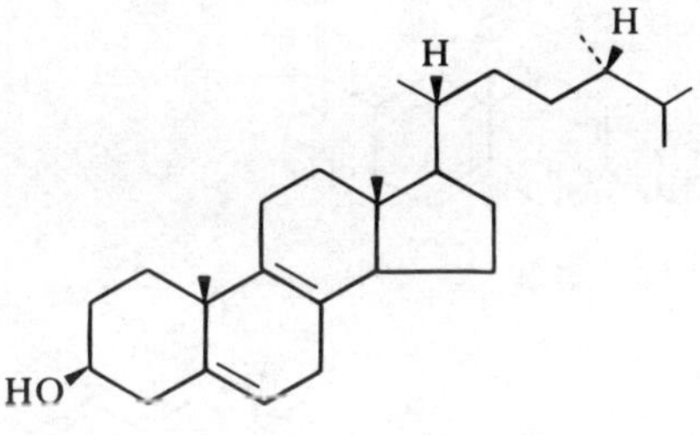

(55)

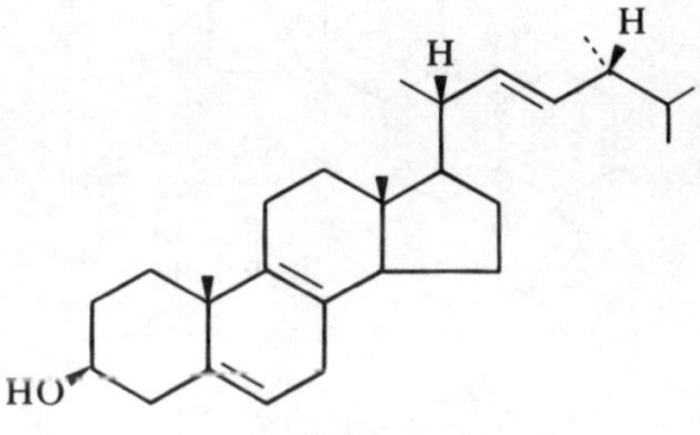

(56)

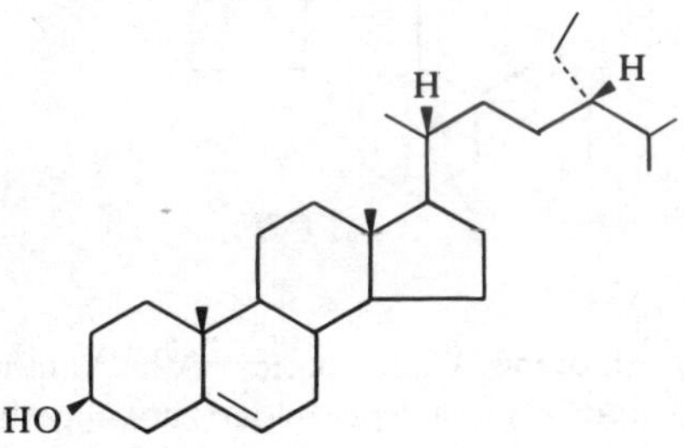

(57)

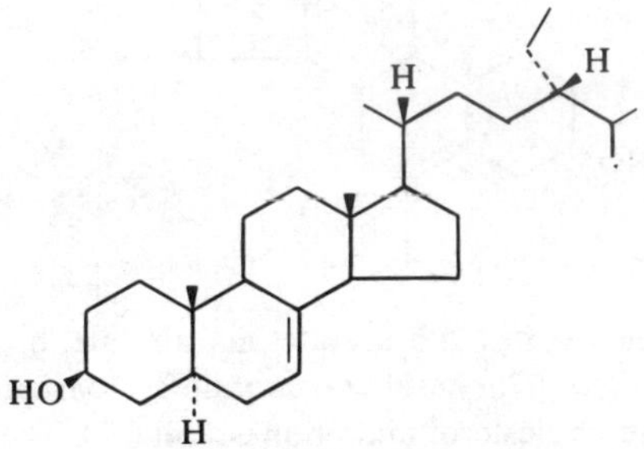

(58)

Another strain of *C. ellipsoidea* is reported to contain, in addition to ergosterol and 5α-ergost-7-enol, two novel sterols; ergosta-5,8-dienol (55) and ergosta-5,8,22-trienol (56) are major components in this strain.[76] The latter compound was also recently found in the lichen *Xanthoria parietina*,[77] and in yeast mutants.[78] Thus it appears

[73] J. T. Chan, G. W. Patterson, S. R. Dutky, and C. F. Cohen, *Plant Physiol.*, 1974, **53**, 244.
[74] G. W. Patterson, P. J. Doyle, L. G. Dickson, and J. T. Chan, *Lipids*, 1974, **9**, 567.
[75] L. B. Tsai, G. W. Patterson, C. F. Cohen, and P. D. Klein, *Lipids*, 1974, **9**, 1014.
[76] G. W. Patterson, M. J. Thompson, and S. R. Dutky, *Phytochemistry*, 1974, **13**, 191.
[77] J. R. Lenton, L. J. Goad, and T. W. Goodwin, *Phytochemistry*, 1973, **12**, 1135.
[78] D. H. R. Barton, J. E. T. Corrie, D. A. Widdowson, M. Bard, and R. A. Woods, *J.C.S. Perkin I*, 1974, 1326.

that C-5,6 dehydrogenation may not require a Δ^7 double-bond in the substrate. Clionasterol (57), a characteristic Δ^5-sterol of dark-grown *Euglena gracilis*, disappears after 3 h illumination of the cultures and is replaced by the Δ^7-sterol chondrillasterol (58).[79] However, no change is observed in the esterified sterols. The rapidity of the change in free sterol content indicates that sterol content and chloroplast development are not related.

Rhodymenia palmata contains cycloartanol (59a) and 31-norcycloartanol (59b) rather than cycloartenol.[80] In addition to the normal sterol distribution found in red algae, 24-methylated sterols and 24-dimethylchola-5,22-dienol (60) were identified. The origin of the latter compound, which appears to be widely distributed in marine invertebrates, remains obscure, since it was not labelled following feeding experi-

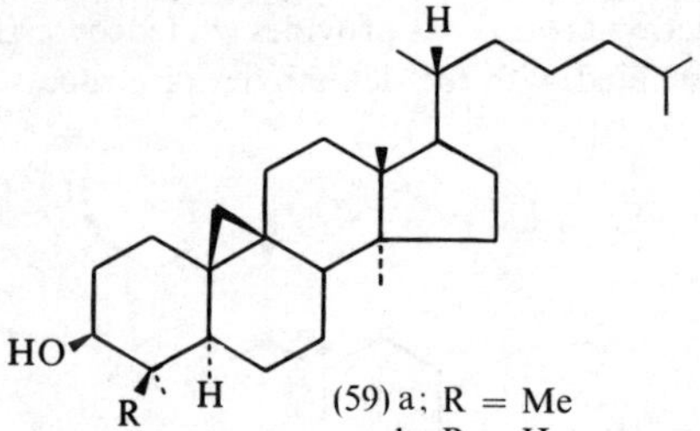

(59) a; R = Me
b; R = H

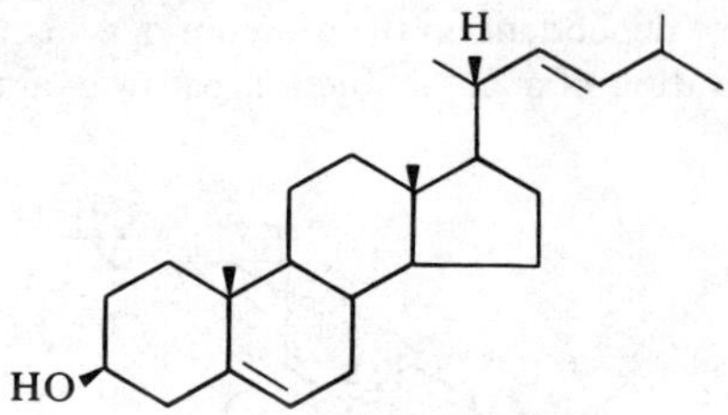

(60)

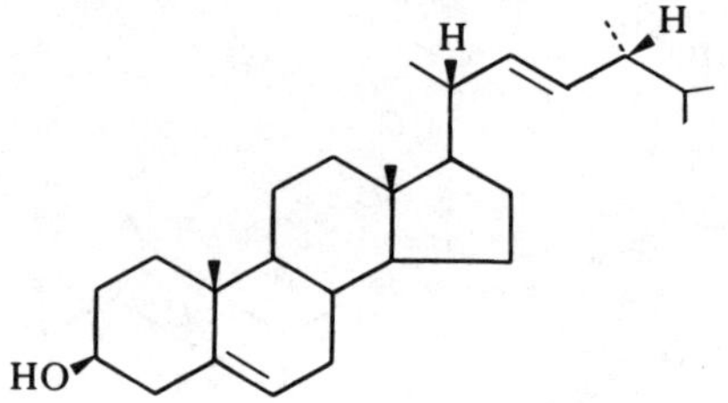

(61)

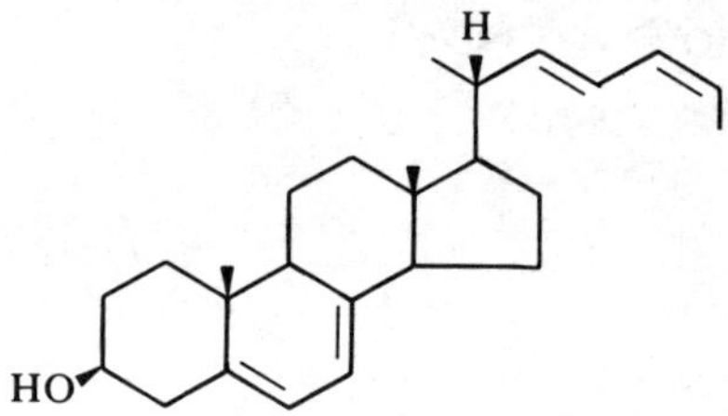

(62)

ments with [^{14}C]acetate, mevalonate, or methionine. Three species of the Xanthophyceae (Chrysophyta) contain cycloartenol and 24-methylenecycloartanol, together with cholesterol and clionasterol (57). The ratio of the two sterols shows no change with age of the cultures.[81] The non-photosynthetic diatom *Nitzschia alba* is reported[82] to produce brassicasterol (61) and clionasterol (57), though no direct evidence is provided for the C-24 configuration. While (61) is present in free and esterified forms, (57) occurs only in a glycosidically bound form.

Fungi. The concept of identifying unit enzyme transformations by the use of blocked mutants is claimed to be a more satisfactory method for elucidation of biosynthetic pathways than more traditional experiments involving feeding of potential precursors.

[79] C. Anding and G. Ourisson, *Physiol. Vegetale*, 1974, **12**, 299.
[80] J. P. Ferezou, M. Devys, J. P. Allais, and M. Barbier, *Phytochemistry*, 1974, **13**, 593.
[81] E. I. Mercer, R. A. London, I. S. A. Kent, and A. J. Taylor, *Phytochemistry*, 1974, **13**, 845.
[82] T. G. Tornabene, M. Kates, and B. E. Volcani, *Lipids*, 1974, **9**, 279.

A number of investigations now support the idea that the conversion of lanosterol into ergosterol in a yeast (*Saccharomyces cerevisiae*) can proceed *via* a matrix of enzymic reactions, some of which demonstrate a low specificity for substrate structures. Detailed analysis of sterol biosynthetic pathways in yeast using this approach has been reported by two groups. The sterols of a series of nystatin-resistant strains, shown to be single-gene mutants, demonstrate metabolic blocks in the processes of C-24 transmethylation, Δ^8–Δ^7 isomerization, C-5,6 dehydrogenation, and C-22,23 dehydrogenation, respectively.[78, 83] Three novel sterols, including cholesta-5,7,22,24-tetraenol (62), were isolated, together with four sterols not previously found in yeast, notably ergosta-5,8,22-trienol (56). A series of mutants which are resistant to different

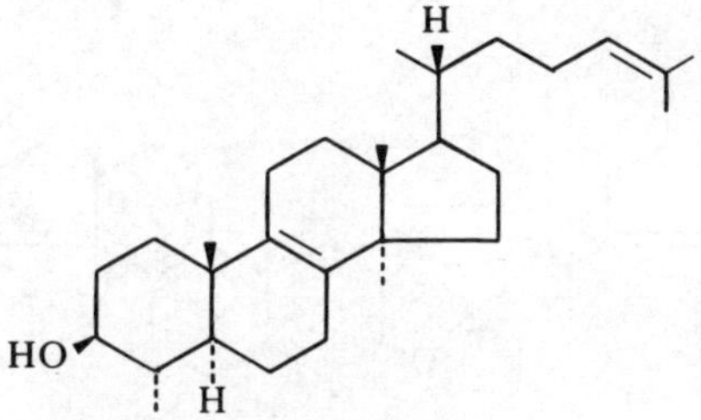

(63)

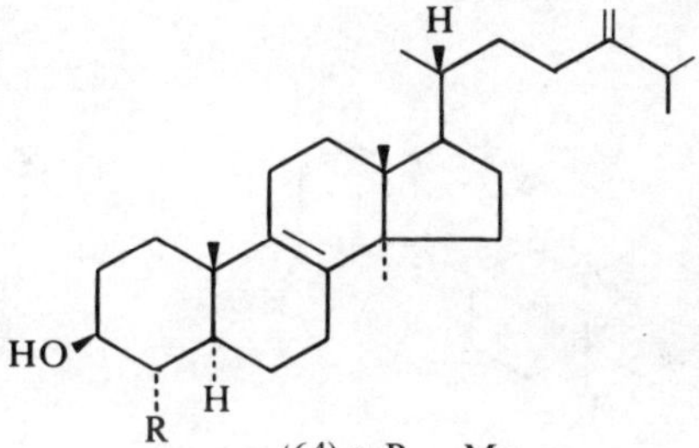

(64) a; R = Me
b; R = H

concentrations of nystatin exhibit loss of C-5,6-dehydrogenase activity in strains amenable only to lower concentrations of the inhibitor, followed by Δ^8–Δ^7-isomerase elimination in strains resistant to higher concentrations.[84] The processes involved in side-chain alkylation and dehydrogenation were not completely blocked in these strains. A further study employing nystatin-resistant strains demonstrated that the C-14 demethylation process can also be blocked.[85, 86] The novel yeast sterols 4,14-dimethylzymosterol (63), obtusifoliol (64a), and 14-methylfecosterol (64b) reveal that C-22,23-dehydrogenase activity is also lost, while demethylation at C-4 is unaffected. A mutant which is incapable of converting squalene into lanosterol has also been reported.[87]

The observation that conversion of lanosterol into zymosterol by a yeast cell-free system is inhibited by carbon monoxide suggests that, as in rat liver, cytochrome P-450 is involved in the later stages of biosynthesis, possibly in the oxidative processes leading to C-14 demethylation.[88] This view is supported by the observation that mutants lacking cytochrome pigments accumulate lanosterol.[89] Two of the strains investigated probably contain lesions in early stages of the porphyrin biosynthetic

[83] D. H. R. Barton, J. E. T. Corrie, D. A. Widdowson, M. Bard, and R. A. Woods, *J.C.S. Chem. Comm.*, 1974, 30.
[84] M. Fryberg, A. C. Oehlschlager, and A. M. Unrau. *Arch. Biochem. Biophys.*, 1974, **160**, 83.
[85] E. G. Gollub, P. Trocha, P. K. Liu, and D. B. Sprinson, *Biochem. Biophys. Res. Comm.*, 1974, **56**, 471.
[86] P. J. Trocha, S. J. Jasne, and D. B. Sprinson, *Biochem. Biophys. Res. Comm.*, 1974, **59**, 666.
[87] F. Karst and F. Lacroute, *Biochem. Biophys. Res. Comm.*, 1974, **59**, 370.
[88] K. T. W. Alexander, K. A. Mitropoulos, and G. F. Gibbons, *Biochem. Biophys. Res. Comm.*, 1974, **60**, 460.
[89] M. Bard, R. A. Woods, and J. M. Haslam, *Biochem. Biophys. Res. Comm.*, 1974, **56**, 324.

pathway, since δ-aminolaevulinic acid and haematoporphyrin restore the formation of sterols from lanosterol. The C-14 demethylation process is thought to result in the introduction and eventual reduction of a Δ^{14} double-bond, since the loss of one hydrogen atom from C-15 of the resulting sterols was demonstrated, using a yeast cell-free preparation.[90] It has now been shown conclusively, by oxidation of the tertiary C-14 carbon atom and stereospecific elimination of the resulting halide, that the 15α-hydrogen of lanosterol is lost and that the 15β-hydrogen occupies the 15α position in the final sterol, as a result of *trans*-hydrogenation of the Δ^{14} bond (Scheme 8).[91] The stereochemical course of the process thus appears identical to that observed in rat liver, in contrast to the different mechanisms apparent for the introduction of the Δ^7 double-bond into these two systems.[92,93]

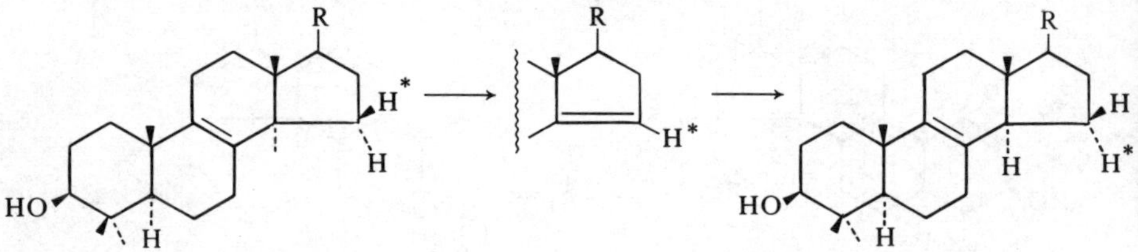

Scheme 8

C-24 Alkylation.—The current state of knowledge on the various mechanisms involved in C-24 alkylation of phytosterol precursors by *S*-adenosylmethionine has been analysed.[65] The involvement of 24-methylene- and 24-ethylidene-sterols in the formation of 24-methyl- and -ethylsterols is reported[65] to have been confirmed by incorporation of [$^2H_3$]methionine into embryos of barley (*Hordeum vulgare*) with retention of two and four deuterium atoms respectively, as determined by mass spectrometry. Earlier use of [$^3H_1$]methionine had given less certain results for the C_{29} sterols of *Zea mays*[94] since the magnitude of possible isotope effects was unknown. Also the observation that both 24*R*- and 24*S*-methyl sterols are found in some higher plants[95] makes the interpretation of tritium-incorporation results even more difficult, since more than one mechanism could be involved. *Withania somnifera* contains 24ξ-methylcholest-5-enol and its 24-ethyl homologue, together with 24ξ-methyl-cholesta-5,22-dienol and isofucosterol.[96] A minor constituent is 24-methylcholesta-5,24-dienol (65a), probably accompanied by the corresponding 24-ethyl derivative (65b); these compounds have been suggested as possible intermediates in one alkylation pathway.[65]

[90] E. Caspi and P. J. Ramm, *Tetrahedron Letters*, 1969, 181.
[91] E. Caspi, J. P. Moreau, and P. J. Ramm, *J. Amer. Chem. Soc.*, 1974, **96**, 8107.
[92] E. Caspi, J. P. Moreau, and P. J. Ramm, *J. Steroid Biochem.*, 1974, **5**, 627.
[93] L. J. Mulheirn and P. J. Ramm, *Chem. Soc. Rev.*, 1972, **1**, 259.
[94] L. J. Goad, A. S. A. Hamman, A. Dennis, and T. W. Goodwin, *Nature*, 1966, **210**, 1322.
[95] L. J. Mulheirn, *Tetrahedron Letters*, 1973, 3175.
[96] W. J. S. Lockley, D. P. Roberts, H. H. Rees, and T. W. Goodwin, *Tetrahedron Letters*, 1974, 3773.

Two species of the Phaeophyta, *Laminaria saccharina* and *Ascophyllum nodosum*, in which fucosterol is predominant, also contain 24-ketocholesterol and saringosterol (66) at concentrations of 1.8 and 6% of total sterol.[97] While the latter compound might be a product of aerial oxidation of fucosterol, its detection in several brown algae implies that it may be a true metabolite in these species. The structure and stereochemistry of the two sterols of *Codium fragile* (Chlorophyceae; Siphonales) have been established as (24*S*)-24-ethylcholesta-5,25-dienol (67a) and the previously unreported methyl homologue (67b) (codisterol).[98] It is thought that the alkylation mechanism in this species is similar to that observed in *Trebouxia* (Chlorococcales)[65] except for the absence of the Δ^{25}-reductase from *C. fragile.*

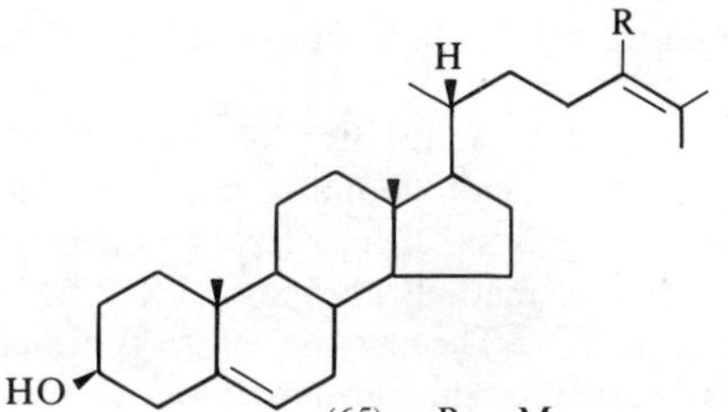

(65) a; R = Me
b; R = Et

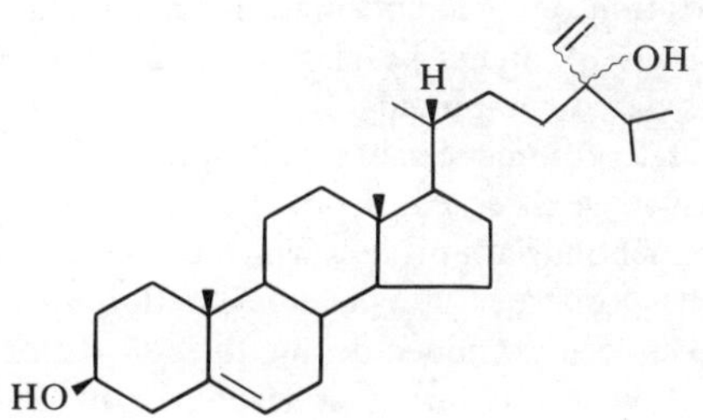

(66)

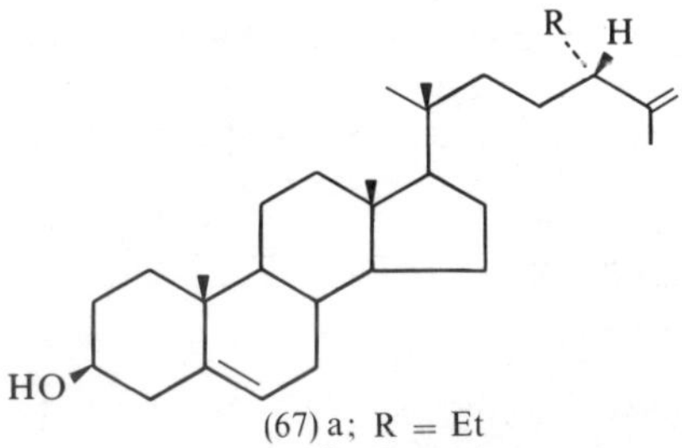

(67) a; R = Et
b; R = Me

A survey of nine other species of green algae reveals that all contain 24-alkylsterols[99] In every case where C-24 stereochemistry could be determined, the 24β (designated by IUPAC/IUB rules as 24*S* in side-chain-saturated sterols, but 24*R* in the presence of a Δ^{22} double-bond) configuration was found. This stereochemistry now appears almost universal in algal sterols. In contrast, however, the major sterol of the diatom *Phaeodactylum tricornutum* (Chrysophyta; Bacillariophyceae) has the 24α configuration.[100] (24*S*)-24-Methylcholesta-5,22*E*-dienol (68) incorporates two deuterium atoms from $[^2H_3]$methionine, indicating the intermediacy of a 24-methylene-sterol. In view of this result, earlier unproven assignments of the 24β

[97] L. M. Safe, C. J. Wong, and R. F. Chandler, *J. Pharm. Sci.*, 1974, **63**, 464.
[98] I. Rubinstein and L. J. Goad, *Phytochemistry* 1974, **13**, 481.
[99] G. W. Patterson, *Comp. Biochem. Physiol.*, 1974, **47B**, 453.
[100] I. Rubinstein and L. J. Goad, *Phytochemistry*, 1974, **13**, 485.

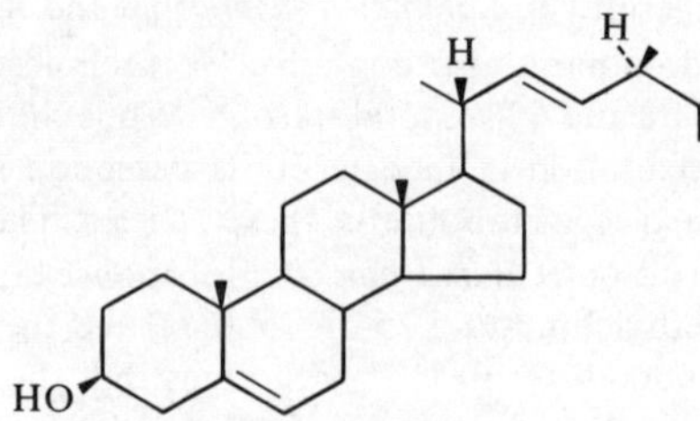

(68)

stereochemistry to sterols of diatoms[82,101] warrant more detailed investigation. The position of algae and diatoms in the marine food-chain makes it probable that the sterols of higher species, such as invertebrates, will include both 24α- and 24β-alkylated compounds.

The commencement of respiration during aerobic adaptation of anaerobically grown yeast results from the development of the promitochondria into functional mitochondria, a process which is accompanied by the onset of ergosterol biosynthesis. The properties of *S*-adenosylmethionine:Δ^{24}-sterol methyltransferase (E.C.2.1.1.41) have been examined during the adaptation process.[102] The enzyme, which is present at low levels during anaerobic growth, is located in the promitochondria–mitochondria, but it is synthesized on cytoplasmic ribosomes and is coded for in nuclear DNA. Inhibition by alkali-metal and ammonium cations is observed *in vitro*.[103] Three different enzyme activities have been discerned, only one of which may be responsible for *in vitro* alkylation of zymosterol. Comparison of Arrhenius kinetics of this enzyme and cytochrome oxidase in wild and mutant strains reveals a lower

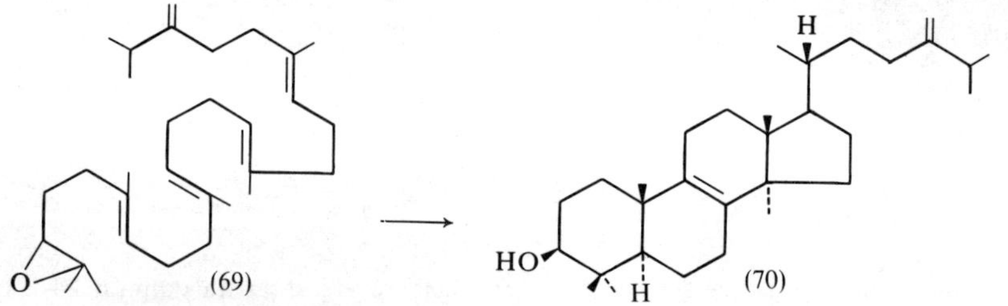

(69) (70)

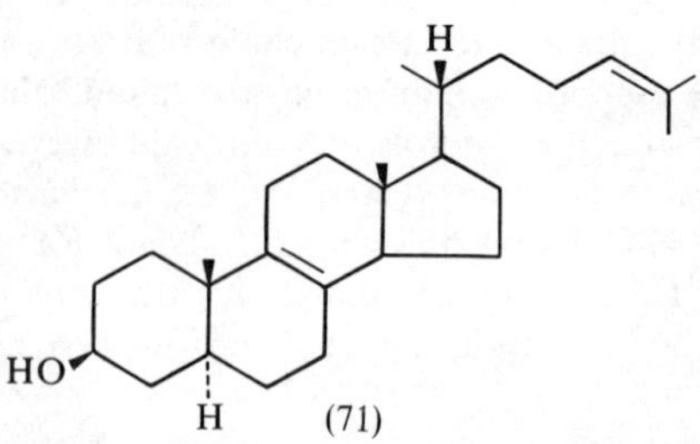

(71)

[101] A. Kanazawa, M. Yoshioka, and S. Teshima, *Bull. Jap. Soc. Sci. Fish.*, 1971, **37**, 899.
[102] E. D. Thompson, R. B. Bailey, and L. W. Parks, *Biochim. Biophys. Acta*, 1974, **334**, 116.
[103] R. B. Bailey, E. D. Thompson and L. W. Parks, *Biochim. Biophys. Acta*, 1974, **334**, 127.

transition temperature for the enzyme in mitochondria of mutant strains, while, in promitochondria, no difference between strains was observed. This may reflect the effects of the normal $\Delta^{5,7}$-sterols and the mutant Δ^{8}-sterols on properties of the mitochondrial membrane and its constituent enzymes.[104] In another study, alkylation of endogenous substrates, following aerobic adaptation, was shown to occur predominantly at the 4-desmethyl-sterol level.[105] No alkylation of 4,4-dimethyl-sterols could be demonstrated, while less than 2% of alkylation involved 4α-methyl-sterols. Thus, in contrast to higher plants, C-24-methylation in yeast occurs predominantly after C-4-demethylation. Interestingly, exogenous 22-methylene-22-dihydrosqualene 2,3-epoxide (69) is cyclized in this organism to 24-methylenedihydrolanosterol (70). Zymosterol (71) and zymost-7-enol constitute the major esterified sterols in the adapting yeast, though a highly reversible esterification process also appears to involve lanosterol and its 14-desmethyl derivatives.

General Aspects.—Particulate enzymes from *Nicotiana tabacum* L. include UDP-glucose:sterol glucosyltransferase.[106] The reaction of sterols with UDP-glucose is stimulated by ATP, while slight inhibition by UDP-glucose was observed. Intact seedlings incorporate labelled sterols into acetylated glycosides more rapidly than into non-acetylated glycosides. Formation of the acetylated derivatives was shown to occur by both acetylation of the glycosides and transfer of the acyl-sugar to the sterol. A study of DL-[2-^{14}C]MVA incorporation into glycosides and fatty acid esters of triterpenes and phytosterols reveals more rapid labelling of cycloartenol and phytosterol esters than of glycosides[107] in the aerial parts of the tobacco plant. However, in the roots, glycosides incorporated MVA faster than the free sterols.

Further Metabolism of Plant Sterols.—Etiolated potato sprouts convert [4-^{14}C]-cholesterol into cholest-4-en-3-one and 26-hydroxycholesterol (72).[108] This first report of the diol in plants provides a possible missing link in sapogenin biosynthesis since it has previously been shown to be metabolized to diosgenin in *Digitalis floribunda*. The first evidence for a pathway from 24-alkylated sterols to sapogenins has been provided by the incorporation of [4-^{14}C,22,23-^{3}H$_4$]sitosterol into diosgenin in 0.92% yield by suspension cultures of *Dioscorea deltoidea*.[109] Under comparable

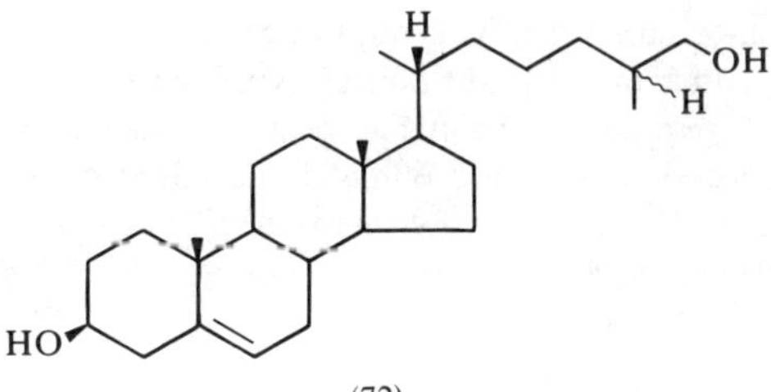

(72)

[104] E. D. Thompson and L. W. Parks, *Biochem. Biophys. Res. Comm.*, 1974, **57**, 1207.
[105] L. W. Parks, C. Anding, and G. Ourisson, *European J. Biochem.*, 1974, **43**, 451.
[106] P. B. Bush and C. Grunwald, *Plant Physiol.*, 1974, **53**, 131.
[107] F. J. Evans, *Planta*, 1974, **116**, 99.
[108] E. Heftmann and M. L. Weaver, *Phytochemistry*, 1974, **13**, 1801.
[109] S. J. Stohs, J. J. Sabatka, and H. Rosenberg, *Phytochemistry*, 1974, **13**, 2145.

conditions *ca.* 7% incorporation of cholesterol is obtained. The ^3H:^{14}C ratio in this experiment dropped from 5.0 to 2.3 in the sapogenin. Thus introduction of oxygen at C-22 proceeds without any loss of hydrogen atoms from C-23, and it is suggested that a mixed-function oxygenase reaction may be involved. However, the observed loss of the 20α hydrogen atom during sapogenin formation from cholesterol[110] implies that a $\Delta^{20(22)}$-dehydrogenation process could also be involved.

The conversion of cholesterol into tigogenin (74) in *Digitalis lanata* has been shown to involve oxidation of the methyl group derived from C-6 of mevalonic acid.[111] Consequently, addition of the hydrogen atom to C-25 of the precursor (73) occurs on the *si*-face of the double bond. The stereochemistry of addition of hydrogen to C-24 during this reduction has now been investigated.[112] The 4-*pro-R* hydrogen of MVA was located in the equatorial 24-*pro-S* position in (74), defining the approach of the incoming proton as being on the *re*-face of the double bond (73). Thus an overall

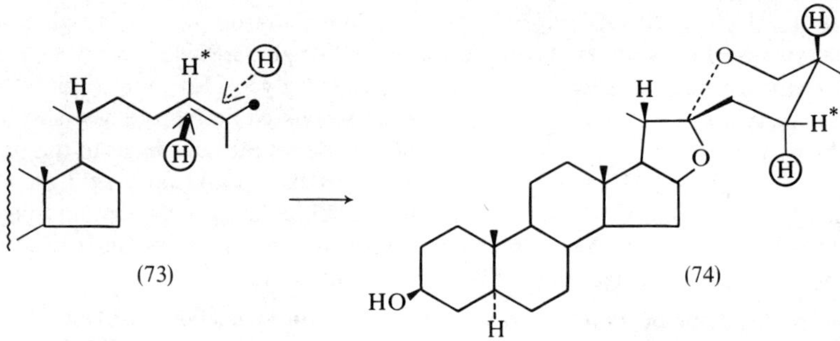

(73) (74)

Scheme 9

trans-reduction process is involved (Scheme 9), in contrast to the *cis*-reduction of Δ^{24} in cholesterol formation by rat liver. It has also been reported that, in *Agave attenuata*, the (25*S*)-sapogenin sarsasapogenin (75) is formed by oxidation of the terminal carbon derived from C-2 of mevalonate.[113] This result demonstrates that the pathways to (25*R*)- and (25*S*)-sapogenins diverge at the point of oxidation of alternative prochiral terminal methyl groups of cholesterol (or the methyl groups of the Δ^{24} precursor) rather than by addition of hydrogen to opposite faces of the Δ^{24} intermediate (73). A parallel situation has been revealed in the formation of the alkaloids solasodine and tomatidine. While the terminal methyl of solasodine (76) is derived from C-2 of MVA,[114] that of tomatidine (77) originates from C-6[115] Once again, the Δ^{24} intermediate in each case has undergone hydrogen addition to C-25 from the *si*-face.

110 A. F. Wickramasinghe, B. P. Burrows, R. K. Sharma, J. B. Greig, and E. Caspi, *Phytochemistry*, 1969, **8**, 1433.
111 R. Joly and C. Tamm, *Tetrahedron Letters*, 1967, 3535.
112 L. Canonica, F. Ronchetti, and G. Russo, *J.C.S. Perkin I*, 1974, 1670.
113 A. G. Gonzalez, C. G. Francisco, R. Freire, R. Hernandez, J. A. Salazar, and E. Suarez, *J.C.S. Chem. Comm.*, 1974, 905.
114 A. R. Guseva and V. A. Paseshnichenko, *Biokhimiya*, 1962, **27**, 721.
115 F. Ronchetti and G. Russo, *J.C.S. Chem. Comm.*, 1974, 785.

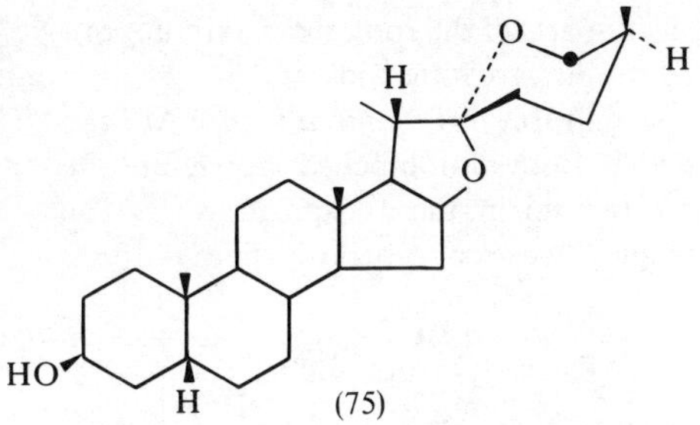

(75)

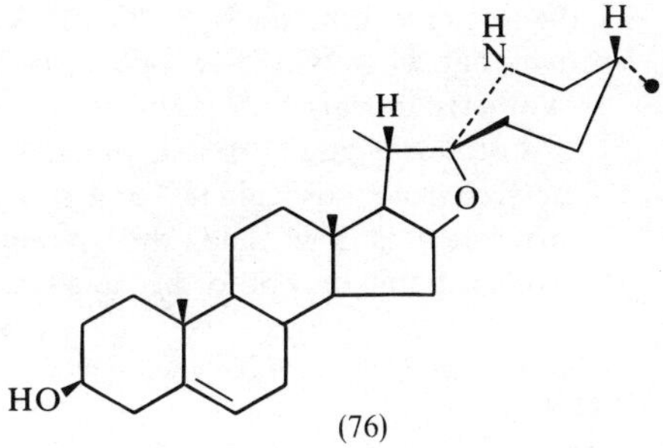

(76)

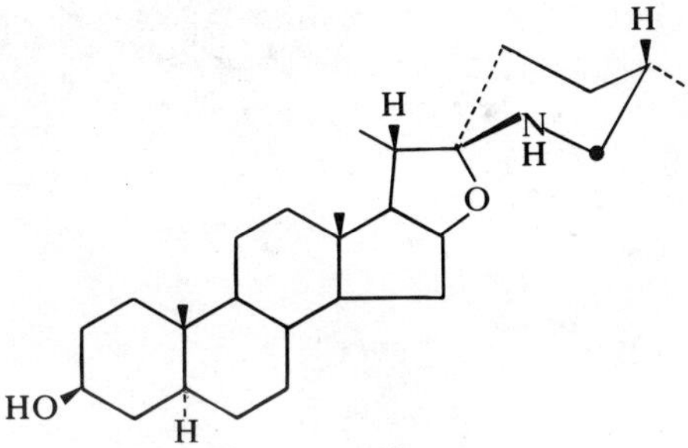

(77)

Synthesis and biogenesis of the steroidal *Solanum* alkaloids have been reviewed in the 1973 Morton Lecture.[116] The formation and breakdown of α-tomatine (78) are discussed in a review of the biochemistry and biology of this steroidal glycoalkaloid.[117]

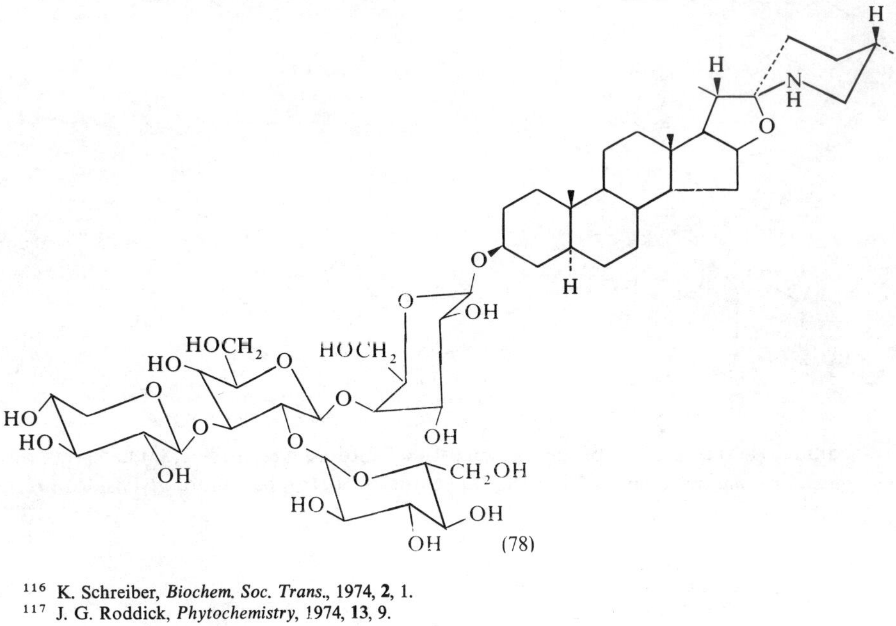

(78)

[116] K. Schreiber, *Biochem. Soc. Trans.*, 1974, **2**, 1.
[117] J. G. Roddick, *Phytochemistry*, 1974, **13**, 9.

Labelling of α-tomatine by [^{14}C]MVA lactone in cultured excised tomato roots has shown that, while synthesis appears to occur in all parts of the root, the maximum rate of synthesis appears to be associated with the rapidly growing root tip.[118]

Prototokorogenin (79), a saponin from tissue cultures of *Dioscorea tokoro* M., can be derived from diosgenin (80) and yonogenin (81). Both radiolabelled sapogenins are incorporated in low yield, with retention of tritium in the 16α-position.[119] The furostan saponin may be formed as a result of high glucose concentration in the culture

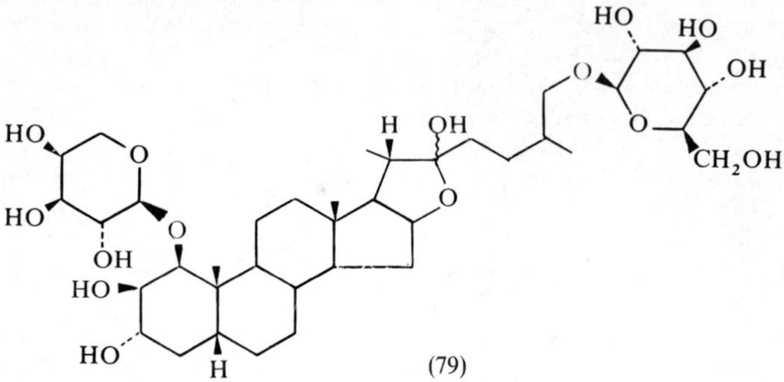

(79)

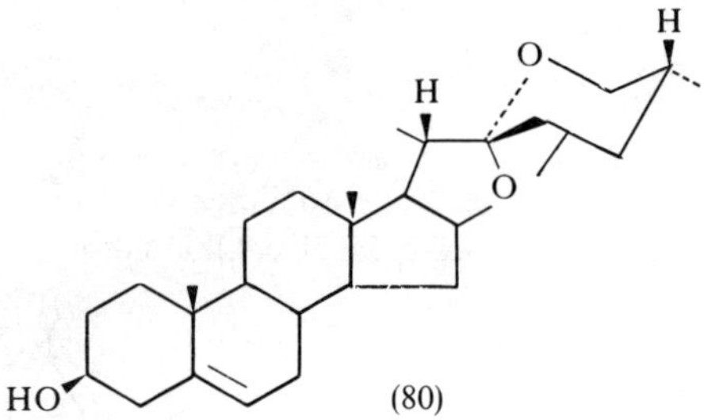

(80)

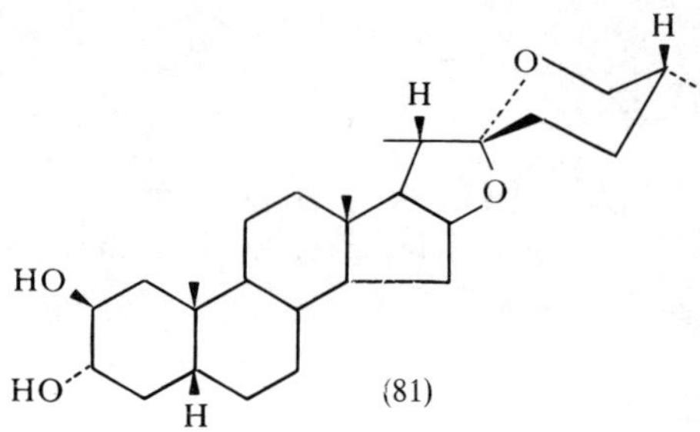

(81)

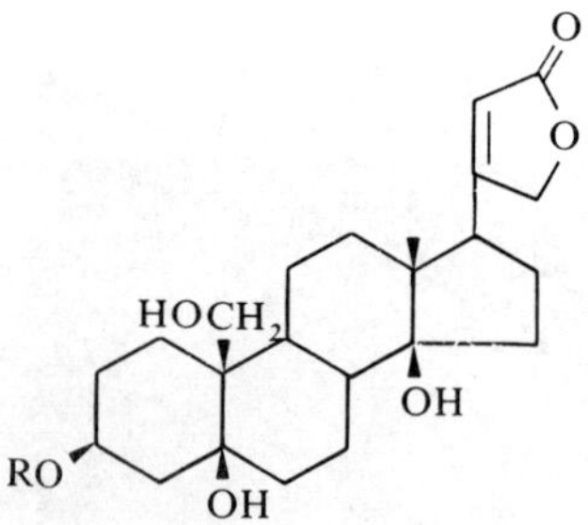

(82) R = rhamnose

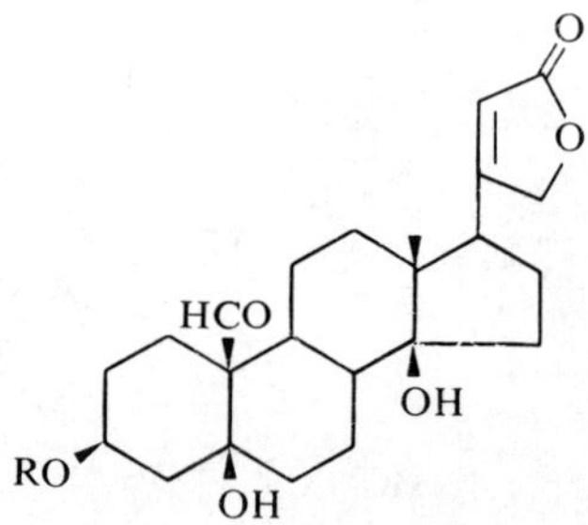

(83) R = rhamnose

medium. The conversion of convallatoxol (82) into convallatoxin (83) in leaves of *Convallaria majalis* suggests that the biosynthesis of strophanthidin glycosides may involve oxidation after glycoside formation.[120]

118 J. G. Roddick, *Phytochemistry*, 1974, **13**, 1459.
119 Y. Tomita and A. Uomori, *Phytochemistry*, 1974, **13**, 729.
120 W. Kubelka, B. Kopp, K. Jentzsch, and H. Ruis, *Phytochemistry*, 1974, **13**, 1805.

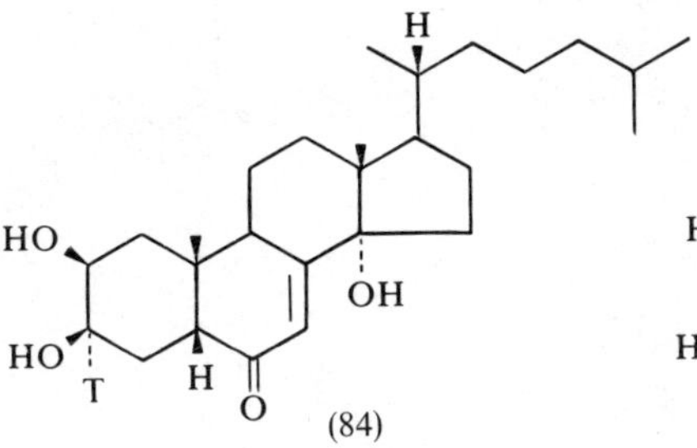

(84)

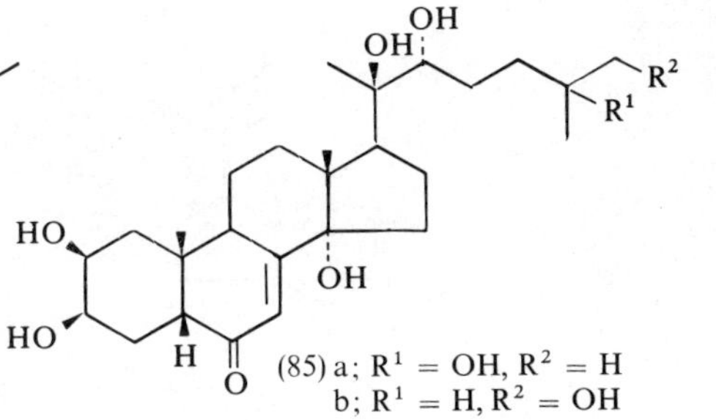

(85) a; R^1 = OH, R^2 = H
b; R^1 = H, R^2 = OH

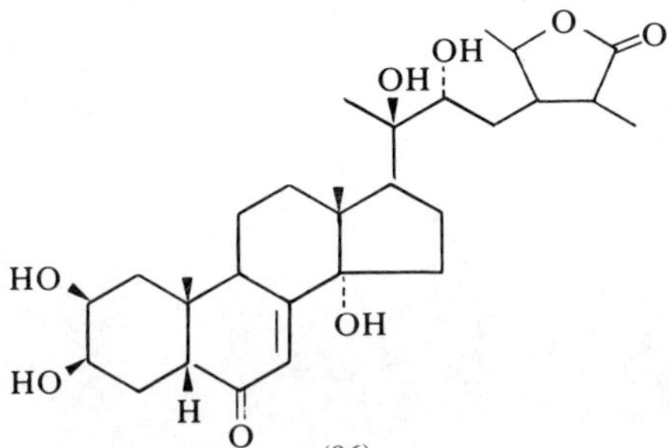

(86)

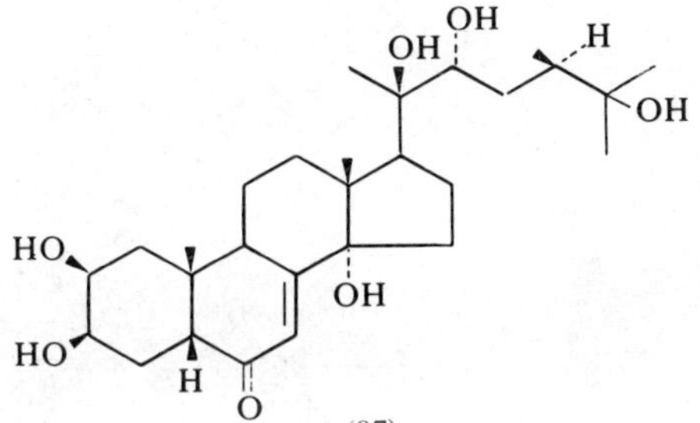

(87)

The ring-oxidized sterol (84), a potential intermediate in ecdysone biosynthesis, has been synthesized and incorporated into ecdysterone (85a) and inokosterone (85b) by young leaves of *Achryanthes fauriei*, indicating that side-chain hydroxylation may occur at a late stage of the sequence.[121] The formation of cyasterone (86) in seedlings of *Cyathula capitata* has been studied by administration of $[2\text{-}^{14}C, (4R)\text{-}4\text{-}^{3}H_1]$-mevalonate.[122] The phytosterols sitosterol and stigmasterol contain only two tritium atoms, presumed to be at C-17 and C-20, while cyasterone retains only one tritium label. If, as appears probable, this is located at C-17, formation of the phytoecdysone must involve either a 24-ethyl-sterol or a $\Delta^{24(28)}$ intermediate which subsequently loses a labelled hydrogen from C-25. The C-24-methylecdysone makisterone A (87) has been shown, by side-chain degradation, to possess the 24*R* configuration,[123] implying derivation of the ecdysone from campesterol.

Biosynthesis of the side-chain of antheridiol (92) (Scheme 10) in the aquatic fungus *Achlya bisexualis* has been examined by feeding and trapping experiments.[124] Fucosterol (88) was shown to be incorporated into the hormone, together with the triene (90) and the acid (91). The acid (89) was not significantly incorporated. Trapping experiments confirmed the conversion of fucosterol into the triene (90) and of the latter into the acid (91). Thus, introduction of the Δ^{22} double-bond precedes side-chain oxidation, while the substrate for oxygenation at C-22/23 may be the acid (91).

High incorporation of acetate into ergosterol is reported during sporulation of yeast.[125] Ergosterol distribution during growth of *Neurospora crassa* has been

121 Y. Tomita and E. Sakurai, *J.C.S. Chem. Comm.*, 1974, 434.
122 R. Boid, H. H. Rees, and T. W. Goodwin, *Biochem. Soc. Trans.* 1974, **2**, 1066.
123 B. Danieli, G. Ferrari, J. Krepinsky, J. Palmisamo, and D. Sardini, *J.C.S. Chem. Comm.*, 1974, 745.
124 C. R. Popplestone and A. M. Unrau, *Canad. J. Chem.*, 1974, **52**, 462.
125 A. K. Bose and J. J. Savarese, *Experientia*, 1974, **30**, 1489.

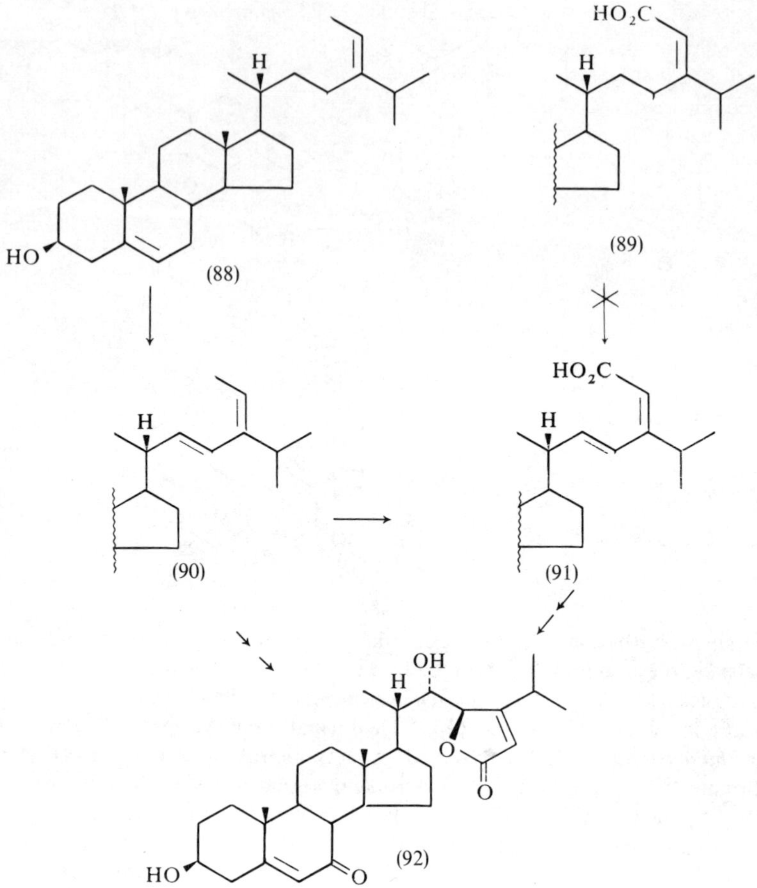

Scheme 10

examined.[126] The free sterol is present in rapidly growing mycelium, while a tightly bound water-soluble derivative predominates during senescence and in conidia. A study of fatty acid esters of sterols, triglycerides, and phospholipids in *Phycomyces blakesleeanus* reveals a non-random esterification process.[127] While ester distribution varies with the age of the culture, free sterol and triterpene levels do not, suggesting that the esters may be involved in a storage process.[128] However, this hypothesis is not supported by starvation experiments in which, while triglyceride levels drop, the sterol ester content is unchanged.[129]

[126] C. G. Elliott, B. A. Knights, and J. A. Freeland, *Biochim. Biophys, Acta.*, 1974, **360**, 339.
[127] E. I. Mercer and K. Bartlett, *Phytochemistry*, 1973, **13**, 1099.
[128] K. Bartlett, M. J. Keat, and E. I. Mercer, *Phytochemistry*, 1974, **13**, 1107.
[129] K. Bartlett and E. I. Mercer, *Phytochemistry*, 1974, **13**, 1115.

7 Triterpenoids and Steroids in Invertebrates

Insects.—Phytophagous insects, Crustacea, and other invertebrates, which do not synthesize sterols, are capable of dealkylating phytosterols to provide the cholesterol required for structural and hormonal functions.[1, 2, 130] The mechanism of dealkylation has been explored by feeding potential intermediates and by the use of inhibitors.

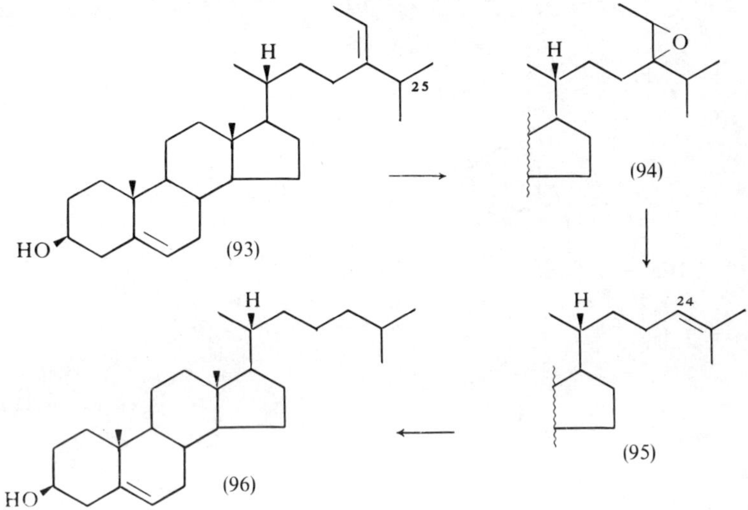

Scheme 11

In the silkworm *Bombyx mori*, dealkylation of sitosterol is thought to involve dehydrogenation to fucosterol (93). Fucosterol 24,28-epoxide (94) is then converted *via* desmosterol (95) into cholesterol (96) (Scheme 11), the final reduction step being susceptible to inhibition by some vertebrate hypocholesterolemic agents. In accord with this mechanism, [25-^{3}H]-24ξ-ethylcholesterol is converted into [24-^{3}H]-desmosterol by *B. mori*.[131] A similar hydrogen transfer has been demonstrated during dealkylation of [25-^{3}H, 26-^{14}C]clionasterol (57) by *Tenebrio molitor*.[132] In the presence of triparanol, desmosterol was isolated, and the tritium atom was located at C-24 by degradation. A mechanism for cleavage of the epoxide that is consistent with the observed hydrogen migration has been proposed (Scheme 12).

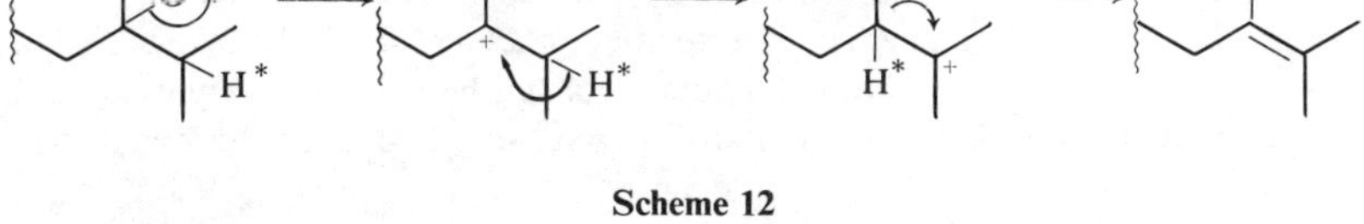

Scheme 12

[130] H. H. Rees and T. W. Goodwin, *Biochem. Soc. Trans.*, 1974, **2**, 1027.
[131] Y. Fujimoto, N. Awata, M. Morisaki, and N. Ikekawa, *Tetrahedron Letters*, 1974, 4335.
[132] P. J. Pettler, W. J. S. Lockley, H. H. Rees, and T. W. Goodwin, *J.C.S. Chem. Comm.*, 1974, 844.

The overall route is supported by studies on the nutritional effects of steroids in *B. mori*.[133] Side-chain-hydroxylated derivatives (97a—d) fail to support growth, as do 24-keto- and 24-hydroxy-cholesterol. Curiously, while campesterol and 24-methylene-cholesterol are active, the epoxide (98) is only partially effective, suggesting that the dealkylation of 24-methyl-sterols may not be analogous to that of the 24-ethyl homologues. 24,28-Iminofucosterol (99a) is reported to disrupt development of *B. mori*

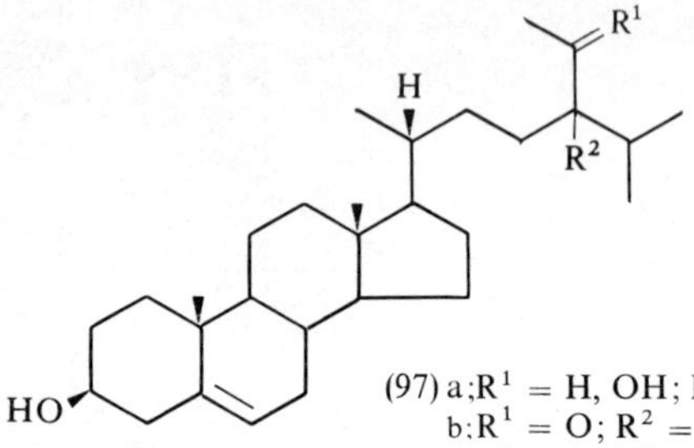

(97) a; R^1 = H, OH; R^2 = H
b; R^1 = O; R^2 = H
c; R^1 = O; R^2 = OH
d; R^1 = H, OH; R^2 = OH

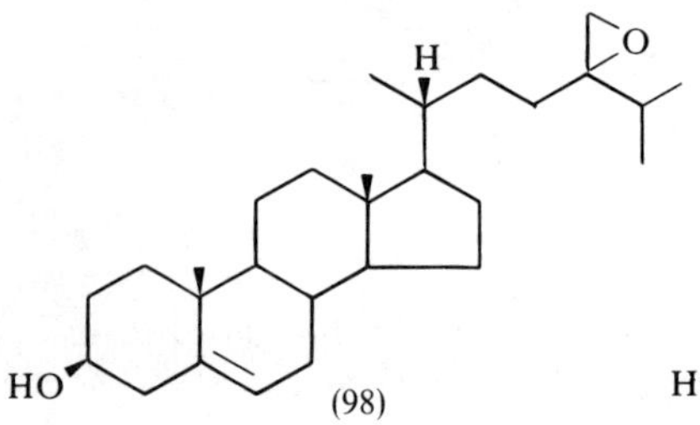

(98)

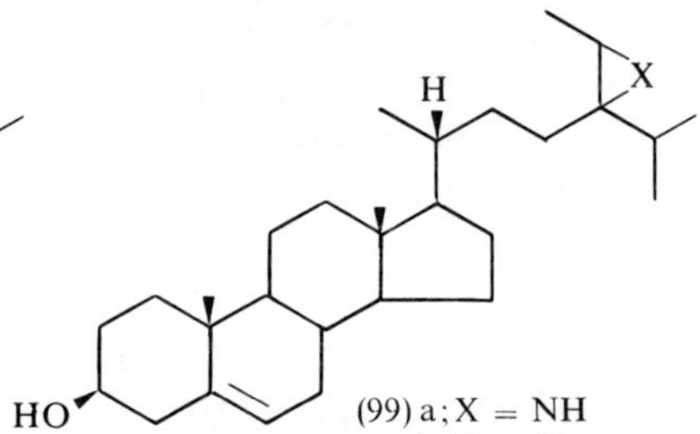

(99) a; X = NH
b; X = O

larvae.[134] However, its mode of action is unclear. Competitive inhibition of the corresponding epoxide seems unlikely, since no accumulation of (99b) was observed. Nor does addition of cholesterol to the diet prevent the inhibitory effect of the aziridine.

Ecdysone biosynthesis from cholesterol in *Calliphora erythrocephala* involves introduction of the Δ^7 double-bond by removal of the 7β- and 8β-hydrogen atoms.[135] The same stereochemistry of elimination is reported for the conversion of cholesterol into cholesta-5,7-dienol by aseptic larvae of this insect, further supporting the intermediacy of the diene in ecdysone biosynthesis.[136] An analogous dehydrogenation of 5α-cholestanol to cholest-7-enol was also observed, though the latter sterol is not further metabolized to the diene. The intermediacy of cholesta-5,7-dienol has also been confirmed in *Bombyx mori*,[137] though cholesterol sulphate, (22*R*)-22-hydroxycholesterol, and 25-hydroxycholesterol were not incorporated, supporting the idea that side-

133 M. Morisaki, H. Ohotaka, N. Awata, and N. Ikekawa, *Steroids*, 1974, **24**, 165.
134 Y. Fujimoto, M. Morisaki, N. Ikekawa, Y. Horie, and S. Nakasone, *Steroids*, 1974, **24**, 367.
135 I. F. Cook, J. G. Lloyd-Jones, H. H. Rees, and T. W. Goodwin, *Biochem. J.*, 1974, **136**, 135.
136 P. Johnson, H. H. Rees, and T. W. Goodwin, *Biochem. Soc. Trans.*, 1974, **2**, 1062.
137 H. Kazuaori and O. Eiji, *J. Insect Physiol.*, 1974, **20**, 2375.

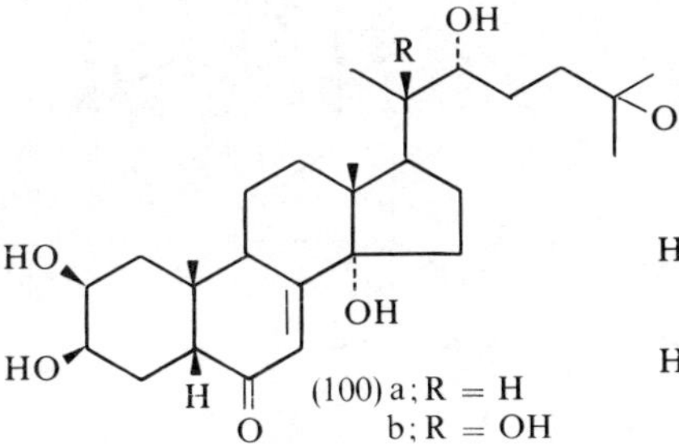

(100) a; R = H
b; R = OH

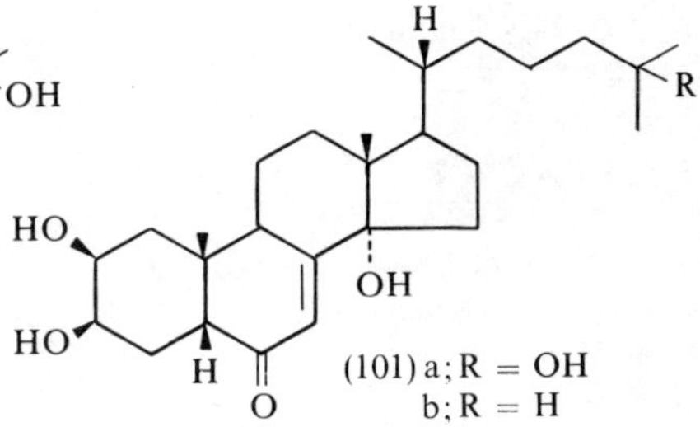

(101) a; R = OH
b; R = H

chain hydroxylation occurs late in the biosynthetic sequence. Evidence for biosynthesis of ecdysone (100a) but not ecdysterone (100b) in the prothoracic glands of *B. mori*[138] and *Manduca sexta*[139] has been obtained using tissue cultures. The suggestion that ecdysone acts as a prohormone is supported by structure–activity tests using cultured wing-discs of *Plodia interpunctella*,[40] and by the observation that target organs slowly convert it into ecdysterone (100b).[141] 3-Epi-ecdysone has been isolated from the meconium of *Manduca sexta*.[142] Midgut preparations from fifth instar larvae contain enzymes capable of converting ecdysone into its 3α-epimer. This dehydrogenase–isomerase activity is located in the supernatant fraction and requires NADH or NADPH for maximal activity.[143] In addition to ecdysone, substrates include ecdysterone (100b) and 22-deoxyecdysone (101a) but not 22,25-dideoxyecdysone (101b) or cholesterol, indicating a specific requirement for the 25-hydroxy-group. In view of the low biological activity of the 3α-epimer (102) and the occurrence of the corresponding 3-ketone in other species, it is postulated[143] that the epimerization process may be a metabolic control point for hormone activity during moulting (Scheme 13).

Other Invertebrates.—*Triterpenoid and Sterol Biosynthesis.* Exogenous cholesterol inhibits biosynthesis of the triterpene tetrahymanol(103) in the protozoan *Tetrahymena pyriformis*.[144] Two major sites of inhibition have been defined in whole cells. Rapid inhibition of acetate conversion into mevalonate is observed, though 3-hydroxy-3-methylglutaryl-CoA reductase, a cytosolic enzyme, is unaffected. Mevalonate conversion into tetrahymanol is blocked at both squalene synthetase and squalene cyclase. While neither cholesterol nor tetrahymanol directly inhibit the enzymes *in vitro*, protein synthesis is regulated. It is suggested that cholesterol affects membrane structure and hence the efficiency of membrane-bound enzymes.

Despite the fact that lanosterol has been identified in the free-living nematode *Panagrellus redivivus*, the organism requires dietary sterol for axenic growth.[145]

138 H. Chino, S. Sakurai, T. Ohtaki, N. Ikekawa, H. Miyazaki, M. Ishibashi, and H. Abuki, *Science*, 1974, **183**, 529.
139 D. S. King, W. E. Bollenbacher, D. W. Borst, W. V. Vedeckis, J. D. O'Connor, P. I. Ittycheriah, and L. I. Gilbert, *Proc. Nat. Acad. Sci. U.S.A.*, 1974, **71**, 793.
140 H. Oberlander, *Experientia*, 1974, **30**, 1410.
141 D. S. King and E. P. Marks, *Life Sci.*, 1974, **15**, 147.
142 M. J. Thompson, J. N. Kaplanis, W. E. Robbins, S. R. Dutky, and H. N. Nigg, *Steroids*, 1974, **24**, 359.
143 H. N. Nigg, J. A. Svoboda, M. J. Thompson, J. N. Kaplanis, S. R. Dutky, and W. E. Robbins, *Lipids*, 1974, **9**, 971.
144 A. S. Beedle, K. W. Munday, and D. C. Wilton, *Biochem. J.*, 1974, **142**, 57.
145 J. D. Willett and W. L. Downey, *Comp. Biochem. Physiol.*, 1973, **45B**, 139.

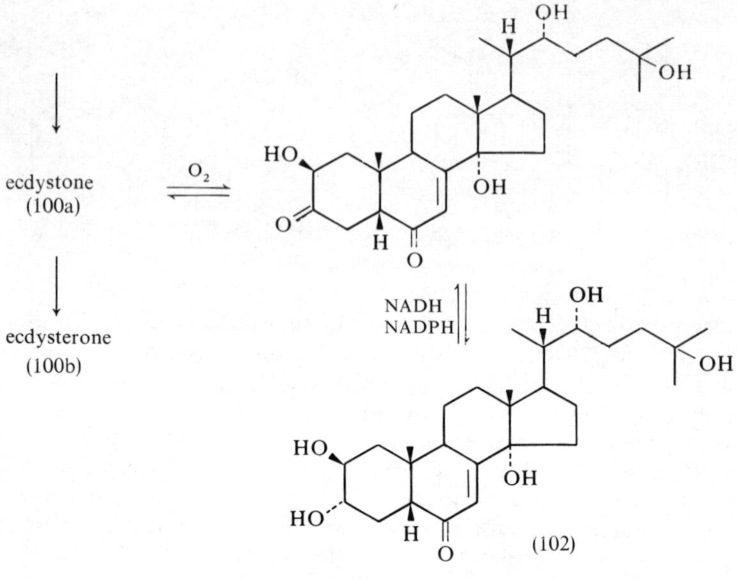

Scheme 13

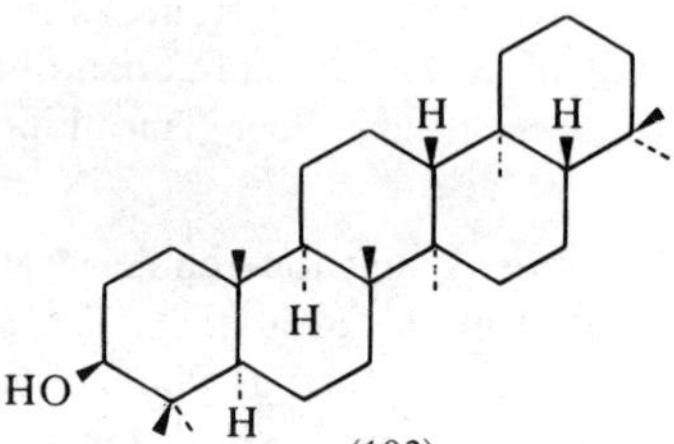

(103)

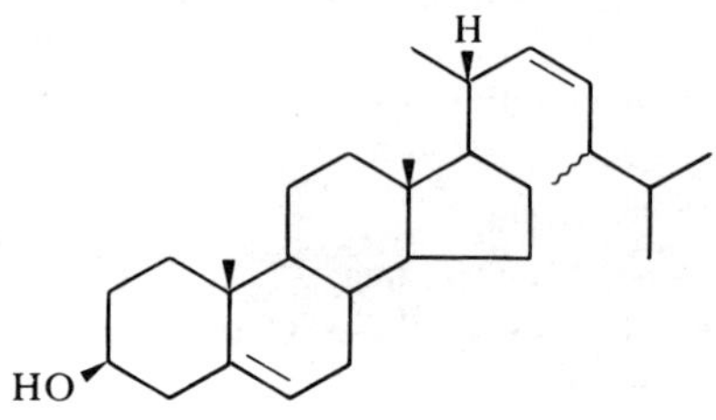

(104)

While neither acetate nor mevalonate is incorporated into sterols, an earlier report noted conversion of mevalonate into non-saponifiable material.[146] Low conversion of [4-^{3}H]squalene 2,3-oxide into lanosterol has now been reported.[147] Thus it appears that the normal pathway of sterol biosynthesis is at least partially operative in this nematode, although metabolism beyond lanosterol is blocked. The sea urchin *Echinus esculentus* incorporates mevalonate into squalene, lanosterol, and desmosterol.[148] The absence of a Δ^{24}-reductase is supported by the conversion of exogenous lanost-8-enol into cholesterol in this organism. No incorporation of mevalonate into C_{26}, C_{28}, or C_{29} sterols was observed, suggesting that these sterols are of dietary origin. Neither dealkylation of sitosterol nor metabolism of cholesterol could be

[146] R. J. Cole and L. R. Krusberg, *Life Sci.*, 1968, **7**, 713.
[147] J. D. Willett and W. L. Downey, *Biochem. J.*, 1974, **138**, 233.
[148] A. G. Smith and L. J. Goad, *Biochem. J.*, 1974, **142**, 421.

demonstrated. A novel sterol isolated from the clam *Tapes philippinarum* has been identified as 22-*cis*-24-methylcholesta-5,22-dienol (104), only the second 22-*cis*-sterol to be isolated from natural sources.[149] Low incorporation of mevalonate into sterols of *Anemonia sulcata* has been achieved.[150] Conversion of acetate and mevalonate into cholesterol and cholest-7-enol is reported to occur in the mollusc *Lepidochitona cinerea.*[151] The mussel *Mytilus edulis* and abalone *Haliotis gurneri* also possess the full pathway of sterol synthesis.[152] However, in the worms *Echinococcus granulosus* and *Taenia hydatigena* the sequence is blocked at farnesol.[153]

Metabolism of Sterols. Further sterols containing novel side-chain structures have been isolated from marine sources. It remains to be seen whether amuresterol (105)

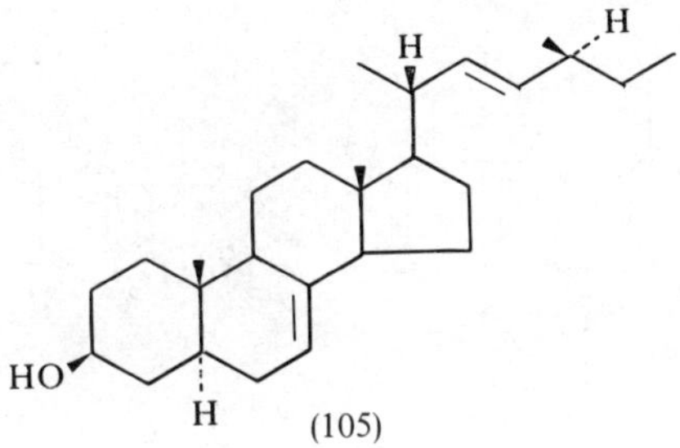

(105)

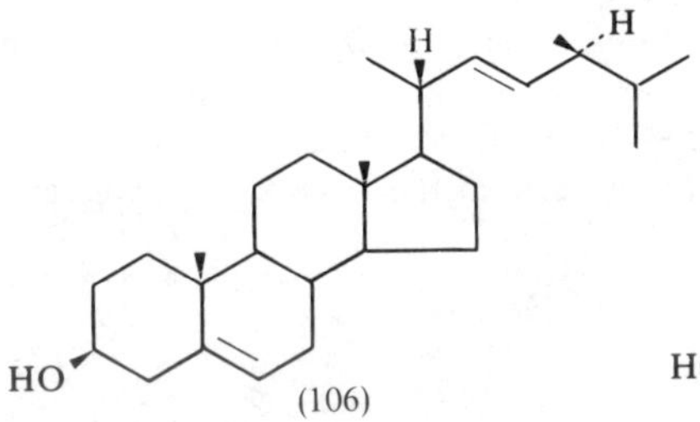

(106)

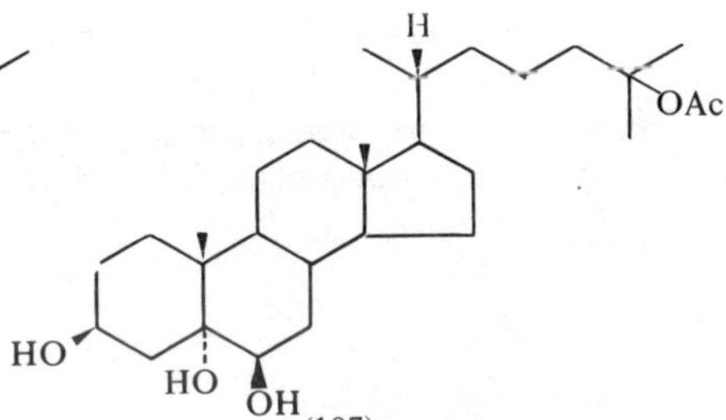

(107)

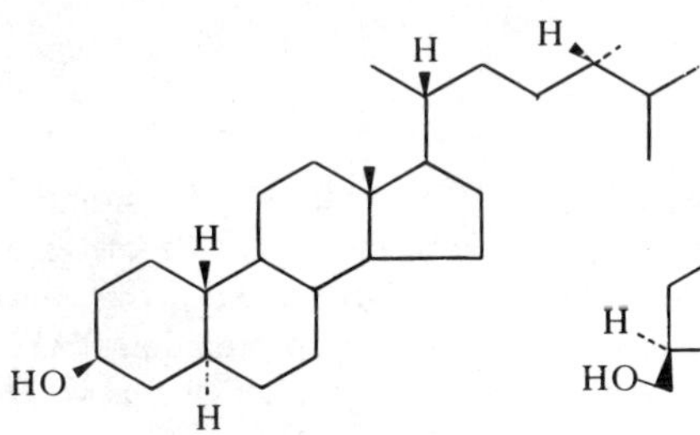

(108)

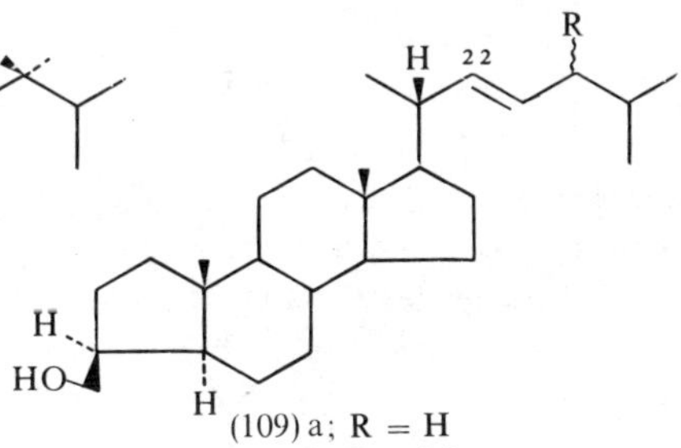

(109) a; R = H
b; R = Me
c; R = Et

[149] S. Teshima, A. Kanazawa, and T. Ando, *Comp. Biochem. Physiol.*, 1974, **47B**, 507.
[150] P. A. Voogt, J. M. van de Ruit, and J. W. A. van Rheenen, *Comp. Biochem. Physiol.*, 1974, **48B**, 47.
[151] P. A. Voogt and J. W. A. van Rheenen, *Comp. Biochem. Physiol.*, 1974, **47B**, 131.
[152] S. I. Teshima and A. Kanazawa, *Comp. Biochem. Physiol.*, 1974, **47B**, 555.
[153] G. J. Frayha, *Comp. Biochem. Physiol.* 1974, **49B**, 93.

from the starfish *Asterias amurensis*,[154] occelasterol (106) from the annelid *Pseudopotamilla occelata*,[155] and the tetrol acetate (107) of the alcyonarian *Sarcophyton elegans*[156] represent novel routes of sterol synthesis or metabolites of more common phytosterols. Novel modifications of the sterol A-ring are evident in the sterols (108) and (109a—c) of the sponges *Axinella polypoides*[157] and *A. verrucosa*,[158] respectively.

Metabolism of [4-^{14}C]progesterone by two species of starfish, *Asterias rubens* and *Marthasterias glacials*, leads to conjugates of 3β-hydroxy-5α-pregnan-20-one (110) and 3β,6α-dihydroxy-5α-pregnan-20-one.[159] The saponin of (111) contains 3β-sulphate and 6α-tetrasaccharide groups.

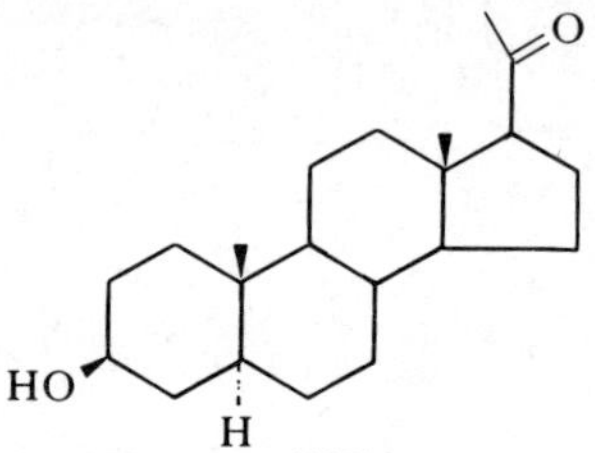

(110)

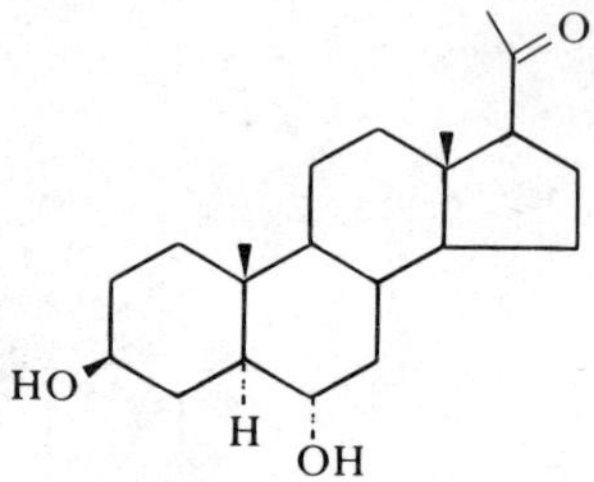

(111)

8 Carotenoids

Phytoene Formation.—Photoinduced carotenogenesis in a *Mycobacterium* species has been shown to involve photoinduction of the enzymes required for phytoene metabolism.[160] This response to light is now reported to include *de novo* synthesis of prephytoene pyrophosphate (112) and an increase in the level of geranylgeranyl

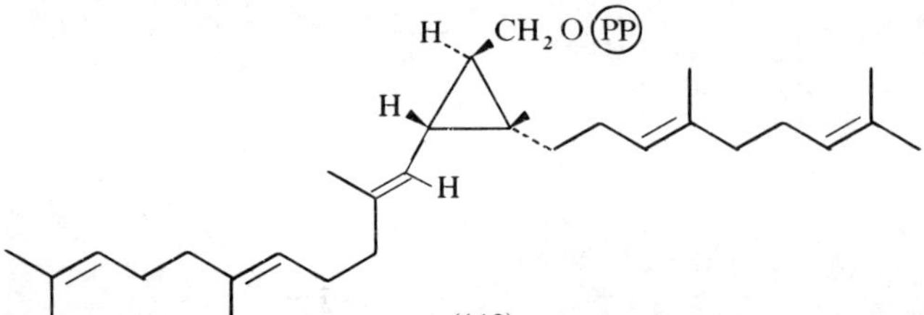

(112)

pyrophosphate synthetase (prenyltransferase),[161] in contrast to the situation in *Neurospora crassa*, in which the earliest known photo-inducible enzymes are the carotene dehydrogenases. Assay of prenyltransferase activity revealed the synthesis of two products, *viz*. a C-35 or C-40 polyprenyl pyrophosphate, production of which is unconnected with carotenogenesis and unaffected by light, and all-*trans*-geranyl-

[154] M. Kobayashi and H. Mitsuhashi, *Tetrahedron*, 1974, **30**, 2147.
[155] M. Kobayashi and H. Mitsuhashi, *Steroids*, 1974, **24**, 399.
[156] J. M. Moldowan, B. M. Tursch, and C. Djerassi, *Steroids*, 1974, **24**, 387.
[157] L. Minale and G. Sodano, *J.C.S. Perkin I*, 1974, 1888
[158] L. Minale and G. Sodano, *J.C.S. Perkin I*, 1974, 2380.
[159] J. Gaffney and L. J. Goad, *Biochem. J.*, 1974, **138**, 309.
[160] H. C. Rilling, *Biochim. Biophys. Acta*, 1964, **79**, 464.
[161] J. H. Johnson, B. C. Reed, and H. C. Rilling, *J. Biol. Chem.*, 1974, **249**, 402.

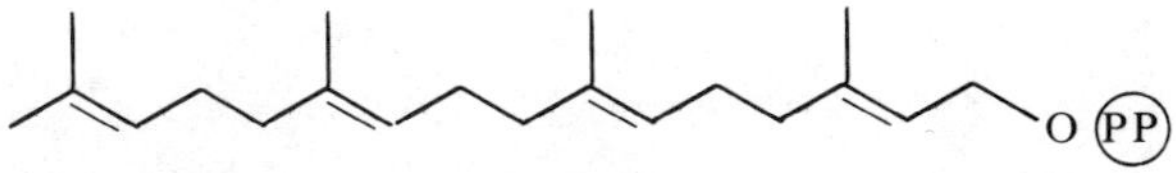

(113)

geranyl pyrophosphate (113). Both induced and un-induced cells synthesize exclusively the all-*trans*-isomer of (113), making it unlikely that the polyprenyl pyrophosphate results from prenylation of *cis,trans,trans*-geranylgeranyl pyrophosphate.

Blue-green mutants of the photosynthetic bacterium *Rhodospirillum rubrum* and a strain of the fungus *Mucor hiemalis* grown anaerobically in light produce both *cis*- and *trans*-phytoene,[162] as had been noted earlier when inhibitors of carotenogenesis were used.[163] One mutant of *R. rubrum* produces the *trans*-isomer predominantly in these conditions, though the *cis*-isomer is the major component when the strain is grown aerobically in the dark. Oxygen is required to permit synthesis of enzymes required for further desaturation of phytoene in *M. hiemalis*. The influence of temperature on light-induced carotenogenesis in *Neurospora crassa* results in variations in the quantity of carotenoids which accumulate in the subsequent dark-grown phase.[164] Optimal synthesis is obtained at 6 °C, while exposure to higher temperatures following illumination results in temperature-dependent synthesis of carotenoids. Sensitivity to inhibitors of protein synthesis is also observed during this period. It is postulated that the light reaction may lead to production of a compound which subsequently induces enzyme synthesis.

Desaturation of Phytoene.—A soluble enzyme system from plastids of red tomato fruits converts *cis*-[^{14}C]phytofluene into the *trans*-isomer, which is subsequently metabolized anerobically to *trans*-ζ-[^{14}C]carotene (114).[165] Both isomers of phytofluene are converted into neurosporene (115), lycopene (116), α-carotene (117), and β-carotene (118) under anaerobic conditions by this system. Maximum conversion of phytofluene into more unsaturated carotenoids requires $NADP^+$ and FAD, is enhanced by Mg^{2+} or Mn^{2+} ions, and shows little pH dependence over the range 7.0—8.5. Two varieties of tangerine tomato fruit also convert *cis*- and *trans*-phytofluene into *trans*-ζ-carotene.[166] Further desaturation leads to proneurosporene (120) and prolycopene (121), in addition to their *trans*-isomers and both α- and β-carotene. The intermediacy of *cis*-ζ-carotene (119) in the formation of the *cis*-carotenoids and its reversible conversion into the *trans*-isomer (114) has also been demonstrated.[167] However, no isomerization of prolycopene (121) to lycopene (116) or further metabolism of this *cis*-isomer could be detected. Thus *trans*-ζ-carotene (114) appears to be the branchpoint for poly-*cis*-carotenoid synthesis (Scheme 14), though a possible route of synthesis *via cis*-phytofluene cannot be ruled out. Of the cyclic carotenoids, only α- and β-carotenes are found in these tangerine tomato strains.[166]

The carotenoids of etiolated seedlings from mutant strains of barley (*Hordeum vulgare*) have been examined.[168] The *xan*^{21} mutant, considered to be blocked in

[162] B. Maudinas, R. Herber, and J. Villoutreix, *Biochim. Biophys. Acta,* 1974, **348**, 357.
[163] R. Herber, B. Maudinas, J. Villoutreix, and P. Granger, *Biochem. Biophys. Acta,* 1972, **280**, 194.
[164] R. W. Harding, *Plant Physiol.*, 1974, **54**, 142.
[165] A. A. Qureshi, A. G. Andrewes, N. Qureshi, and J. W. Porter, *Arch. Biochem. Biophys.*, 1974, **162**, 93.
[166] A. A. Qureshi, M. Kim, N. Qureshi, and J. W. Porter, *Arch. Biochem. Biophys.*, 1974, **162**, 108.
[167] A. A. Qureshi, N. Qureshi, M. Kim, and J. W. Porter, *Arch. Biochem. Biophys.*, 1974, **162**, 117.
[168] O. F. Nielsen and S. Gough, *Physiol. Plantarum.*, 1974, **30**, 246.

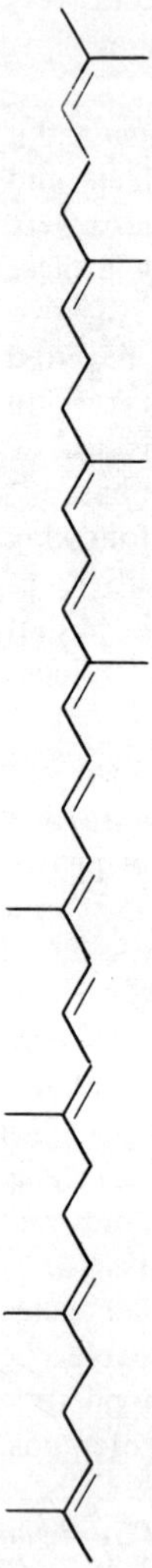

(114)

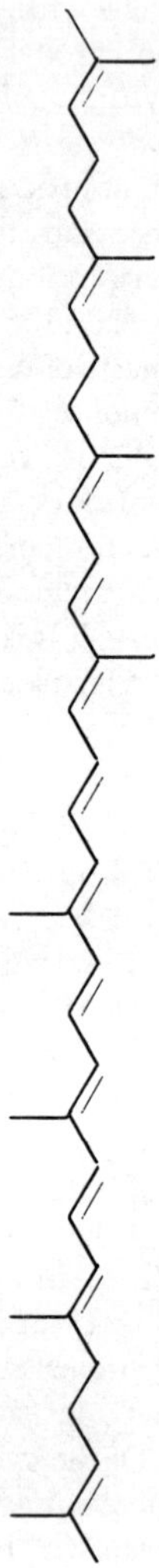

(115)

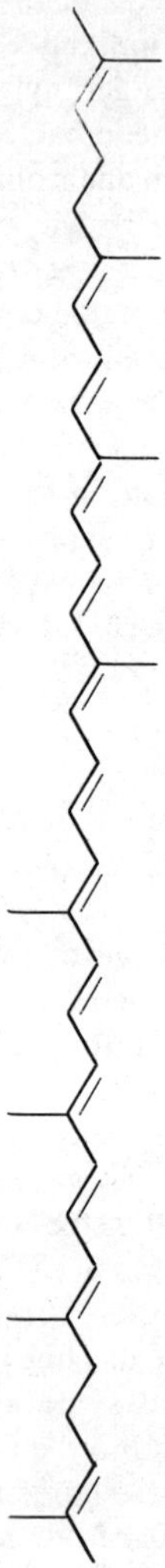

(116)

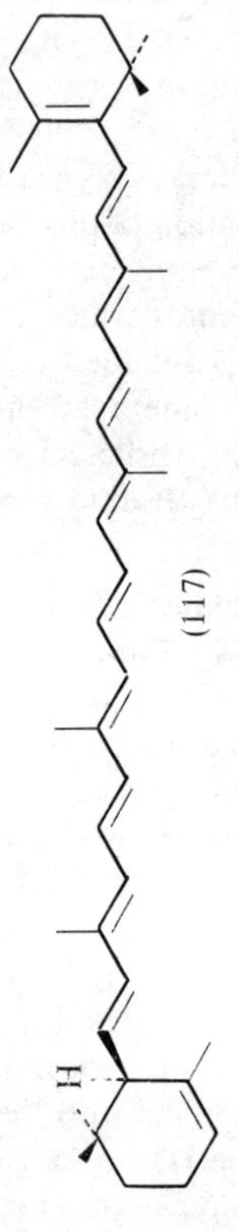

(117)

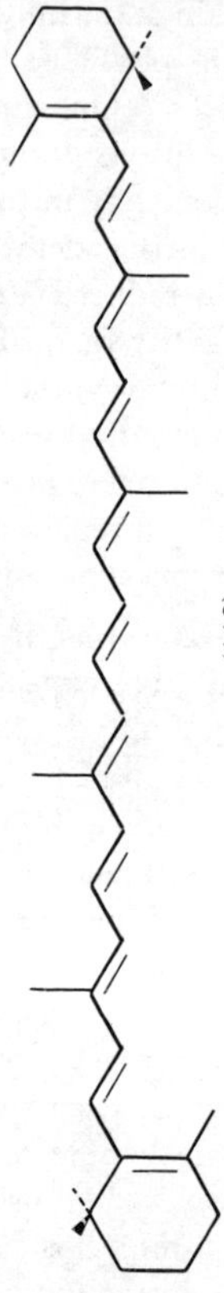

(118)

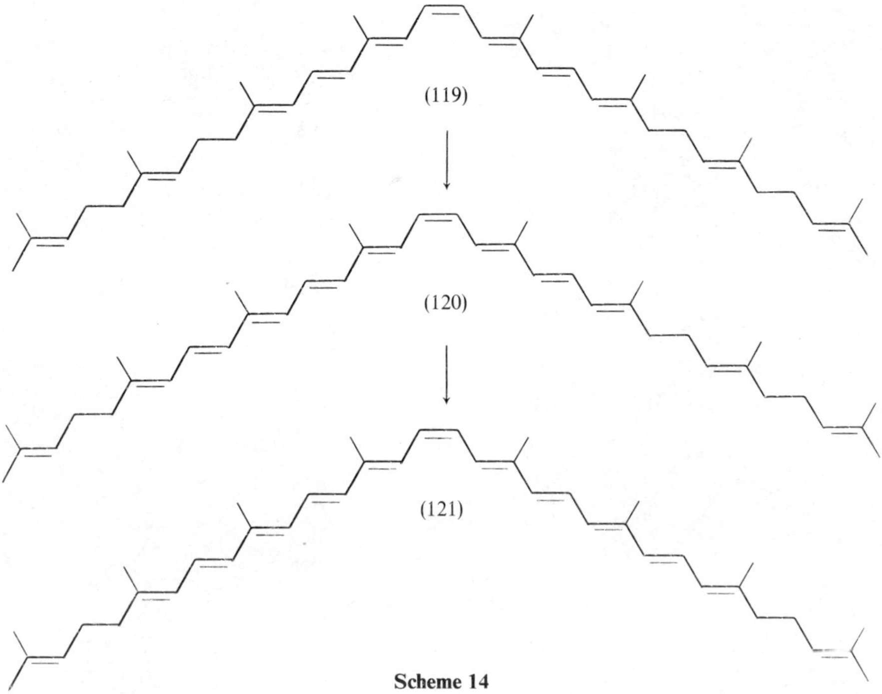

(119)

(120)

(121)

Scheme 14

protochlorophyllide synthesis, is also blocked in β-carotene synthesis in the dark and accumulates phytofluene, proneurosporene (120), poly-*cis*-lycopenes, neolycopene, and lycopene, rather like the tangerine tomato. Illumination with blue light appears to promote conversion into the *trans*-isomers, and β-carotene synthesis can be detected. Four other mutants, thought to be blocked in the control mechanisms of chlorophyll biosynthesis, retain the ability to synthesize β-carotene but also accumulate various other intermediates, *e.g.* ζ-carotene (*tig-b*^{23}), a lycopene-like pigment (*tig-n*^{32}), and lycopene (*tig-o*^{34}). Attenuated ability to synthesize the later carotenoids in etioplasts appears to be correlated with specific changes in plastid membrane structure and, in some mutants, with blocks in the protochlorophyllide pathway.

Three new carotenoids (122)—(124a) have been isolated from the purple non-sulphur bacterium *Rhodopseudomonas globiformis*, together with the ketone (124b).[169] Inhibition of dehydrogenation by diphenylamine gave three further novel ketones (125)—(127), suggesting the biosynthetic relationship outlined in Scheme 15. The pattern of metabolites resembles that found in purple sulphur bacteria except for the absence of the cyclization steps.

C_{30} Carotenoids. Further details have been reported on the novel diapocarotenoids of *Streptococcus faecium*.[170] 4,4′-Diapophytoene is found predominantly as the 12-*cis*-isomer (128), though a small amount of the all-*trans*-compound was detected. The

[169] K. Schmidt and S. Liaaen-Jensen, *Acta Chem. Scand.*, 1974, **27**, 3040.
[170] R. F. Taylor and B. H. Davies, *Biochem. J.*, 1974, **139**, 751.

(122)

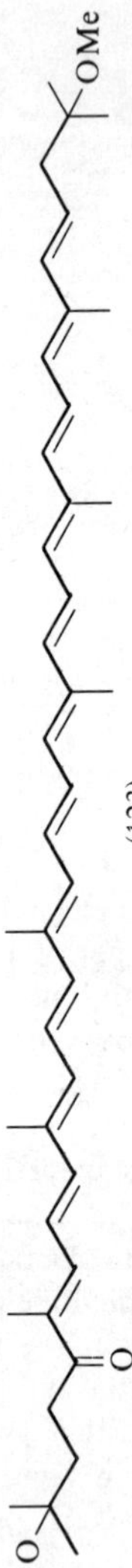

(123)

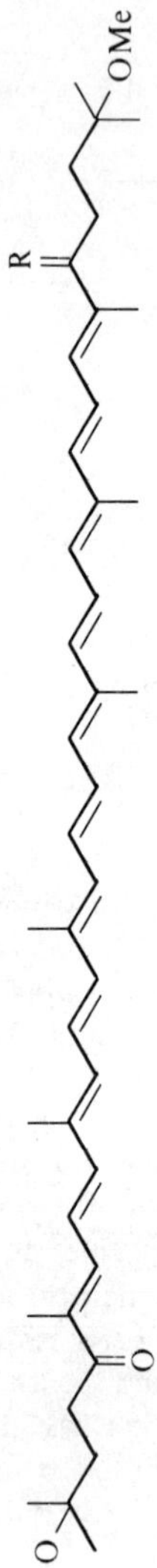

(124) a; R = O
b; $R = H_2$

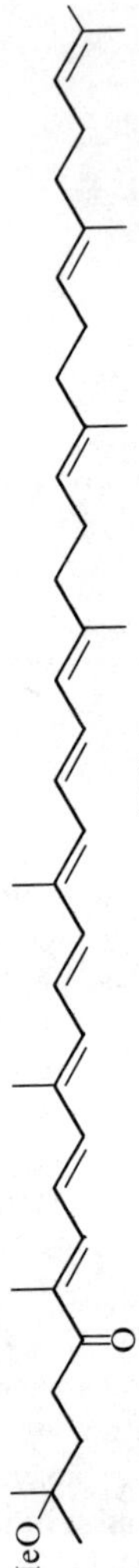

(125)

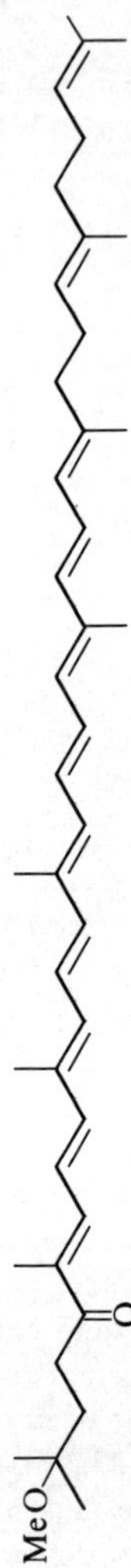

(126)

(127)

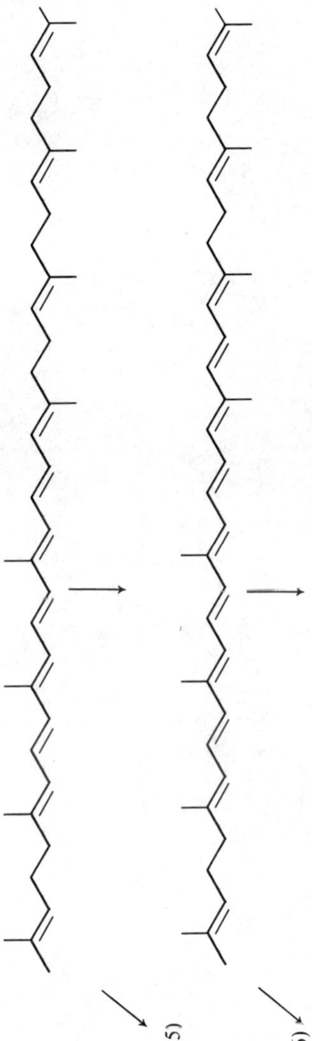

(125) → (126) → (127)

→ (122)–(124)

Scheme 15

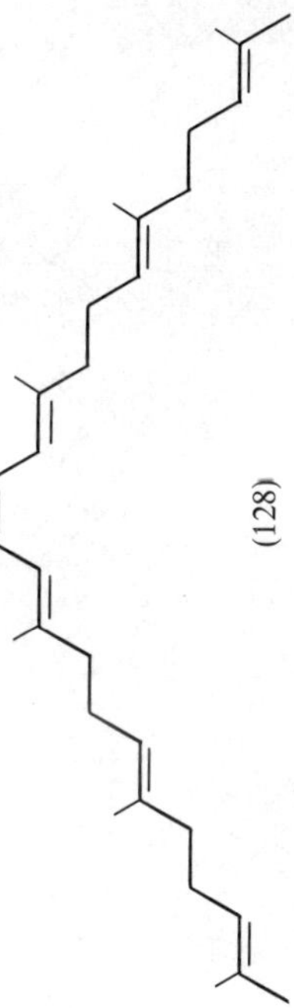

(128)

(129)

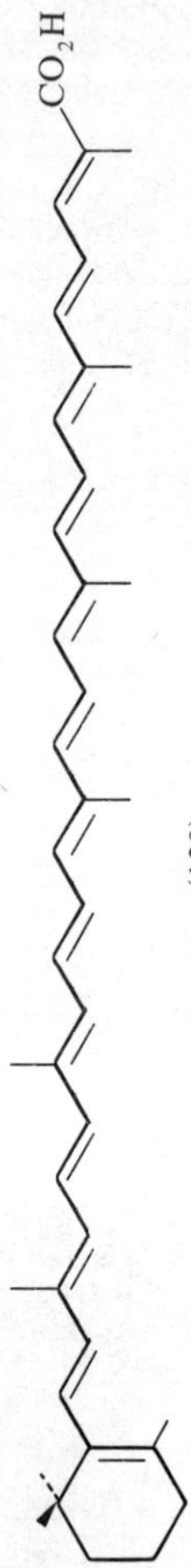

(130)

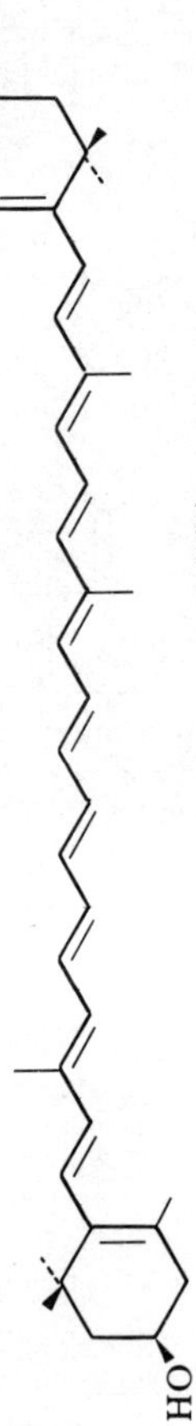

(131)

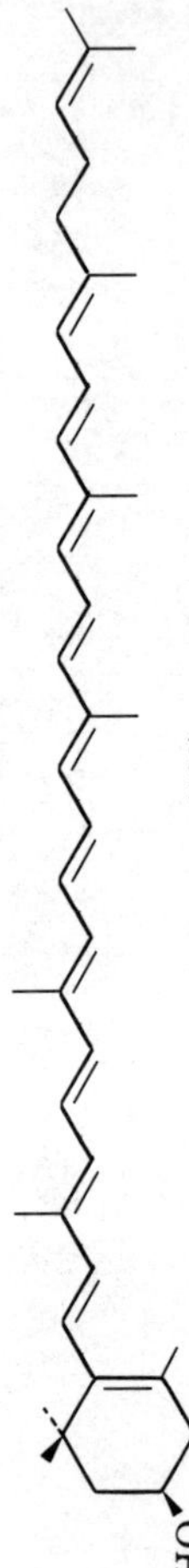

(132)

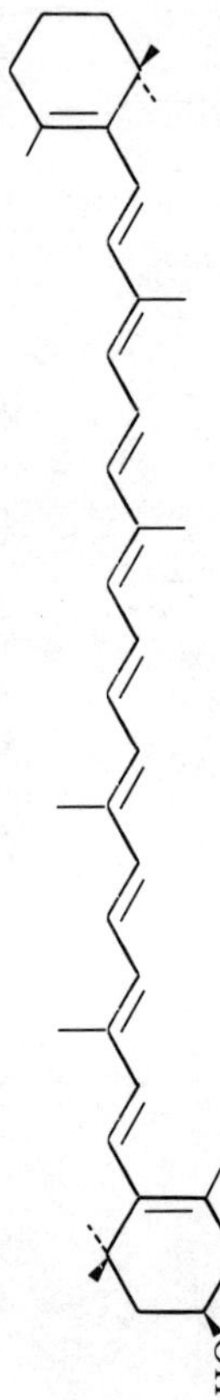

(133)

more unsaturated compounds all have *trans*-double-bonds with the exception of diaponeurosporene, in which some *cis*-isomers were detected.[170] Two further metabolites have been identified as 4-hydroxy-4,4′-diaponeurosporene (129) and its glucoside.[171]

Cyclization and Further Metabolism.—The herbicide CPTA [2-(4-chlorophenylthio)-triethylamine hydrochloride], which has been shown to cause lycopene accumulation in *Blakeslea trispora* and grapefruit,[3] produces a similar effect in cotyledons of *Cucurbita moschata*, together with decreased chlorophyll synthesis and cotyledon growth.[172] However, it remains to be seen whether lycopene accumulation results from cyclase inhibition, as appears to be the case in lower organisms, or from its position as the rate-limiting intermediate in enhanced carotene production, stimulated by blocking of other biosynthetic pathways. A similar effect is observed in *Rhodotorula glutinis*, though here, while the β-carotene level dropped on treatment with the herbicide, the level of γ-carotene remained constant.[173] Conversion of α-carotene into torularhodin (130) may also be unaffected. Ergosterol and fatty acid content was reduced by CPTA. With the exception of fatty acid production, all effects of the inhibitor were reversed following its removal. The action of the herbicide Sandoz 6706 [4-chloro-5-(dimethylamino)-2-(α,α,α-trifluoro-*m*-tolyl)-3(2*H*)-pyridazinone] on carotenogenesis in wheat seedlings has also been examined.[174] Inhibition of β-carotene synthesis and production of phytoene epoxides appear to be the major effects.

Time-course studies on carotenoid production and mycelial growth of *Phycomyces blakesleeanus* mutants reveal the lycopene concentration to be almost maximal after two days, as β-carotene synthesis is commencing.[175] The later accumulation of α- and β-carotene and the constant lycopene level indicated that the normal pathway to β-carotene may be circumvented in later stages of growth, with β-zeacarotene and α-carotene being intermediates on the major pathway. However, the presence in the mutants of defective cyclases, having reduced substrate specificity, could also explain the observed shift in product distribution. Vitamin A stimulates β-carotene production in wild-type strains of *P. blakesleeanus*.[176] *Car* B and *Car* R mutants, which accumulate phytoene and lycopene respectively, also show stimulation by vitamin A. In *Car* A mutants, which are normally unable to synthesize carotenoids, the vitamin partially counters the blockage and re-establishes β-carotene synthesis. A similar effect is reported to be caused by β-ionone. The increased carotenoid level results from protein synthesis, as judged by the lack of stimulation in the presence of cycloheximide, and this indicates that the active compounds interact in some way with a *Car* A gene product. It is known that, in *P. blakesleeanus*, mating results in increased carotenogenesis, stimulated by the trisporic acid hormones. This effect is also prevented by cycloheximide. Thus vitamin A and β-ionone may partly mimic the effect of the hormones in unmated strains.

Further details have been reported on the cyclization of lycopene in a *Flavobacterium* species and the inhibition of this process by nicotine.[177] Lycopene can be cyclized to

[171] R. F. Taylor and B. H. Davies, *Biochem. J.*, 1974, **139**, 761.
[172] J. Fortino and W. E. Splittstoesser, *Plant Cell Physiol.*, 1974, **15**, 59.
[173] E. P. Hayman, C. O. Chichester, and K. L. Simpson, *Phytochemistry*, 1974, **13**, 1123.
[174] A. Ben-Aziz and E. Koren, *Plant Physiol.*, 1974, **54**, 916.
[175] W.-J. Hsu, D. C. Ailion, and M. Delbruck, *Phytochemistry*, 1974, **13**, 1463.
[176] A. P. Eslava, M. I. Alvarez, and E. Cerda-Olmedo, *European J. Biochem.*, 1974, **48**, 617.
[177] J. C. B. McDermott, D. J. Brown, G. Britton, and T. W. Goodwin, *Biochem. J.*, 1974, **144**, 231.

β-carotene anaerobically and to zeaxanthin (131) under aerobic conditions. The *in vivo* hydroxylation of β-carotene to zeaxanthin was also demonstrated, confirming that cyclization precedes oxidation. An alternative pathway to zeaxanthin was detected since rubixanthin (132) was converted into *β*-cryptoxanthin (133) and zeaxanthin (131). The results obtained using nicotine as an inhibitor are interpreted in terms of two similar enzyme sites which can desaturate and cyclize substrates. In the presence of low inhibitor concentrations, abnormal biosynthetic routes may result from inhibition of one of the sites of cyclization.

The absolute configuration of zeaxanthin has been confirmed, as a result of an *X*-ray study on the allene (134) obtained from the grasshopper *Romalea microptera* and as a degradation product of fucoxanthin.[178] Since the hydroxy-groups of fucoxanthin and neoxanthin have been shown to have the same configuration as those in

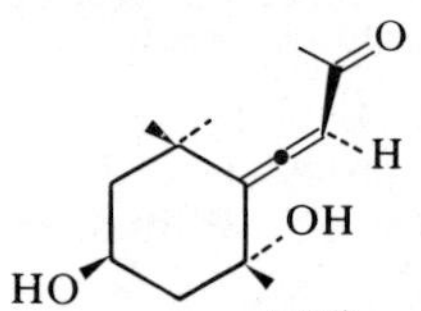

(134)

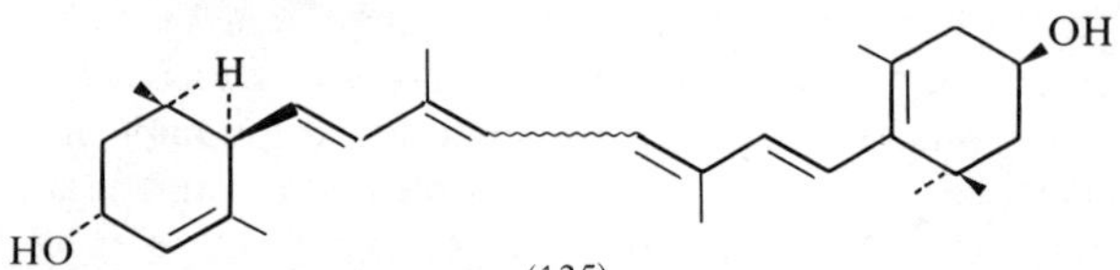

(135)

zeaxanthin, the configuration of the latter is defined as (3*R*,3′*R*) (131). The C-3′ configuration of lutein (135) has now been firmly established. Earlier assignment of the (3′*S*) stereochemistry was at variance with n.m.r. spectroscopic evidence obtained from model compounds. The (3′*R*) configuration is supported by the observation that base-catalysed isomerization of lutein gives (136), the optically inactive *meso*-isomer of zeaxanthin.[179] This result has been confirmed by detailed spectroscopic and degradative studies on lutein and model compounds which establish lutein as the (3*R*,3′*R*,6′*R*)-isomer (135).[180,181] In view of this result and the assignment of the (3*S*,3′*S*) configuration to astaxanthin (137),[182] the reported conversion of lutein into astaxanthin in goldfish[183] warrants re-examination. The stereochemistry of lutein epoxide has been defined as (3*S*,5*R*,6*S*,3′*R*,6′*R*) (138).[184] The major pigment of *Caltha palustris* has been identified as the 3′*S*-epimer of lutein and called calthaxanthin (139).[185]

The previous tentative assignments of stereochemistry to *β* ε-caroten-2-ol (140), ββ-caroten-2-ol, and ββ-carotene-2,2′-diol from the green alga *Trentepohlia iolithus* have been confirmed by a degradative sequence.[186] All hydroxy-groups have the (*R*) configuration. Also, β ε-caroten-2-ol has the (6′*R*) stereochemistry, identical to that of lutein in higher plants. Two new epoxides have been isolated from *T. iolithus* and assigned structures (141) and (142).[187] The absolute configuration of the C_{50}

178 J. R. Hlubucek, J. Hora, S. W. Russell, T. P. Toube, and B. C. L. Weedon, *J.C.S. Perkin I*, 1974, 848.
179 A. G. Andrewes, *Acta Chem. Scand.* (*B*), 1974, **28**, 137.
180 R. Buchecker, P. Hamm, and C. H. Eugster, *Chimia* (*Switz.*), 1971, **25**, 192.
181 R. Buchecker, P. Hamm, and C. H. Eugster, *Helv. Chim. Acta*, 1974, **57**, 631.
182 A. G. Andrewes, G. Borch, S. Liaaen-Jensen, and G. Snatske, *Acta Chem. Scand.* (*B*), 1974, **28**, 730.
183 W. J. Hsu, D. B. Rodriguez, and C. O. Chichester, *Internat. J. Biochem.*, 1972, 3, 333.
184 H. Cadosch and C. H. Eugster, *Helv. Chim. Acta*, 1974, **57**, 1466.
185 A. G. Dabbagh and K. Egger, *Z. Pflanzenphysiol.*, 1974, **72**, 177.
186 R. Buchecker, C. H. Eugster, H. Kjosen, and S. Liaaen-Jensen, *Acta Chem. Scand.* (*B*), 1974, **28**, 449.
187 G. Nybraaten and S. Liaaen-Jensen, *Acta Chem. Scand.* (*B*), 1974, **28**, 483.

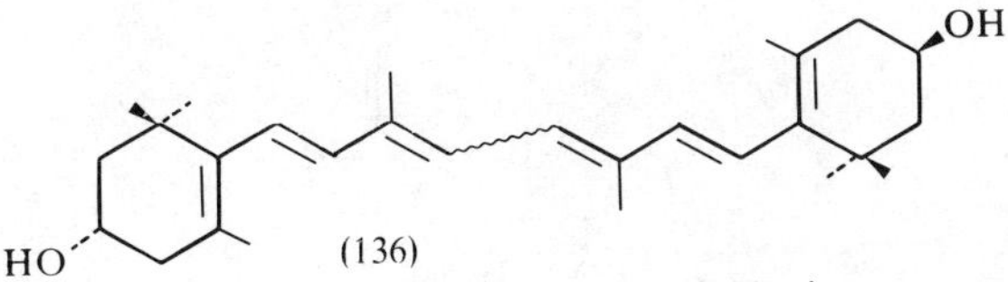

(136)

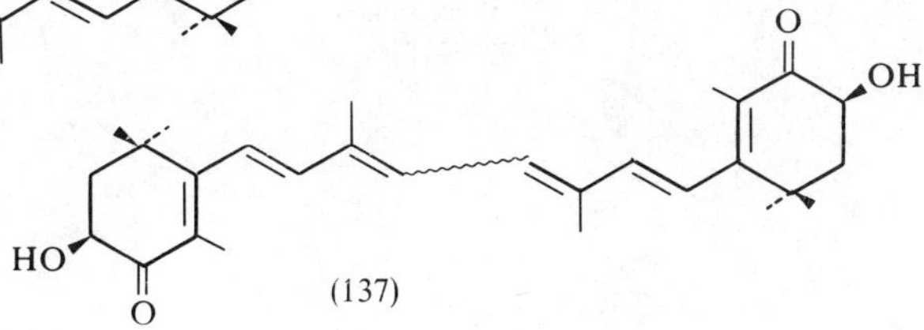

(137)

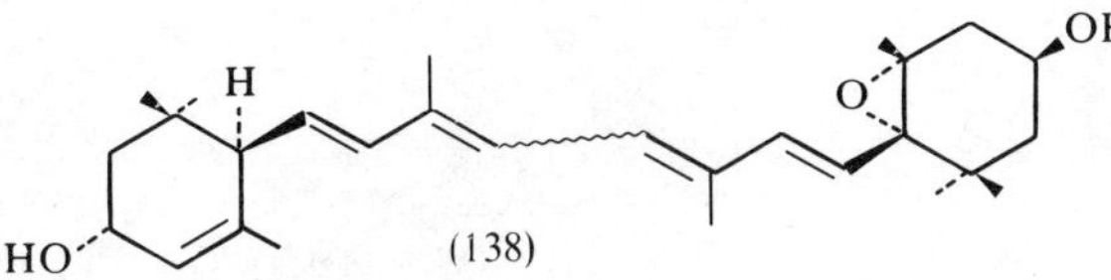

(138)

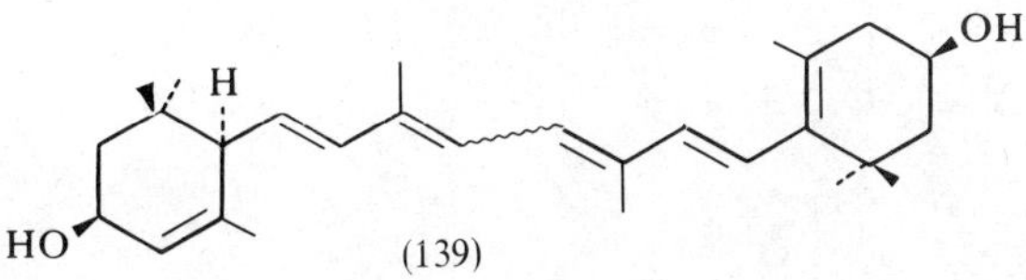

(139)

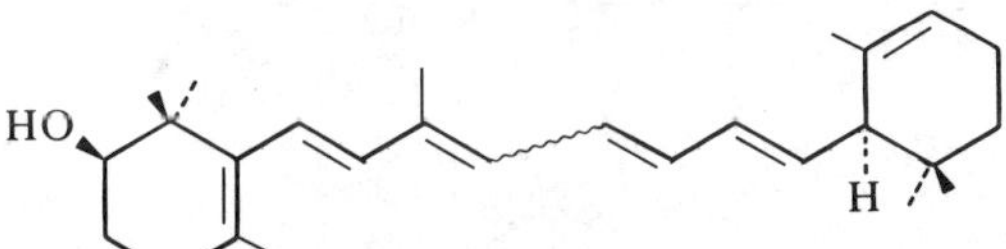

(140)

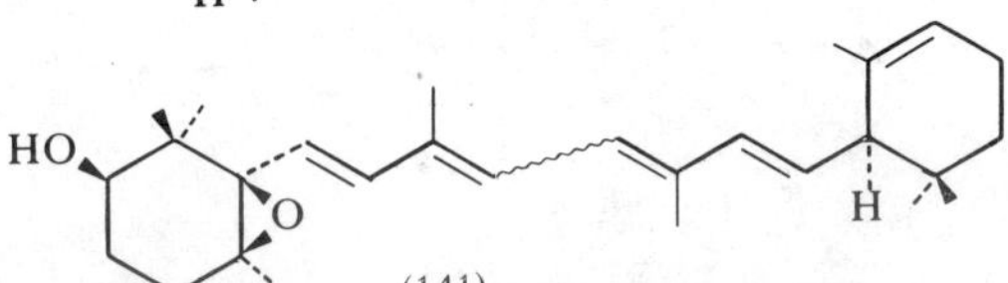

(141)

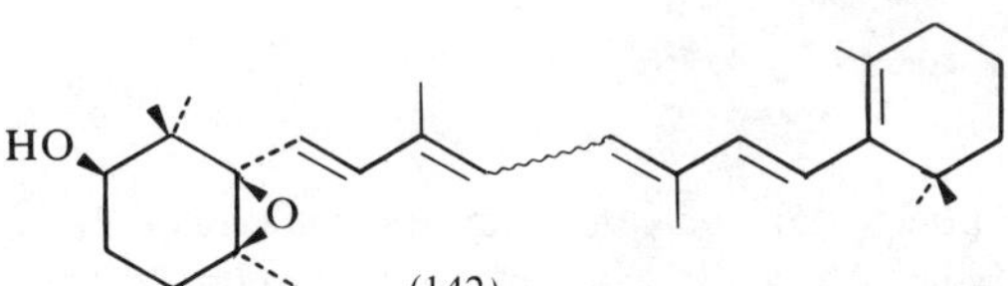

(142)

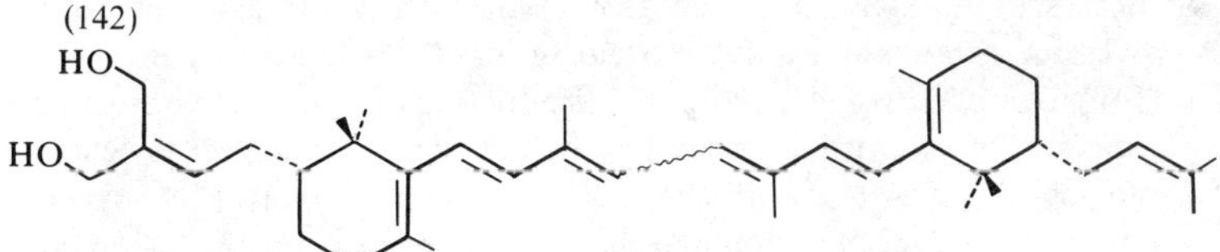

(143)

carotenoid Cp. 450 has been determined and its structure defined as (2*R*,2′*R*)-2-(4-hydroxy-3-hydroxymethyl-2-butenyl)-2′-(3-methyl-2-butenyl)-ββ-carotene (143).[188]

Additional evidence for an alternative route to astaxanthin from β-carotene in goldfish has been obtained by feeding potential intermediates to carotenoid-depleted

[188] A. G. Andrewes, S. Liaaen-Jensen, and G. Borch, *Acta Chem. Scand.* (*B*), 1974, **28**, 737.

β-carotene (118)

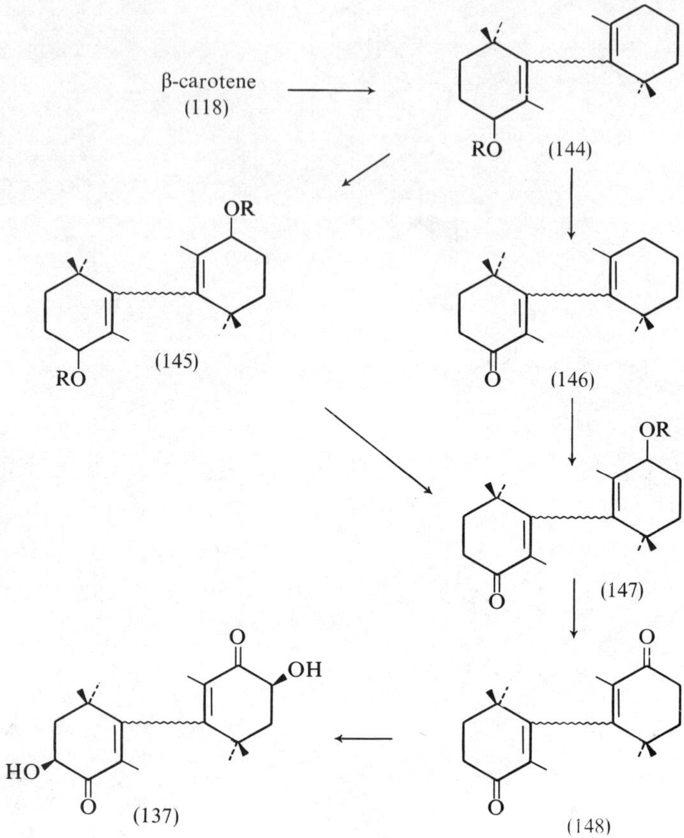

Scheme 16

fish.[189] Isocryptoxanthin (144), echinenone (146), and canthaxanthin (148) markedly increased the astaxanthin content of the fish, while isocryptoxanthin (144) raised levels of echinenone (146), isozeaxanthin (145), and 4-hydroxy-4′-keto-β-carotene (147). Echinenone (144) promoted production of (147) and canthaxanthin (148). Thus astaxanthin biosynthesis may involve the sequence shown in Scheme 16. However, the position of isozeaxanthin is unclear since it is converted into the monoketone (147) only at high concentration and does not increase the astaxanthin content.

The predominant carotenoid of several species of photosynthetic dinoflagellates of the Dinophyceae is perdinin (149a). Additional members of this class of C_{37} carotenoids have been identified, including perdininol (149b), pyrrhoxanthin (150a), and pyrrhoxanthinol (150b), while the structure of dinoxanthin has been proposed to be (151).[190]

[189] D. B. Rodriguez, K. L. Simpson, and C. O. Chichester, *Internat. J. Biochem.*, 1974, **5**, 157.
[190] J. E. Johansen, W. A. Svec, S. Liaaen-Jensen, and F. T. Haxo, *Phytochemistry*, 1974, **13**, 2261.

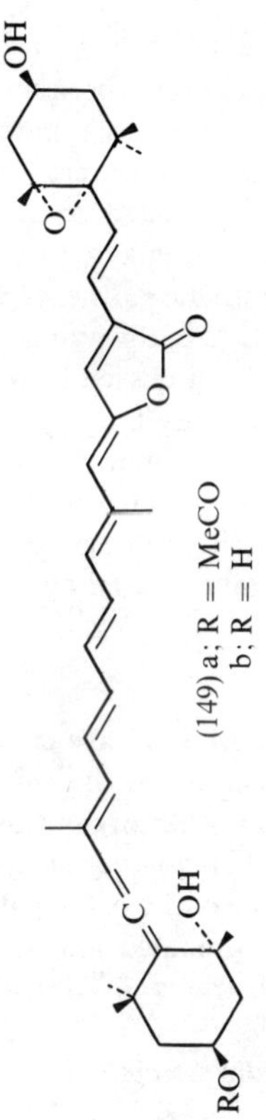

(149) a; R = MeCO
b; R = H

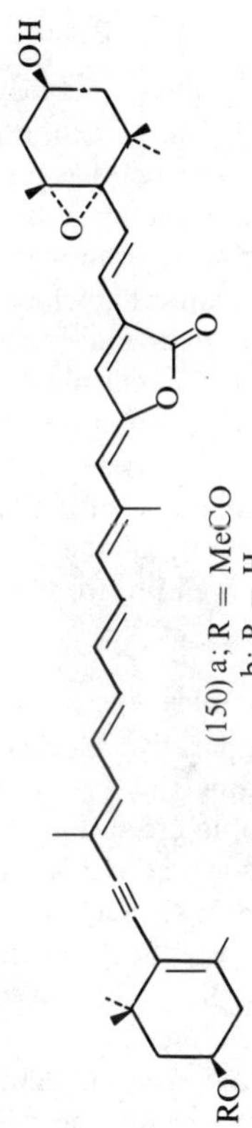

(150) a; R = MeCO
b; R = H

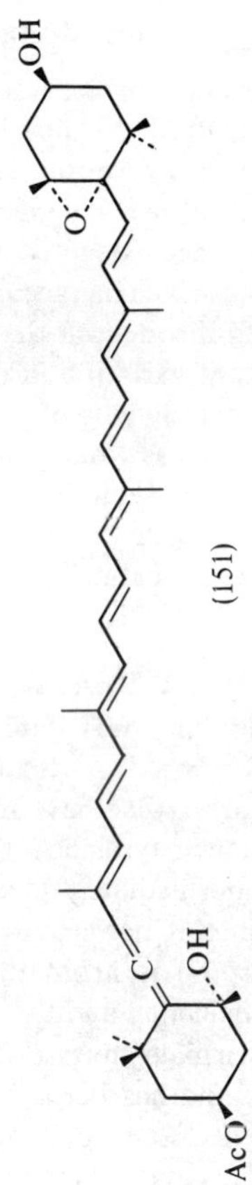

(151)

4
Phenolic Compounds Derived from Shikimate

BY J. B. HARBORNE

1 Introduction

In terms of progress in understanding phenolic biosynthesis, 1974 has been a year of consolidation. No new pathways to rare or exotic phenolic compounds have been opened up by feeding experiments. The emphasis instead has been on enzymology and several new enzymes of secondary metabolism have been detected. Much work has also been devoted to the continued examination of known enzymes. Further evidence which has accumulated indicates that whereas biosynthetic pathways may be uniform throughout the plant kingdom, the enzymes of any one pathway may show significant variation in properties in different plant groups. As discussed below, this is particularly true of the enzymes of the shikimic acid pathway itself.

Other areas where significant advances have been made are in lignin biosynthesis and in the assembly of flavan oligomers. Mention should also be made of the novel discovery in a number of higher plants of the presence of the homogentisic acid pathway for catabolism of tyrosine, well known in micro-organisms and animals.

2 Phenols and Phenolic Acids

The Shikimic Acid Pathway.—Comparative aspects of this pathway have continued to be explored, particularly regarding the enzymes and their mode of regulation. Stenmark *et al.*[1] have found that blue-green algae, unlike all other organisms, synthesize their tyrosine (1) from prephenate (2) by a unique route involving pretyrosine (3) as intermediate. Indeed, it is a simple switch in the last two enzymes of the pathway, since in this new route (Scheme 1) prephenate is first transaminated and then the product (3) is aromatized. This pathway has been demonstrated in *Agmenellum quadruplicatum* and five other related algae.

Chorismate mutase has been purified some 2200-fold from *Streptomyces aureofaciens* and has been found to be unusually thermostable.[2] The K_m for chorismate mutase is 5.2×10^{-4} mol l^{-1}. It is a single protein with the one activity and differs from the same enzyme as isolated from other bacteria. In *E. coli*, for example, chorismate mutase forms a functional unit with prephenate dehydrogenase and, in *Aerobacter aerogenes*, with prephenate dehydratase. The *streptomyces* enzyme also differs from bacterial enzymes in having a mol. wt. of 60000 and being derived from four subunits of mol. wt. 14500; the bifunctional enzymes by contrast have two subunits of mol. wt. 40000.

[1] S. L. Stenmark, D. L. Pierson, R. A. Jensen, and G. I. Glover, *Nature*, 1974, **247**, 290.
[2] H. Görisch and F. Lingens, *Biochemistry*, 1974, **13**, 3790.

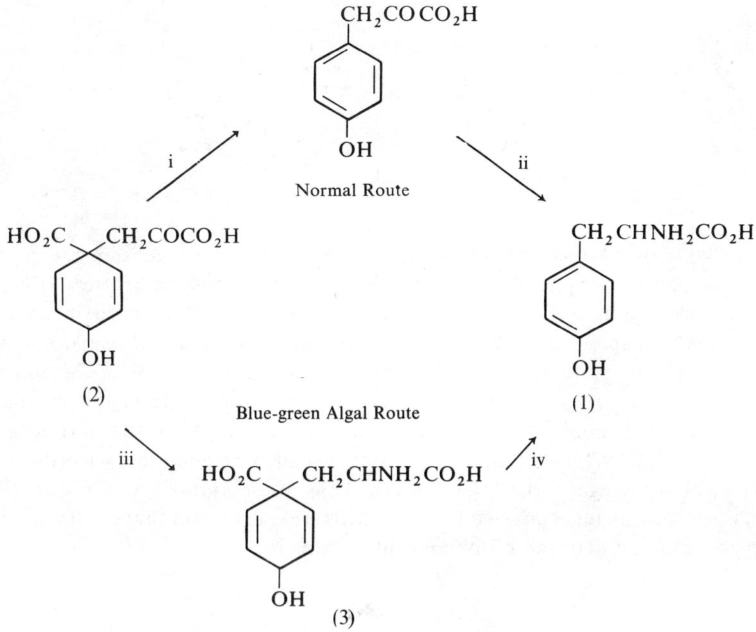

Enzymes: i, prephenate dehydrogenase, NAD^+; ii, 4-OH phenylpyruvate transaminase, pyridoxal 5′-phosphate; iii, prephenate transaminase, pyridoxal 5′-phosphate; iv, pretyrosine dehydrogenase, NAD^+

Scheme 1

Another bifunctional enzyme of the shikimic acid pathway has recently been described, this time from higher plants. It combines the activities of 5-dehydroquinate hydrolyase and shikimate:$NADP^+$-oxidoreductase[3]. Extensive purification using Sephadex, DEAE-cellulose, hydroxylapatite, and isoelectric focusing failed to separate the two enzyme activities. This bifunctionality holds for enzyme isolated from eight angiosperms, one gymnosperm (*Ginkgo biloba*) and one fern (*Scolopendrium officinale*). By contrast, previous workers have shown that the same two enzyme activities are readily distinguished from each other when obtained from prokaryotes and from mosses. This is further evidence that the pathway may be regulated in different ways in different plants. The link between these two particular enzymes in higher plants may be related to the relatively high concentrations of quinic acid in these organisms, which might well interfere with the pathway if the enzymes were not linked.

An alternative to the shikimate pathway for the production of phenolic acids has been noted in *Corynebacterium cyclohexanicum*, an organism which can convert 4-ketocyclohexane carboxylic acid (4) directly to *p*-hydroxybenzoic acid (5) in three stages.[4] The intermediates in this aromatization have yet to be characterized. The only other similar reaction recorded previously appears to be the conversion of the CoA ester of (4) to benzoate in guinea pig liver mitochondria.[5]

[3] A. M. Boudet and R. Lécussen, *Planta*, 1974, **119**, 71.

[4] T. Kaneda, *Biochem. Biophys. Res. Comm.*, 1974, **58**, 140.

[5] C. Mitoma, H. S. Posner, and F. Leonard, *Biochim. Biophys. Acta*, 1958, **27**, 156.

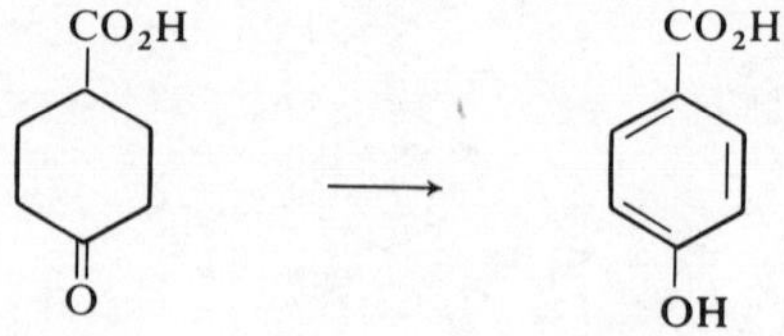

(4) (5)

Chloramphenicol Biosynthesis.—It is not yet known at what step the route to the antibiotic chloramphenicol (6) diverges from the main shikimate pathway. Emes *et al.*[6] set out to solve this problem by feeding several labelled intermediates to *Streptomyces* sp. 3022a but failed, because neither labelled chorismic nor prephenic acids could be persuaded to enter the cell. Fortunately, some of the shikimic acid fed did get in and the results of feeding showed unambiguously that the *pro*-6*R*-hydrogen was eliminated during ring aromatization. Also, the antibiotic labelled from [*G*-^{14}C, 6*S*-^{3}H]shikimate (7) was chemically degraded to 2,4,6-tribromoaniline (8) without change of the ^{14}C:^{3}H ratio. Thus, these authors were able to show that (7) is specifically incorporated into the phenyl ring of (6) and that the tritium is present at positions *ortho* to the propanoid side-chain.

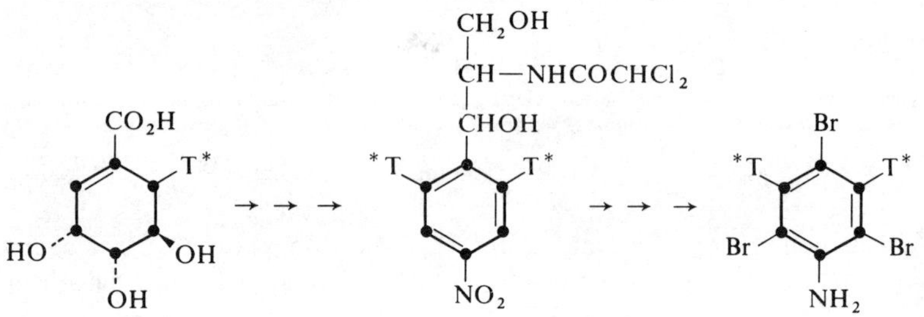

(7) (6) (8)

Isoprenylated Phenols.—Laboratory experiments relevant to the biosynthesis of such compounds as the resorcinol-based cannabindiols have been reported. The regio-specificity of the pyridine-catalysed reaction of α-unsaturated aldehydes and acetals with phenols is discussed.[7] Further experiments of phenol–terpene condensations in aqueous acid media have also been described.[8]

3 Phenylpropanoids

Chlorogenic Acid Biosynthesis.—In spite of the fact that 3-*O*-caffeyl-D-quinic acid (9) is ubiquitous in higher plants, there is still controversy about its mode of biosynthesis. According to Nagels and Parmentier,[9] caffeic acid, a likely precursor of this

[6] A. Emes, H. G. Floss, D. A. Lowe, D. W. S. Westlake, and L. C. Vining, *Canad. J. Microbiol.*, 1974, **20**, 347.
[7] D. G. Clarke, L. Crombie, and D. A. Whiting, *J.C.S. Perkin I*, 1974, 1007.
[8] K. L. Stevens, L. Jurd, and G. Manners, *Tetrahedron*, 1974, **30**, 2075.
[9] L. Nagels and F. Parmentier, *Phytochemistry*, 1974, **13**, 2759.

ester, does not play a major role in chlorogenic acid biosynthesis in *Cestrum poeppigi* (Solanaceae). [3-^{14}C]Cinnamic acid was fed to this plant at different stages of growth and the incorporation of activity into (9) was found to vary considerably. No detectable amounts of (9) were found in trapping experiments with *p*-coumaric acid, so this acid apparently has a regulating effect on the biosynthesis or turnover of the caffeic ester. That caffeic acid is not a direct intermediate was apparent from kinetic experiments in which the specific activity of caffeic from added cinnamic acid was lower than that of (9), except at one point at 17 h, throughout a 26 h feeding period.

The above results do not rule out a direct relationship between caffeic acid and (9) if the caffeic acid is present in an active bound form, such as the coenzyme A ester (10). This could be synthesized *in vivo* directly from cinnamyl–CoA *via p*-coumaryl–CoA, without the intermediacy of the free acids. Evidence in favour of such a route has been provided by Stöckigt and Zenk,[10] who obtained an enzyme preparation from cell suspension cultures of *Nicotiana alata* which catalyses the formation of (9) from (10) and quinic acid. The crude enzyme, which contained hydroxycinnamic acid CoA-ligase activity, produce the depside in the presence of caffeic acid, quinic acid, CoA, and ATP. Subsequently, the same preparation gave labelled chlorogenic in the presence of synthetic (10) and [^{14}C]quinic acid.

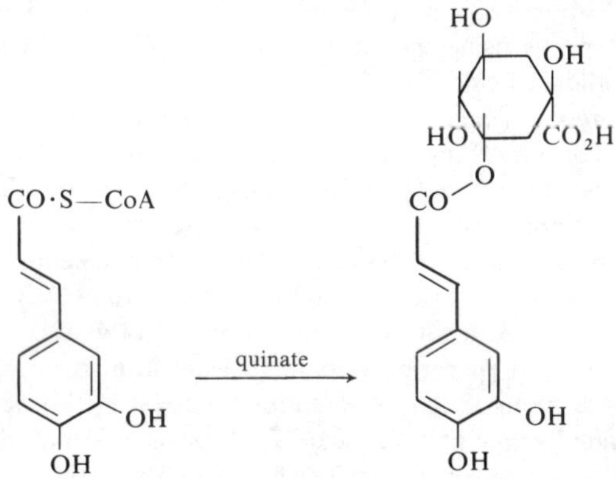

(10) (9)

Additional evidence that CoA esters of cinnamic acids are important intermediates in phenolic biosynthesis has been obtained by Suetfeld and Wiermann,[11] who noted that *p*-coumarate CoA-ligase activity in anther extracts from *Lilium* is correlated with PAL (phenylalanine ammonia lyase) activity. While *p*-coumaric, caffeic, and ferulic acids were converted into the CoA esters in the presence of this enzyme, cinnamic, isoferulic, and sinapic acids were not. Thus, the enzyme specificity is restricted to those acids required for flavonoid synthesis and not those involved in lignin production. When light is shone on cell cultures of soyabean, activities of *p*-coumarate CoA-ligase, PAL, and cinnamic 4-hydroxylase all increase at the same rate prior to the stationary phase of growth. However, light has no effect on the concentration of two

[10] J. Stöckigt and M. H. Zenk, *F.E.B.S. Letters*, 1974, **42**, 131.
[11] R. Suetfeld and R. Wiermann, *Z. Pflanzenphysiol.*, 1974, **72**, 163.

enzymes of lignin biosynthesis (*O*-methyltransferase and cinnamyl alcohol dehydrogenase), indicating again separate regulation of flavonoid and lignin syntheses.[12] The whole topic of regulation of phenylpropanoid biosynthesis in plants has recently been reviewed.[13]

Considerable attention has been given during the year to the enzymes which catalyse the successive oxidations of cinnamic to *p*-coumaric to caffeic acids. Fritz *et al.*[14] have demonstrated that the hydroxyl oxygens of caffeic acid are derived from molecular oxygen, carrying our appropriate experiments in *Helianthus annuus* grown in light in an atmosphere enriched with $^{18}O_2$. The activation of 4-hydroxycinnamic acid hydroxylase in *Sorghum bicolor* has been examined by Stafford,[15] who reports that the lag period in the reaction *p*-coumaric → caffeic acid can be eliminated by adding catalytic amounts of caffeic acid or by increasing enzyme concentrations. Freezing an incubation mixture containing substrate and ascorbate also activated the enzyme.

A soluble phenolase capable of catalysing the oxidation of *p*-coumaric to caffeic has also been isolated from *Polyporus hispidus* cultures, where it apparently mediates in the biosynthesis of the phenylpropanoid hispidin (see Vol. 3[16]). Two forms with mol. wts. of 185000 and 45000 were isolated, the higher mol. wt. form having the additional ability to further oxidize caffeic acid to the corresponding quinone. Both isozymes oxidized bisnoryangonin to hispidin. They were both highly specific, since they had no effect on tyrosine, *p*-hydroxyphenylpropionic, *p*-hydroxybenzoic, or *p*-hydroxyphenylmandelic acids.[17]

The enzymic methylation of caffeic to ferulic acid *in vivo* must be closely controlled since isoferulic acid (*i.e.* 3-hydroxy-4-methoxycinnamic acid) is rarely found as a plant constituent. The existence of an *O*-methyltransferase which specifically methylates caffeic acid only in the *m*-position has not been reported in Pampas grass, *Cortaderia selloana.*[18] Interestingly enough, when presented with protocatechuic acid as substrate, the enzyme catalyses methylation at a much lower rate (1/10th of that for caffeic acid) but methylates both hydroxy-groups so that a mixture of vanillic and isovanillic is formed. Unlike a similar preparation obtained earlier from parsley (see ref. 13), the enzyme catalyses the methylation of *m*-coumaric to *m*-methoxycinnamic and of isoferulic to 3,4-dimethyoxycinnamic acid. The above evidence suggests that the Pampas grass enzyme has separate activities for *m*- and *p*-methylation; this was not, however, confirmed because the enzyme became increasingly unstable on purification.

Studies on the biosynthesis of caffeic and chlorogenic acids in plants are inevitably complicated by the competing pathways that exist for the precursor phenylalanine and by the further turnover of these cinnamic acids in plant metabolism. For these

[12] J. Ebel, B. S. Hekeler, K. H. Knobloch, E. Wellman, H. Grisebach, and K. Hahlbrock, *Biochim. Biophys. Acta*, 1974, **362**, 417.
[13] H. Grisebach and K. Hahlbrock, *Recent Adv. Phytochem.* 1974, **8**, 22.
[14] G. J. Fritz, R. W. King, and R. C. Dougherty, *Phytochemistry*, 1974, **13**, 1473.
[15] H. A. Stafford, *Plant Physiol.*, 1974, **54**, 686; *Recent Adv. Phytochem.*, 1974, **8**, 53.
[16] J. B. Harborne, in 'Biosynthesis', ed. T. A. Geissman (Specialist Periodical Reports), The Chemical Society, London, 1975, Vol. 3, p. 98.
[17] C. P. Vance, E. B. Tregunna, A. M. D. Nambudiri, and G. H. N. Towers, *Biochim. Biophys. Acts*, 1974, **343**, 138; A. M. D. Nambudiri, C. P. Vance, and G. H. N. Towers, *ibid.*, p. 148; G. H. N. Towers, C. P. Vance, and A. M. D. Nambudiri, *Recent Adv. Phytochem.*, 1974, **8**, 81.
[18] B. J. Finkle and S. H. Kelly, *Phytochemistry*, 1974, **13**, 1719.

reasons, Woodhead and Swain[19] found it difficult to correlate the effect of light on increasing caffeic acid synthesis in dark-grown *Amaranthus caudatus* seedlings with any other factor. Although the rate of increase was similar to that of the betalain pigment amaranthin, also competing for phenylalanine and PAL activity, there was no obvious relationship with the rate of formation of three phenolic acids present, vanillic, *p*-hydroxybenzoic, and gentisic. The only clearcut result from the series of experiments was the sequential formation of ferulic acid from caffeic in this plant.

Caffeic and ferulic acids are also produced together in suspension cultures of carrot cells. Biosynthesis is highest in the early logarithmic phase and declines during the stationary phase.[20] The concentrations of these two acids were found to be inversely related to the amount of 2,4-D auxin in the medium. When cells were grown in cultures containing 0.05 p.p.m. 2,4-D, the content of caffeic acid rose 17 times on a wt. basis and 27 times on a cell basis compared to growth in the presence of 1 p.p.m. 2,4-D. Under the same conditions, the incorporation of $[U\text{-}^{14}C]$phenylalanine into protein remained the same. Thus both PAL activity and lignin formation can be considerably affected by varying auxin content, without an effect necessarily being observed in protein synthesis.

Allyl- and Propenyl-phenols.—The fact that mixtures of allyl- and propenyl-phenols, or their methyl derivatives, sometimes occur together in the same plant, e.*g*. eugenol with isoeugenol, suggests that these two series of isomeric phenylpropanoids are biosynthetically interconvertible, since they only differ in the positions of the double bond in the C_3-sidechain. Indeed, they are chemically interconvertible and Birch and Slaytor[21] have suggested a plausible route by which they are formed directly from a common C_6–C_3 precursor, the pyrophosphate of the corresponding cinnamyl alcohol. Geissman and Crout[22] have, however, indicated that such a prototropic isomerization is highly improbable in plants and prefer to consider that the allyl-propenyl pairs are formed by independent routes. Experimental evidence supporting this latter view has now been published. Feeding experiments in *Ocimum basilicum* indicate that the allyl componds, eugenol (11). estragole, and chavicole, are formed by one route,[23] while those in *Pimpinella anisum*[24] show that the propenyl derivative anethole (12) is produced by a second pathway (Scheme 2).

In *Ocimum*, L-phenylalanine and cinnamic acid (but not tyrosine) were acceptable precursors of eugenol. However, from feeding $[1\text{-}^{14}C]$phenylalanine, it was clear that the carboxy-carbon was lost during biosynthesis, probably at the ferulic acid stage, an extra C_1-unit being added later in biosynthesis. Labels from $[2\text{-}^{14}C]$- and $[3\text{-}^{14}C]$-phenylalanine were specifically incorporated in the expected positions (see Scheme 2), as proved by degradation of the labelled product, so that no rearrangement of the sidechain apparently occurred. In *Pimpinella*, L-phenylalanine was incorporated into anethole synthesis, with retention of all three sidechain carbon atoms. DL-Phenylalanines ^{14}C-labelled in the sidechain in all three positions were

[19] S. Woodhead and T. Swain, *Phytochemistry*, 1974, **13**, 953.
[20] N. Sugano, R. Iwata, and A. Nishi, *Phytochemistry*, 1974, **13**, 1653.
[21] A. J. Birch and M. Slaytor, *Chem. and Ind*., 1956, 1524.
[22] T. A. Geissman and D. H. G. Crout, 'Organic Chemistry of Secondary Plant Metabolism', Freeman-Cooper, San Francisco, 1969, p. 151.
[23] P. Manitto, D. Monti, and P. Gramatica, *J.C.S. Perkin I*, 1974, 1727.
[24] P. Manitto, D. Monti, and P. Gramatica, *Tetrahedron Letters*, 1974, 1567.

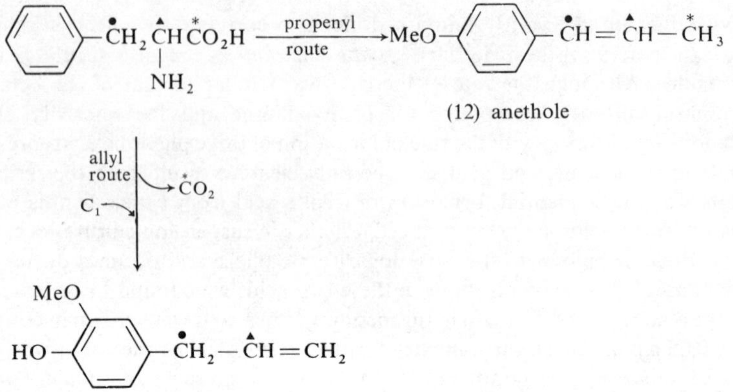

(12) anethole

(11) eugenol

Scheme 2

mixed with tritiated phenylalanine and fed variously to *P. anisum* plants. After 20 days, work up gave active anethole in all three experiments. The % incorporation was practically the same in all three cases: 0.13% for [3-^{14}C]-, 0.14% for [2-^{14}C]-, and 0.12% for [1-^{14}C]-phenylalanine. No loss of the terminal carbon could, thus, have occurred.

The stage is now set for the crucial experiment to prove the existence of two independent pathways, *i.e.* to feed the three side-chain labelled phenylalanines separately to a plant which contains both pairs of isomer. One such plant is the nutmeg *Myristica fragans*, which is known to have both myristicin (allyl isomer) and iso-myristicin (propenyl isomer). Biosynthetic studies in this plant would also be of interest, since oxidative coupling of these monomers occurs in the fruit, with production of a range of diarylpropanoids, such as dehydrodi-isoeugenol.[25]

Coumarin Biosynthesis.—The only developments in this area appear to be at the enzymic level, with the discovery of two key enzymes. The first of these is required for the biosynthesis of coumarin itself, and carries out the *o*-hydroxylation of cinnamic acid. While preparations which will catalyse the introduction of a *p*-hydroxyl into cinnamate have often been described, one which will bring about *o*-hydroxylation has been searched for in vain. At last, Gestetner and Conn[26] have detected this activity in *Melilotus alba* seedlings, principally through using chloroplast preparations as the source of enzyme. This phenolase converted *trans*-[3-^{14}C]cinnamic acid into 2-hydroxy-*trans*-[3-^{14}C]-cinnamic and the product was verified by recrystallization to constant specific activity. The pH optimum was 7.0 and the enzyme activity increased 4-fold when glucose 6-phosphate was added as a source of reducing power. One of the remaining problems with the *o*-hydroxylase is that it appears to be bound to the lamellar membrane of the chloroplast; sonication of the crude preparation gave a 50% increase in activity.

The second new enzyme was discovered in suspension cultures of young leaves of

[25] J. E. Forrest, R. A. Heacock, and J. P. Forrest, *J.C.S. Perkin I*, 1974, 205.
[26] B. Gestetner and E. E. Conn, *Arch. Biochem. Biophys.*, 1974, **163**, 617.

Ruta graveolens[27], and it catalysed the reaction between unbelliferone (7-hydroxycoumarin) and dimethyl-allyl pyrophosphate to give demethylsuberosin (13), the first key intermediate in the biosynthesis of linear furanocoumarins. The enzyme is quite specific in its requirements and abilities: it fails to use 7-methoxycoumarin (herniarin) as substrate and also it only produces substitution in the 6-, and not the 8-position. It has a requirement for Mn^{2+} which can be partly replaced by Fe^{2+} or Zn^{2+} but not by Mg^{2+}. The fact that the enzyme is particle-bound in the particulate fraction may explain earlier failures to find this important biosynthetic enzyme.

Lignin Biosynthesis.—Following the demonstration reported in Volume 3[28] of enzymes capable of carrying out the reduction of cinnamic acids (as their CoA esters) to the corresponding aldehydes and then alcohols, several papers have appeared giving further details of these important enzymes of lignin biosynthesis. The most detailed work has been on a specific NADP (H) dependent cinnamyl alcohol oxidoreductase, which has been purified 600-fold from *Forsythea suspensa* tissues.[29] It is the first alcohol dehydrogenase found in plants with an absolute specificity towards cinnamyl substrates. It catalyses the reaction aldehyde ↔ alcohol in both directions, in the presence of the appropriate form of the cofactor (Scheme 3). The mol. wt. is 80000. The pH optimum for aldehyde reduction is 7.6 with the enzyme being very unstable below 6.8. In keeping with its role in lignin biosynthesis, it is much more active with cinnamaldehydes substituted in the 3- and 4- positions than with any others. Thus K_m values for cinnamaldehyde and its 2- and 4-nitro-derivatives were between 97 and 114 × 10^{-6} mol l^{-1} whereas they were 17 × 10^{-6} mol l^{-1} for *p*-coumarylaldehyde and coniferylaldehyde.

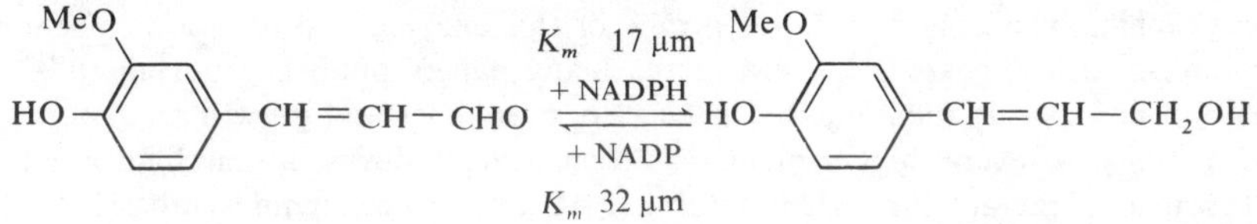

Scheme 3

Cinnamyl alcohol dehydrogenase activity has been sought for and found in a wide range of plant tissues.[29] Relatively low levels were recorded in bryophytes and pteridophytes; there were contrasting high levels in gymnosperms, which are naturally rich in lignin. Medium levels were found generally in angiosperms, with most activity in cambial samples and least activity in leaf or tissue-culture samples.

The reduction of *p*-coumaric and ferulic acids by a crude enzyme mixture from swede-root tissue, *Brassica napobrassica*, has also been reported.[30] The preparation reduced the acids only in the presence of CoA, ATP, Mg^{2+}, and a reduced pyridine nucleotide. Curiously, the reduction of *p*-coumaric specifically required $NADPH_2$, whereas that of ferulic acid occurred in the presence of either $NADPH_2$ or $NADH_2$

[27] B. E. Ellis and S. A. Brown, *Canad. J. Biochem.* 1974, **52**, 734.
[28] J. B. Harborne, in 'Biosynthesis', ed. T. A. Geissman (Specialist Periodical Reports), The Chemical Society, London, 1975, Vol. 3, p. 94.
[29] R. L. Mansell, G. G. Gross, J. Stöckigt, H. Franke, and M. H. Zenk, *Phytochemistry*, 1974, **13**, 2427.
[30] M. J. C. Rhodes and L. S. C. Wolltorton, *Phytochemistry*, 1974, **13**, 107.

In this latter case, $NADPH_2$ is 2.5 times more active than $NADH_2$. This is interesting as a possible control mechanism for determining the ratio of *p*-coumaryl to ferulyl residues to be condensed into lignin.

A comparison of the relative abilities of gymnosperms and angiosperms to reduce ferulic and sinapic acids to the corresponding alcohols was carried out by Nakamura *et al.*,[31] using the simple technique of feeding sliced xylem tissues with the free acids. While tissue from gymnosperms only reduced ferulic, similar tissue from angiosperms reduced both acids equally effectively. In young (less differentiated) xylem tissue or callus tissue of angiosperms, the ability to reduce sinapic acid is markedly lower than in fully differentiated xylem. These results indicate that the reductive enzymes or the CoA-ligase could provide the control mechanism whereby only guaiacyl residues appear in gymnosperm lignins and both guaiacyl and syringyl residues are absorbed into angiosperm lignin.

The last stage in lignification is polymerization of the cinnamyl alcohols and Nakamura *et al.*[31] found that peroxidases from both gymnosperms and angiosperms were equally capable of producing lignin-like material from mixtures of coniferyl and sinapyl alcohols. Lignins, although very stable in living plants, are degraded by fungi and bacteria in dead tissue; little is known about the pathways of metabolism. Ciaramitaro and Steelink[32] have, therefore, carried out some model oxidation experiments with H_2O_2 and peroxidase on synthetic 4-hydroxybenzylphenyl ethers and acetates to see which side-groups are cleaved and how quickly depolymerization occurs. Interestingly, syringyl moieties were oxidized more readily than guaiacyl residues, hinting that lignins of angiosperms may be more rapidly broken down in Nature than those of gymnosperms.

Phenylalanine Ammonia Lyase.—A review of this enzyme and its role in phenolic metabolism has appeared.[33] A few of the many papers published during 1974 on PAL are now briefly mentioned. The effect of herbicides and growth regulators on PAL activity in several higher plants has been studied, during a search for a herbicide which might act through inhibiting PAL and hence lignin synthesis.[34] The induction of PAL in pods of *Pisum sativum* by adding various polyamines and basic polypeptides has been described by Hadwiger *et al.*[35] PAL activity was increased from 84 to 604 nmol cinnamic acid (g fresh wt.)$^{-1}$ after 18 h in the dark at 22 °C in the presence of added protamine; a similar increase in the isoflavonoid pisatin was also noted.

Action spectra and phytochrome control of PAL in *Hordeum vulgare* shoots have been reported.[36] The biochemical properties of PAL, especially its kinetics, have been exhaustively examined in the enzyme isolated from gherkin hypocotyls, *Cucumis sativa.*[37] Two isozymes of PAL have been detected in strawberry leaf discs, only one of which is light sensitive.[38] There is increasing evidence from several laboratories

31 Y. Nakamura, H. Fushiki, and T. Higuchi, *Phytochemistry*, 1974, **13**, 1777.
32 D. Ciaramitaro and C. Steelink, *Phytochemistry*, 1975, **14**, 543.
33 L. L. Creasy and M. Zucker, *Recent. Adv. Phytochem.*, 1974, **8**, 1.
34 N. O. Jangaard, *Phytochemistry*, 1974, **13**, 1765, 1769.
35 L. A. Hadwiger, A. Jafri, S. Broembsen, and R. Eddy, *Plant Physiol.* 1974, **53**, 52.
36 J. W. McClure, *Phytochemistry*, 1974, **13**, 1065, 1071.
37 S. E. Iredale and H. Smith, *Phytochemistry*, 1974, **13**, 575.
38 P. P. Wong, M. Zucker, and L. L. Creasy, *Plant Physiol.* 1974, **54**, 659.

that the regulation of PAL in plants is controlled through an inactivating system, but the nature of the inactivator still awaits structural elucidation.[39]

4 Flavonoids

The Chalcone Stage.—Although labelling experiments have shown the intervention of the CoA esters of acetate, malonate, and cinnamate as precursors of chalcones in plants, much remains to be learnt about the enzymology of these first stages in flavonoid biosynthesis. A cell-free extract has now been obtained from 2—3-day-old buckwheat seedlings, which catalyzes at optimal pH of 7.5 the incorporation of the carboxy-carbon of [1-^{14}C]acetyl–CoA into ring A of quercetin.[40] Unfortunately, the enzyme activity is not only in the particulate fraction but also is very labile, so it may be some time before it can be completely characterized.

The A-ring of most naturally occurring flavonoids is phloroglucinol-derived and although the early experiments of Neish back in 1957 indicated that phloroglucinol itself was not on the pathway, conclusive evidence of its non-intervention has not been available. This question has been re-investigated by Ali and Kagan[41] who fed 2,4,6-[^{14}C]phloroglucinol to buckwheat seedlings and allowed metabolism to take place over 3—73 h. The isolated rutin at all times was labelled, but most of the activity was in the sugar moiety, with the remainder being approximately equally distributed over both A- and B-rings of quercetin. The phloroglucinol had thus clearly undergone extensive degradation before incorporation and was not therefore a natural precursor. The hypothesis of a biological 'Fries rearrangement' of phoroglucinyl cinnamate (14) to 2′,4′,6′-trihydroxychalcone was also tested and eliminated following a comparison of the efficiency of incorporation of labelled (14) with those of labelled phloroglucinol and cinnamic acid.

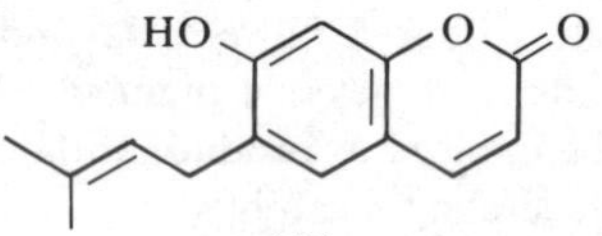

(13)

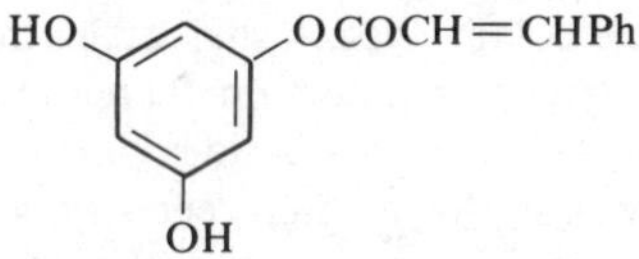

(14)

The role of chalcones as intermediates (but not of their glucosides) in flavonoid synthesis is now well accepted. Endress[42] has shown that a number of chalcone 4′-*O*-glucosides are incorporated into anthocyanins in *Petunia hybrida.* Feeding these chalcone glucosides also causes increased incorporation of labelled acetate into anthocyanins; this effect is apparently on those biosynthetic enzymes which require divalent ions as cofactors.[43] A review of the circumstantial evidence supporting a role for α-hydroxychalcones in the biosynthesis of 3-hydroxyflavonoids has been published.[44]

[39] J. A. Attridge, C. B. Johnson, and H. Smith, *Biochim. Biophys. Acta* 1974, **343**, 440; J. A. Attridge and H. Smith, *ibid.*, 452; L. L. Creasy, *Phytochemistry*, 1974, **13**, 1391.
[40] K. Saito, *Biochim. Biophys. Acta*, 1974, **343**, 392.
[41] M. A. Ali and J. Kagan, *Phytochemistry*, 1974, **13**, 1479.
[42] R. Endress, *Z. Pflanzenphysiol.* 1974, **74**, 179.
[43] R. Endress, *Phytochemistry*, 1974, **13**, 421, 599.
[44] D. G. Roux and D. Ferreira, *Phytochemistry*, 1974, **13**, 2039.

Anthocyanin Biosynthesis.—A simple but elegant approach to the problem of determining the position of gene blocks along the pathway to anthocyanins has been developed by Stickland and Harrison.[45] These workers have found that, following imbibition of dihydroflavonol or flavanone through the stems, petals of recessive white phenotypes can often be persuaded to synthesize the anthocyanins of the dominant colour type. Thus, feeding dihydroquercetin to flowers of the albino *niv* form of *Antirrhinum majus* produces cyanidin 3-rutinoside, indicating that the gene blockage is at the stage of assembly of the C_{15}-skeleton. A second white form (*inc*) synthesizes some aurone, showing that the blockage must be after chalcone formation; since feeding dihydroquercetin to *inc* gives anthocyanin, the gene blockage must be before the dihydroflavonol stage (Scheme 4). Flowers of a third phenotype, *pal*, do not produce anthocyanin on feeding, indicating that the interruption in synthesis must occur after the dihydroflavonol is formed.

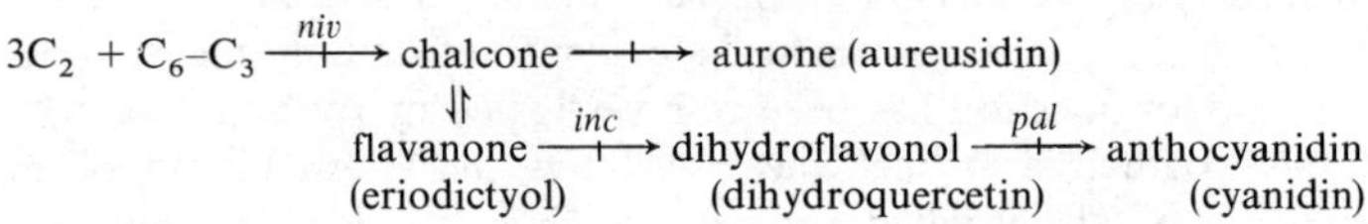

Scheme 4

These experiments indicate that all the enzymes for anthocyanin production beyond the gene block are either present in white flowers or else are induced by feeding the precursor. The system is such that it is possible to make white flowers produce pigments not normally present in the plant. Thus, if dihydromyricetin is fed to albino petals of *Antirrhinum*, delphinidin glycoside is produced *de novo*. In the natural coloured flowers, flavonol glycosides accompany anthocyanins in the dominant colour types, whereas in the artificially coloured flowers, flavonols are absent. Precursor induction therefore only extends to the cyanic pigments. White flowers of other species besides *Antirrhinum* can be induced to produce anthocyanin in this way, so this method seems to be of fairly general applicability.

Developmental studies of anthocyanin synthesis in flowers of *Impatiens*, *Petunia*, and *Primula* have indicated that there is a progression from simple to complex anthocyanins and that the enzymes responsible for methylation, acylation, and extra glycosylation only exert their full effect in the mature petal.[46] In *I. balsamina*, for example, buds contain pelargonidin 3-glucoside while the fully opened flower has pelargonidin [*i.e.* 3-(*p*-coumarylglucoside) 5-glucoside]. An opposite developmental sequence has now been noted in *Lathyrus* and *Pisum*.[47] Here, as the flower ages, the enzymes controlling methylation and glycosylation become inactive, so that the pigments in the mature flower are simpler than those of the young petal; *e.g.* in *L. odoratus* young petals contain petunidin and malvidin 3-rhamnoside 5-glucosides, while older blooms have increasing amounts of delphinidin 3-rhamnoside 5-glucoside and malvidin 3-rhamnoside. A similar situation may occur in bracts of *Euphorbia pulcherrima* since pelargonidin glycosides are formed later than cyanidin glycosides;

[45] R. G. Stickland and B. J. Harrison, *Heredity*, 1974, **33**, 108; B. J. Harrison and R. G. Stickland, *ibid.*, p. 112; R. G. Stickland and B. J. Harrison, *J. Roy. Hort. Soc.* 1974, 526.
[46] H. Reznik, *Flora*, 1961, **150**, 454; C. W. Hagen, *Amer. J. Botany*, 1966, **53**, 54.
[47] C. M. Statham and R. K. Crowden, *Phytochemistry*, 1974, **13**, 1835.

the proportion of pelargonidin increases from 2 to 55% during development. There may be a shortage of oxidases in the mature bracts.[48]

The photo-control of anthocyanin synthesis in plants continues to receive much attention.[49] Anthocyanin synthesis in dark-grown cabbage seedlings can be stimulated by reagents such as n-propanol, which affect membrane permeability; addition of growth regulators can also increase anthocyanin production.[49,50] Knowledge of flavonoid biosynthesis has been applied to the production of new colour forms in ornamental plants. Selection among kaempferol-producing lines of *Petunia hybrida* has yielded, for the first time, a flower pigmented by pelargonidin.[51]

Biosynthesis of Proanthocyanidins.—Although a range of proanthocyanidins (or flavolans) of increasing molecular size is known to occur in Nature, most attention so far has been given to the simplest molecules of this type, the dimeric procyanidins. These are of two types: simple 6(or 8)-4′ linked dimers, such as (15) where several stereoisomers are known; and doubly linked dimers of the type (16). Preliminary feeding experiments[52] suggested that these compounds were formed by oxidative dimerization of two similar flavan-3-ol (or catechin) molecules. More detailed examination of the biosynthesis of (−)-epicatechin and related procyanidins in *Aesculus carnea* and *Rubus* sp. has now shown that they are of 'mixed' biosynthetic origin, being formed by condensation of monomers from different metabolic pools.[53]

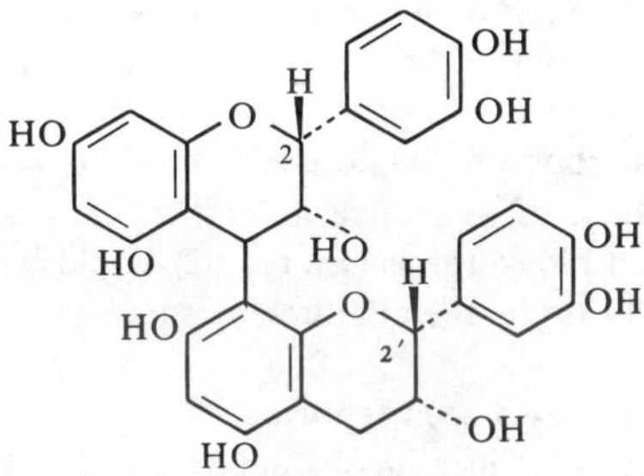

(15) Procyanidin B2

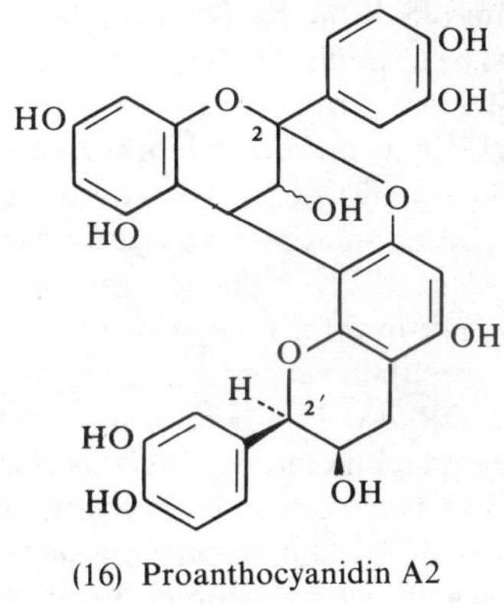

(16) Proanthocyanidin A2

Doubly labelled cinnamic acid was administered to embryo horsechestnut fruits and (−)-epicatechin, procyanidin B2 (an epicatechin dimer), and proanthocyanidin A2 were isolated after 6 days. After purification to constant activity, measurement of ^{3}H–^{14}C ratios (following degradation to veratric acid and the corresponding flav-2-ene) showed that both labels in (−)-epicatechin were at the predicted 2-position to an extent greater than 90%. Degradation of procyanidin B2 with toluene-α-thiol gave (−)-epicatechin (from the lower half) and thioether (17) from the upper half.

[48] J. Billot, *Phytochemistry*, 1974, **13**, 2673.
[49] R. C. Pecket and T. A. H. Bassim, *Phytochemistry*, 1964, **13**, 815, 1395.
[50] E. Lachele, F. Lenner, and V. Sietz, *Z. Pflanzenphysiol.*, 1975, **71**, 75.
[51] A. Cornu, M. Paynot, and H. Touvin, *Phytochemistry*, 1974, **13**, 2022.
[52] E. Haslam, *J. Chem. Soc. (C)*, 1969, 1824.
[53] D. Jacques and E. Haslam, *J.C.S. Chem. Comm.*, 1974, 231.

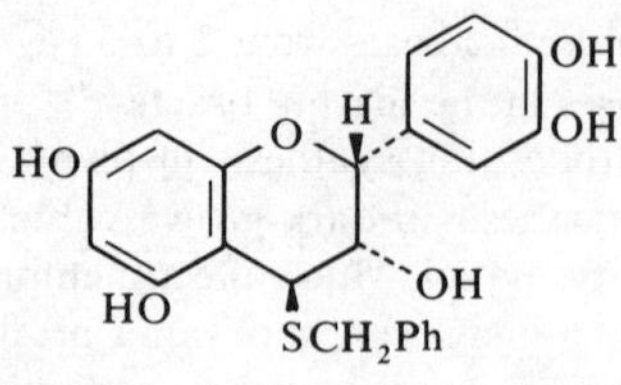

(17)

Again, the labels were clearly located (> 85%) at C-2 in both halves. However, measurements of specific activities always showed incorporation into the upper half to be 3–4 times greater than in the lower half. Although degradation of proanthocyanidin A2 was not attempted, similar measurements of $^{3}H:^{14}C$ ratios indicated that incorporation took place at C-2 and C-2′ and that there was the same difference of incorporation in upper and lower halves. Analogous measurements made on the biosynthesis of (−)-epicatechin and procyanidin B4 [a (+)-catechin-(−)-epicatechin dimer] in *Rubus fruticosus* and the additional feeding of [*G*-^{14}C]-epicatechin showed that the label is almost exclusively located in the lower half of the dimer in this plant.

On the basis of this and other work, Jacques and Haslam[53] have proposed a general scheme for proanthocyanidin biosynthesis which could account for the above findings (Scheme 5). The proposal also encompasses the production of anthocyanidins (*e.g.* cyanidin), which are often formed in proanthocyanidin-containing tissue, *e.g.* in fruits during ripening and in leaves during senescence. The postulated intermediate (18) in its protonated form is equivalent to the 4-carbonium ion derived from a flavan-3,4-diol. The earlier failure to detect flavan-3,4-diols in vegetative tissues, which was used as an argument against their intervention in biosynthesis (see ref. 52), could be alternatively explained as due to their reactivity or to their failure to survive the conditions for their isolation.

Synthetic experiments supporting Scheme 5 have also been reported.[54] Treatment of 2*R*, 3*S*, 4*S*-4-benzylthioflavan-3,5,7,3′4′-pentaol (17) in aqueous acidic dioxan at room temperature with (−)-epicatechin gave a 75% yield of procyanidin B2, together with two isomers in which the interflavan bond is between ring A (C-6 or C-8) and C-4 of the two epicatechin units. Both these new isomers were detected at the same time in natural sources. Furthermore, acid-catalysed equilibration of any one of the three dimers based on (−)-epicatechin gave a mixture of all three in the same proportions. Natural mixtures were also found to occur in the very same proportions. Natural and synthetic dimers based on (+)-catechin behaved in the same way. These experiments strongly suggest that the condensation of flavan units to yield proanthocyanidins is under thermodynamic rather than enzymic control in plants. The situation is thus fairly similar to that in lignin biosynthesis. Oxidative enzymes (*e.g.* peroxidases) may participate but they have little control over the products formed.

Indirect evidence supporting the intermediate role of catechins in proanthocyanidin synthesis has been obtained by Das,[55] who noted little turnover of [*U*-^{14}C]-

[54] E. Haslam, *J.C.S. Chem. Comm.*, 1974, 594.
[55] N. P. Das, *Planta Med.*, 1974, **25**, 274.

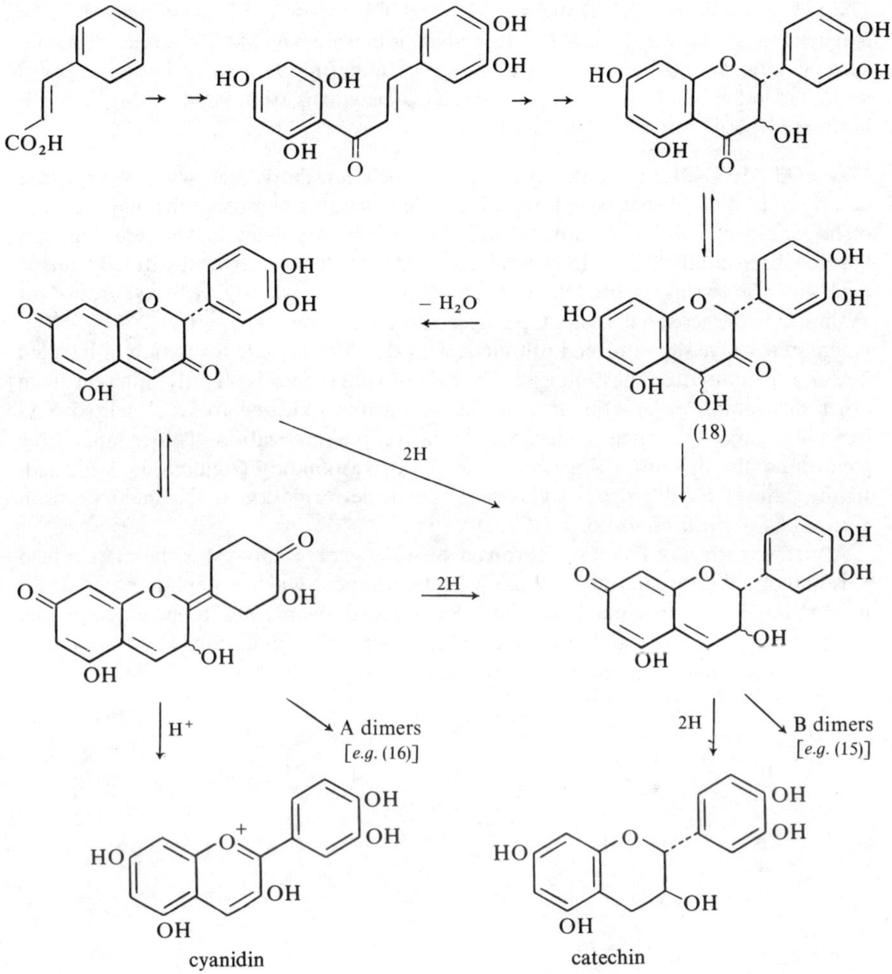

Scheme 5

catechin when it was fed to young leafy shoots of *Uncaria gambir*, a tannin-producing plant. A method of obtaining [^{14}C]- labelled epicatechin and procyanidins by injecting seed shells of horsechestnut fruits *in situ* with [1-^{14}C]acetate has been described by Eastmond and Gardner.[56] Incorporation of label into recrystallized [^{14}C]-epicatechin after a fortnight's growth on the tree was 0.4% and into procyanidin 0.1%.

Glycosylation of Flavonoids.—The first xylosyl transferase of flavonoid synthesis has been detected, in petal extracts of *Melandrium album*.[57] It catalyses the transfer of xylose from UDP-xylose to the 7-hydroxy-group of isovitexin(6-*C*-glucosyl-apigenin).

[56] R. Eastmond and R. J. Gardner. *Phytochemistry*, 1974, **13**, 1477.
[57] J. van Brederode and G. van Nigtevecht, *Phytochemistry*, 1974, **13**, 2763.

The pH optimum is 7.0—7.5 and the K_m for UDP-xylose is 1.0×10^{-3} mol l^{-1}, for isovitexin it is 0.04×10^{-3} mol l^{-1}. Its activity is inhibited by $MnCl_2$. A relatively non-specific glucosyltransferase, isolated from germinating *Phaseolus aureus* seeds,[58] is similar in being affected by metal ions. This enzyme is stimulated by Mg^{2+} but is inhibited by Zn^{2+} at 10×10^{-3} mol l^{-1}.

Flavonoid Metabolism.—A useful review on the degradative pathways of aromatic compounds in plants has been published.[59] The turnover of most of the major classes of flavonoid by cell suspension cultures derived from parsley and several legumes has now been studied.[60–62] In general, the B-ring is recovered as a substituted benzoic acid and the A-ring is broken down to CO_2. [α-^{14}C]-6,4′-Dihydroxyaurone, for example, was recovered as [^{14}C]-*p*-hydroxybenzoic acid and $^{14}CO_2$.[60] [*U*-^{14}C]-Kaempferol gave the same end products, with the intermediate formation of labelled 2,3,5,7,4′-pentahydroxyflavanone.[61] The fate of isoflavones is slightly different from other flavonoids.[62] 7,4′-Dihydroxyisoflavone is first oxidized to 7,3′,4′-trihydroxy-isoflavone and this then undergoes oxidative polymerization. Under anaerobic conditions, the oxidation is arrested and the corresponding *O*-glucoside is formed. In the case of 6,7-dihydroxyisoflavones, glucoside formation is the main reaction even under aerobic conditions.

A further aspect of flavonoid turnover of some general interest is the metabolism of isoflavonoid phytoalexins by those fungal pathogens which stimulate phytoalexin production in the original host plant. Such metabolism leads to products which are both less fungitoxic and more accessible to further degradation. The metabolism of the isoflavonoid phaseollin (19) *in vitro* has been examined in two laboratories

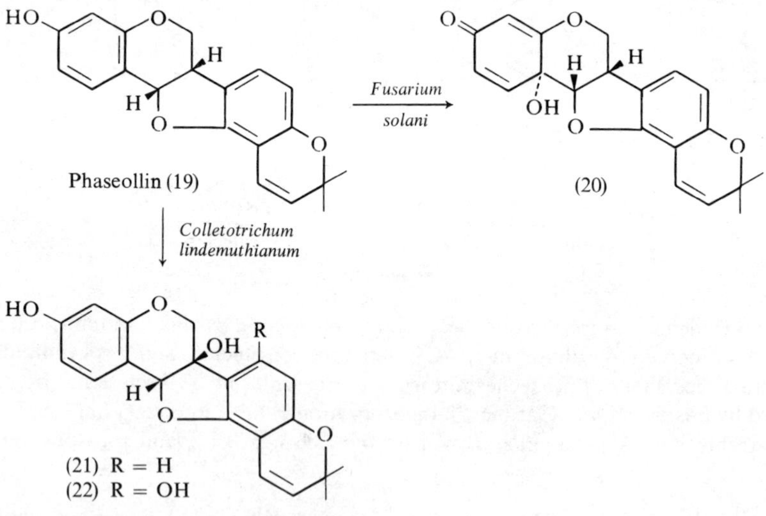

Phaseollin (19)

(20)

(21) R = H
(22) R = OH

[58] D. L. Storm and W. Z. Hassid, *Plant Physiol.*, 1974, **54**, 840.
[59] B. E. Ellis, *Lloydia*, 1974, **37**, 168.
[60] W. Barz, F. Mohr, and E. Teufel, *Phytochemistry*, 1974, **13**, 1785.
[61] W. Hösel, G. Frey, and W. Barz, *Phytochemistry*, 1975, **14**, 417.
[62] J. Berlin. P. Kiss, D. M. Enoch, D. Gierse. W. Bara. and B. Janistyn, *Z. Naturforsch.*, 1974, **29c**, 374.

and it has been found that different fungi produce different metabolites. Unexpectedly, the bean pathogen *Fusarium solani* converts it into the quinonoid product (20).[63] On the other hand, the pathogen *Colletotrichum lindemuthianum* changes it to the two hydroxy-derivatives (21) and (22):[64] evidence that these two latter substances are the only metabolites was obtained by feeding labelled (19) to the fungus and recovering the radio-activity exclusively in these metabolites, to the extent of 18 and 17% respectively, the remainder being lost as $^{14}CO_2$. Labelled phaseollin used in these experiments was produced by inoculating bean hypocotyls with tobacco necrosis virus in the presence of [U-^{14}C]NaOAc.

5 Naphthoquinones

The *o*-Succinylbenzoic Acid Pathway.—One of the best established pathways to naphthoquinones in plants is that involving incorporation of shikimate and the intermediacy of *o*-succinylbenzoic acid, or 4-(2′-carboxyphenyl)-4-oxobutyric acid (23).[65] However, although (23) has been fed to plants and been shown to be incorporated into naphthoquinone, it had, until recently, still to be identified as a natural plant substance. Grotzinger and Campbell,[66] in a study of lawsone (24) biosynthesis in *Impatiens balsamina* cuttings, have now done this. Furthermore, on feeding [U-^{14}C]-glutamate to the plant, (23) became labelled and was then effectively incorporated (0.43%) into lawsone. By using a combined g.c.–m.s. proportional counter scanning procedure, the same authors were able to detect all metabolites derived from the labelled precursor. In this case, a second acid (25) was also found to be labelled. From its dilution value, it was assumed to be a catabolic product of lawsone, formed by oxidative cleavage of the 1,2-bond (Scheme 6).

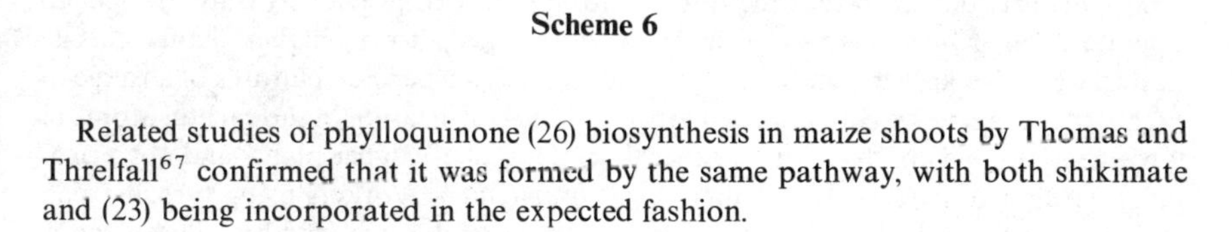

(23) (24) lawsone (25)

Scheme 6

Related studies of phylloquinone (26) biosynthesis in maize shoots by Thomas and Threlfall[67] confirmed that it was formed by the same pathway, with both shikimate and (23) being incorporated in the expected fashion.

[63] J. van den Heuvel, H. D. van Etten, J. W. Serum, D. L. Coffen, and T. Williams, *Phytochemistry*, 1974, **13**, 1129.

[64] R. S. Burden, J. A. Bailey and G. G. Vincent, *Phytochemistry*, 1974, **13**, 1789.

[65] R. Bentley, in 'Biosynthesis', ed. T. A. Geissman (Specialist Periodical Reports), The Chemical Society, London, 1975, Vol. 3, p. 181.

[66] E. Grotzinger and I. M. Campbell, *Phytochemistry*, 1974, **13**, 923.

[67] G. Thomas and D. R. Threlfall, *Phytochemistry*, 1974, **13**, 807.

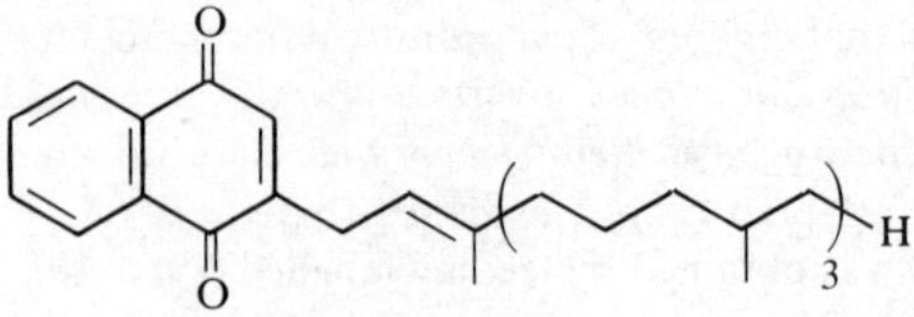

(26)

The Homogentisic Acid Ring Cleavage Pathway.—In the course of studying the biosynthesis of plumbagin (27) in 16 Droseraceae species, Durand and Zenk[68] discovered that the most efficient precursor was [β-^{14}C]tyrosine, which was incorporated to the extent of 20%. However, chemical degradation of (27) showed that the label was incorporated indirectly; in fact, the pattern of labelling in both plumbagin and 7-methyljuglone in *Drosophyllum lusitanicum* showed that they must be formed *via* the acetate–polymalonate path. The reason why tyrosine is such an excellent precursor is simply that it is broken down *in vivo* to acetate (see Scheme 7) and this is

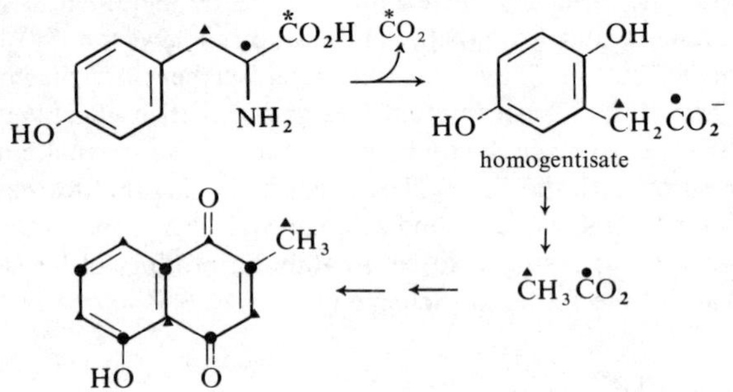

Scheme 7

then incorporated into plumbagin in the usual alternate fashion. Exhaustive tests in *Drosophyllum* confirmed that the homogentisic pathway, previously only thought to operate in micro-organisms and animals, does take place in higher plants. Sterile tissues were also able to catabolize tyrosine, and cell suspension cultures of a range of higher plants also possess the same ability.[69] The re-utilization of the carbon atoms of tyrosine in this way may thus be a general property of all higher plants and not a property peculiar to carnivorous plants like *Drosophyllum.* Workers using tyrosine as a precursor in phenolic biosynthesis must now be careful to see whether catabolism is occurring and, if so, then appropriate corrections must be made in their results.

[68] R. Durand and M. H. Zenk, *Phytochemistry*, 1974, **13**, 1483.
[69] R. Durand and M. H. Zenk, *F.E.B.S. Letters*, 1974, **39**, 218.

5
Biosynthesis of Alkaloids

BY E. LEETE

1 Introduction

This Report follows the same format as previous years, and attempts to review all material concerned with the biosynthesis of alkaloids and related compounds published during 1974. This year some nitrogen-containing microbial products have been included, which, in the eyes of the Reporter, resemble the more 'conventional' alkaloids. The nitrogenous compounds formed by the *Claviceps* species have long been regarded as alkaloids, and no apology is required for including them. Experiments involving the use of labelled precursors are again summarized in an extensive Table. I have attempted to be critical of work which is not thought to be significant, and deplore the practice of some authors who publish a preliminary account of biosynthetic experiments as a communication, and then several years later, publish a full paper with little, if any, additional data.

Carbon-13, fully discussed by Tanabe in Chapter 7 of this Volume and in previous volumes of this series, is being used to an increasing extent in biosynthetic studies.[1] Most of the experiments have been concerned with the origin of microbial products, which are often formed from ^{13}C-labelled precursors with good specific incorporations (3—10%). The positions enriched with ^{13}C are then readily detected in the ^{13}C-n.m.r. spectrum. This year this method has been applied for the first time to natural products formed in higher plants, in studies on the biosynthesis of camptothecin[2] and colchicine.[3]

Several reviews relating to alkaloid biosynthesis and metabolism have appeared in 1974.[4,5] Floss *et al.* have written an excellent account of the regulatory control mechanisms in alkaloid biosynthesis, especially those relating to the formation of the ergot alkaloids.[6] A lengthy and somewhat irreverent discussion of the biosynthesis of the indole alkaloids has been written by Cordell.[7] He has collected a vast amount of information (413 refs.) relating to both *in vivo* and *in vitro* studies on the origin and transformation of these alkaloids; it is perhaps not surprising, in view of his academic background, that he favours the Manchester School in the controversy

[1] U. Séquin and A. I. Scott, *Science*, 1974, **186**, 101.
[2] C. R. Hutchinson, A. H. Heckendorf, P. E. Daddona, E. Hagaman, and E. Wenkert, *J. Amer. Chem. Soc.*, 1974, **96**, 5609; Correction: *ibid.*, 1975, **97**, 1988.
[3] A. R. Battersby, P. W. Sheldrake, and J. A. Milner, *Tetrahedron Letters*, 1974, 331.
[4] R. B. Herbert, in 'The Alkaloids', ed. E. Saxton (Specialist Periodical Reports), The Chemical Society, London, 1974, Vol. 4, p. 1.
[5] K. Schreiber, *Biochem. Soc. Trans.*, 1974, **2**, 1.
[6] H. G. Floss, J. E. Roberts, and P. F. Heinstein, in 'Recent Advances in Phytochemistry', ed. V. C. Runeckles and E. E. Conn, Vol. 8, Academic Press, 1974, p. 141.
[7] G. A. Cordell, *Lloydia*, 1974, **37**, 219.

regarding the transformation products of stemmadenine in refluxing acetic acid.[8,9] T. Robinson[10] has discussed the metabolism and function of alkaloids in plants. The present Reporter does not agree with his remarks relating to rate of turnover of nicotine in the tobacco plant. Robinson's calculations suggest that 10% of the photosynthetically fixed carbon dioxide is directly involved in the biosynthesis of nicotine Recent work on the metabolism of nicotine in *Nicotiana glauca*[11] has indicated that 60% of the administered nicotine was recovered either as nicotine, or closely related alkaloids (nornicotine and myosmine) after 3 days. Feeding experiments with $^{14}CO_2$ also indicate that only a very small fraction of the fixed carbon dioxide ends up in the alkaloids of the opium poppy[12] or tobacco.[13] After studying the biosynthesis of alkaloids for more than 20 years I am willing to concede that *most* alkaloids do not play an important biological role in the plants which produce them; however there is much still to be learned regarding the biochemical relationship of alkaloids to other plant components, and this is still a viable area of research.

Little progress has been made on the isolation of enzymes responsible for the formation of alkaloids, contrary to the Reporter's earlier predictions.[14]

2 Some Highlights of 1974

Biosynthesis of the Pyrrolidine Ring of Nicotine examined by Short-term $^{14}CO_2$ Exposure.—In the year under review Rapoport and co-workers have described additional experiments on the exposure of tobacco seedlings (*Nicotiana glutinosa*) to $^{14}CO_2$ for varying lengths of time and different light intensities.[15] The resultant radioactive nicotine (1) was degraded to determine the distribution of activity. A new method for estimating the activity at C-2′ was developed, and is apparently more reliable than previous methods. In some of the experiments unsymmetrical labelling of the pyrrolidine ring of nicotine occurred. Unsymmetrical labelling of the pyrrolidine ring is incompatible with almost all the tracer experiments which have been carried out in the past 20 years on the origin of this ring. It is generally agreed that the pyrrolidine ring of nicotine is derived from the *N*-methyl-Δ^1-pyrrolinium salt (2), which is formed from ornithine and related compounds by the routes illustrated in Scheme 1. Enzymes have been isolated from tobacco roots which catalyse the formation of (2) from ornithine.[16] It is expected that any unsymmetrical labelling of the pyrrolidine ring of nicotine occurred. Unsymmetrical labelling of putrescine in the biosynthetic sequence.

The experiments involving the exposure of tobacco plants (almost all *N. glutinosa*) to $^{14}CO_2$ are recorded in Table 1. Experiments *1* to *5* are Rapoport's earlier results,[17]

[8] A. A. Qureshi and A. I. Scott, *Chem. Comm.*, 1968, 945.
[9] R. T. Brown, G. F. Smith, J. Poisson, and N. Kunesch, *J. Amer. Chem. Soc.*, 1973, **95**, 5778.
[10] T. Robinson, *Science*, 1974, **184**, 430.
[11] E. Leete and M. R. Chedekel, *Phytochemistry*, 1974, **13**, 1853.
[12] H. Rapoport, F. R. Stermitz, and D. R. Baker, *J. Amer. Chem. Soc.*, 1960, **82**, 2765.
[13] W. L. Alworth and H. Rapoport, *Arch. Biochem. Biophys.*, 1965, **112**, 45.
[14] E. Leete, *Adv. Enzymol.*, 1969, **32**, 373.
[15] M. L. Rueppel, B. P. Mundy, and H. Rapoport, *Phytochemistry*, 1974, **13**, 141.
[16] *cf.* E. Leete, in 'Biosynthesis', ed. T. A. Geissman (Specialist Periodical Reports), The Chemical Society, London, 1975, Vol. 3, p. 129.
[17] A. A. Liebman, B. P. Mundy, and H. Rapoport, *J. Amer. Chem. Soc.*, 1967, **89**, 664.

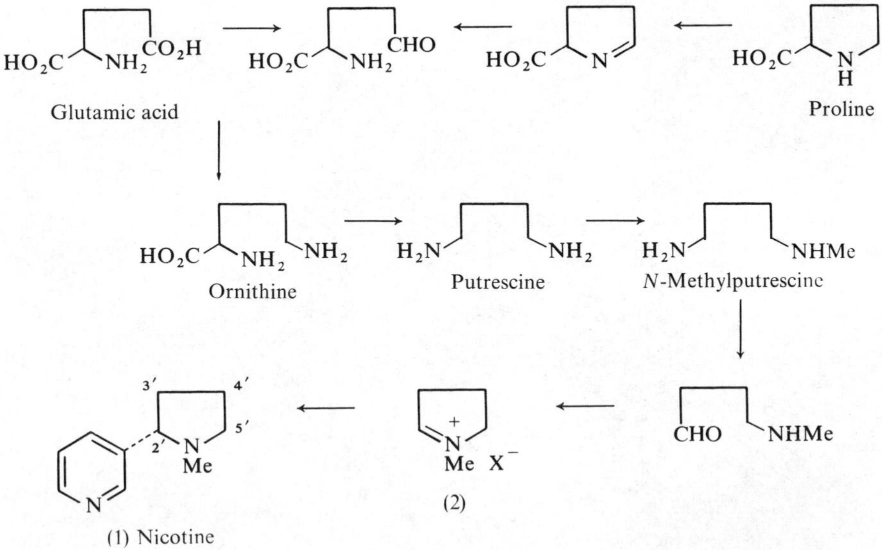

Scheme 1 *Precursors of the pyrrolidine ring of nicotine*

experiments *6* to *11* the recent work,[15] and experiments *12* to *14* the work of Byerrum and co-workers.[18] The distribution of activity in various parts of the nicotine molecule (pyridine ring, *N*-methyl group, pyrrolidine ring) is shown. The percentage distribution of activity within the pyrrolidine ring has also been calculated from the original data on activities at the individual carbons of the ring. In all these experiments, more than average (50%) activity enters the pyridine ring. This result is consistent with the more rapid biosynthesis of the precursor of the pyridine ring, namely nicotinic acid, which is formed from 3-phosphoglyceraldehyde and aspartic acid.[19] It has been established that these molecules are rapidly labelled on short-term exposure of tobacco and other plants to $^{14}CO_2$.[20]

The level of activity (2.6—11.8%) in the *N*-methyl group does not seem to correlate with any of the variables such as duration of feeding, light intensity, or the age or part of the plant.

The distribution of activity in the pyrrolidine ring has been the major source of controversy. In his earlier work Rapoport claimed that C-2′ had in general a lower activity than C-5′. However in the recent work a higher activity was found at C-2′ than at C-5′ (exp. *6*) In a very short feeding time (exp. *7*, 30 min) a higher activity was found at C-5′ than at C-2′. In experiments *8* to *11*, carried out under very similar experimental conditions, essentially uniform labelling was found. Variations in the % of CO_2 in the growth chamber seemed to have little effect on the general pattern of labelling. Byerrum found essentially uniform labelling in the pyrrolidine ring in experiments involving both *N. glutinosa* and *N. rustica*.

[18] H. R. Zielke, R. U. Byerrum, R. M. O'Neal, L. C. Burns, and R. E. Koeppe, *J. Biol. Chem.*, 1968, **243**, 4757.
[19] H. R. Zielke, C. M. Reinke, and R. U. Byerrum, *J. Biol. Chem.*, 1969, **244**, 95.
[20] J. L. Hess and N. Tolbert, *J. Biol. Chem.*, 1966, **241**, 5705.

Table 1. *Distribution of activity in nicotine after short-term exposure of* Nicotiana glutinosa *plants to* $^{14}CO_2$

Exp. no.	*Source of nicotine*	*Duration of feeding/h*	*Activity in pyridine ring(%)*	*Activity in N Me (%)*	*Activity in pyrrolidine Ring (%)*	*Distribution of activity in the pyrrolidine ring (%)* C-2′	C-3′	C-4′	C-5′
1	70-day-old aerial parts	6	79.5	2.6	19.9	12	10	←78→	
2	85-day-old roots	6	75.8	5.0	19.2	14	16	←70→	
3	85-day-old aerial parts	6	79.4	6.6	14.0	11	22	32	34
4	53-day-old roots	6	73.0	7.8	19.2	7	31	32	30
5	6-day-old seedlings	6	73.1	6.5	20.4	17	29	26	28
6	6-day-old seedlings (6200 lux, CO_2: 0.18 → 0.02%)	2	66.9	11.8	21.7	77	11	3	8
7	6-day-old seedlings (17000 lux, CO_2: 0.03 → 0.02%)	0.5	88.2	9.5	2.3	22	21	20	37
8	6-day-old seedlings (17000 lux, CO_2: 0.04 → 0.05%)	2	76.0	6.0	18.0	26	27	25	23
9	6-day-old seedlings (17000 lux, CO_2: 0.04 → 0.05%)	0.5	82.0	5.6	12.4	24	26	24	25
10	6-day-old seedlings (5500 lux, CO_2: 0.05 → 0.05%)	2	90.2	3.4	6.4	25	25	23	27
11	6-day-old seedlings (5500 lux, CO_2: 0.22 → 0.22%)	2	86.4	5.3	8.3	23	25	24	28
12	plants 20—25 cm tall	6	77.8	5.0	17.2	25	25	27	24
13	plants 20—25 cm tall	3	90.4	5.4	4.2	26	27	23	24
14	*Nicotiana rustica*, 20—25 cm tall	6	74.6	6.9	18.5	24	26	25	25

It is the Reporter's opinion that the unsymmetrical labelling of the pyrrolidine ring observed by Rapoport is due to errors in assay of the radioactive degradation products of the pyrrolidine ring; *e.g.*, the nicotine obtained in experiment 7 afforded degradation products having extremely low activities (C-2′ assayed as benzoic acid, 4 d.p.m. mg^{-1}; C-3′ benzoic acid, 4 d.p.m. mg^{-1}; C-4′ barium carbonate, 2 d.p.m. mg^{-1}; and C-5′ formaldehyde dimedone, 2.6 d.p.m. mg^{-1}). It was earlier suggested that a source of error in the degradation was the production of formaldehyde from the *N*-methyl groups of *NN*-dimethylglycine on oxidation with lead tetra-acetate.[21] However, it was subsequently claimed[22] that the degradation was radiochemically valid. In experiment 6, the high value at C-2′ was obtained using an older degradative scheme which Rapoport himself admits was not reliable.

Perhaps this problem can be solved by carrying out feeding experiments with $^{13}CO_2$, and then determining the enrichments by ^{13}C-n.m.r., thus avoiding any ambiguous degradations.

Camptothecin—a Compound Derived from a Terpenoid Indole Alkaloid.— This alkaloid (9) containing a pyrrolo[3,4-*b*]quinoline unit has been shown to be derived from the β-carboline alkaloid isovincoside (3) (also called strictoside). Preliminary work[2, 23] with the plant *Camptotheca acuminata* indicated that established precursors (tryptophan, mevalonic acid, secologanin *etc.*) of the terpenoid indole alkaloids[24] were incorporated into camptothecin. It was then discovered that isovincoside was a precursor of camptothecin. Hutchinson *et al.*[2] considered that this indole alkaloid was incorporated *via* the lactam (4) produced by cyclization of N-4 with the carbomethoxy group (C-22). The 18,19-dihydroisovincoside lactam (5) is then oxidized at the indole double bond affording the di-oxo-compound (6) which undergoes a Camps reaction[25] yielding the quinolone (7). Oxidation leads to the α-pyridone (8) and then further unexceptional steps yield camptothecin (Scheme 2). A similar hypothesis was suggested by Cordell.[7] Experiments with ^{14}C- and ^{3}H-labelled isovincoside lactam indicated that this compound was indeed an excellent precursor (2% incorporation) of camptothecin. [5-^{13}C]Isovincoside lactam was incorporated and the resultant labelled camptothecin was shown to have an enriched ^{13}C peak at C-5 in its ^{13}C-n.m.r. Good incorporation of (5) probably indicates that the vinyl group is reduced at an early stage in the biosynthetic sequence. Hutchinson *et al.*[26] were able to accomplish part of this sequence *in vitro*. Oxidation of modified (5) (with glucose hydroxy-groups acetylated) with sodium metaperiodate in methanol afforded (6)-tetra-acetate, which cyclized on treatment with triethylamine in ethanol to the quinolone (7)-tetra-acetate. The corresponding 3β-isomer (18,19 dihydrovincoside lactam tetra-acetate) underwent the same oxidation with periodate, however at a much slower rate and in a lower yield. This result probably has some biological significance since vincoside lactam was not a precursor of camptothecin *in vivo*.

[21] E. Leete, *J.C.S. Chem. Comm.*, 1971, 1524.
[22] A. A. Liebman, B. P. Mundy, M. L. Rueppel, and H. Rapoport, *J.C.S. Chem. Comm.*, 1972, 1022.
[23] H. Rapoport, unpublished work, referred to in ref. 2.
[24] A. R. Battersby, in 'The Alkaloids', ed. E. Saxton (Specialist Periodical Reports), The Chemical Society, London, 1971, Vol. 1. p. 31.
[25] R. Camps, *Ber.*, 1899, **22**, 3228.
[26] C. R. Hutchinson, G. J. O'Loughlin, R. T. Brown, and S. B. Fraser, *J.C.S. Chem. Comm.*, 1974, 928.

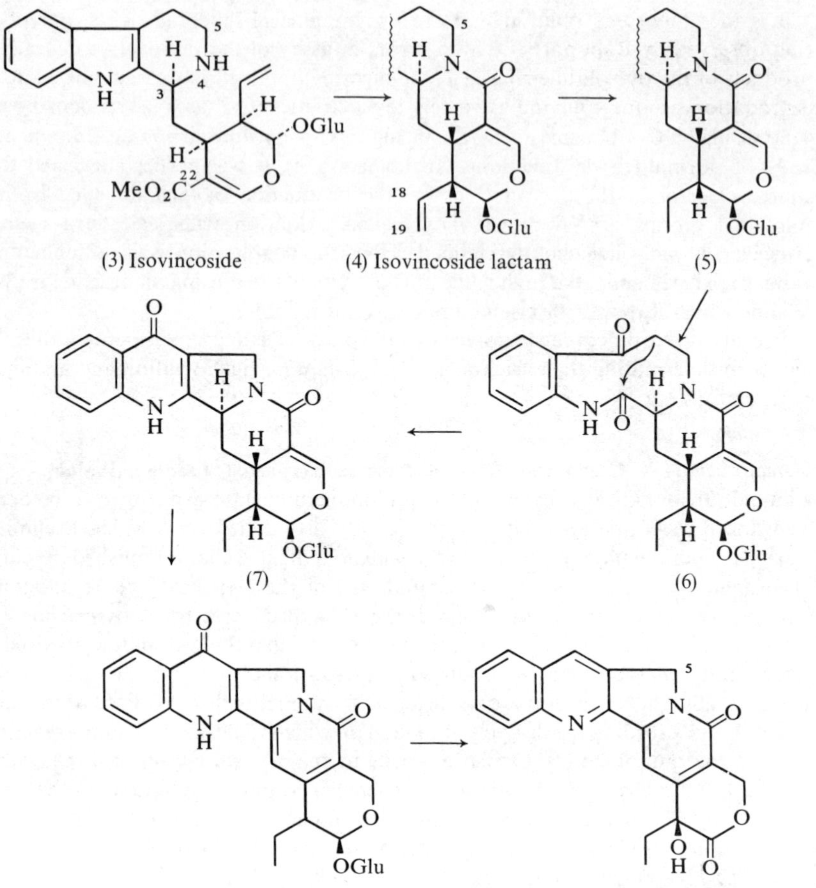

(3) Isovincoside (4) Isovincoside lactam (5) (6) (7) (8) (9) Camptothecin

Scheme 2 *Biosynthesis of camptothecin*

The Ergot Alkaloids—Mechanism of the Formation of Ring D.—Floss *et al.* have presented considerable evidence[27] which favours the biosynthetic pathway illustrated in Schemes 3 and 4. These rationalize the intermolecular transfer of a hydrogen at C-9 in chanoclavine-I (12), and the *cis-trans* isomerization of the C-8–C-9 double bond which occurs during the formation of ring D of agroclavine (19) and elymoclavine (20). The fate of the isotopes originally present in 3*R*-[2-^{14}C,3′, 4-$^{2}H_5$]-mevalonic acid (10) is depicted in the schemes. The early stages in the biosynthesis *via* 4-dimethylallyltryptophan (11) are well authenticated and have been reviewed extensively.[28—30] However, it should be pointed out that the mechanisms of the

[27] H. G. Floss, M. Tcheng-Lin, C.-j. Chang, B. Naidoo, G. E. Blair, C. I. Abou-Chaar, and J. M. Cassady, *J. Amer. Chem. Soc.*, 1974, **96**, 1898.
[28] R. Voigt, *Pharmazie*, 1968, **23**, 285, 353, 419.
[29] E. Ramstad, *Lloydia*, 1968, **31**, 327.
[30] H. G. Floss, *Abhandl. Deut. Akad. Wiss, Berlin, Kl. Chem., Geol., Biol.*, 1972, 395.

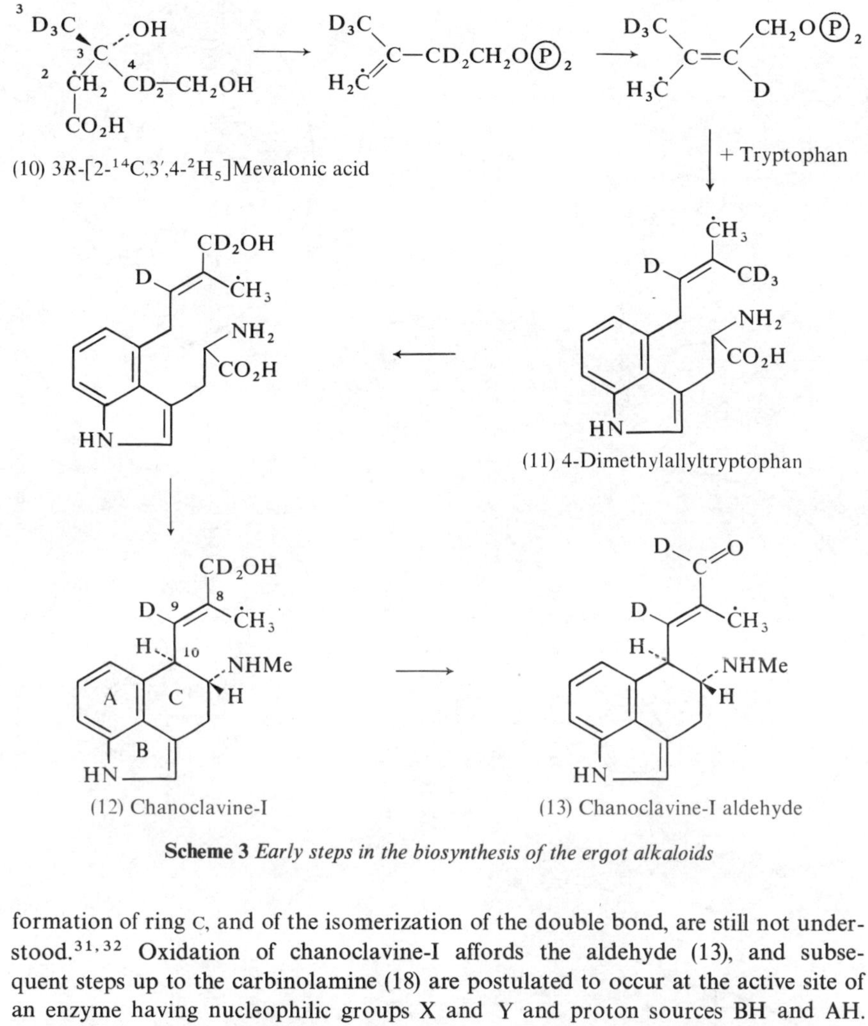

Scheme 3 *Early steps in the biosynthesis of the ergot alkaloids*

formation of ring C, and of the isomerization of the double bond, are still not understood.[31,32] Oxidation of chanoclavine-I affords the aldehyde (13), and subsequent steps up to the carbinolamine (18) are postulated to occur at the active site of an enzyme having nucleophilic groups X and Y and proton sources BH and AH. It is suggested that the group Y binds the aldehyde (13) by the formation of the alcohol, group AH providing a proton [see (14)]. The second nucleophilic group X then adds to the C-8–C-9 double bond, group BH providing a proton at C-9 (15). The substrate molecule is then free to rotate about the C-8–C-9 single bond to yield (16), in which the NHMe group then displaces group Y and a proton is accepted by A to give (17). The carbinolamine (18) is then released from the active site by regeneration of the C-8–C-9 double bond, group B removing a different hydrogen from C-9 (in Scheme 4, a D atom). The active site will then contain a deuterium atom attached to B which will be transferred to the next molecule of chanoclavine-I to enter the site. In the subsequent reduction of the carbinolamine to agroclavine and

[31] M. Seiler, W. Acklin, and D. Arigoni, *Chem. Comm.*, 1970, 1394; *Chimia*, 1970, **24**, 449.

[32] C. I. Abou-Chaar, H. F. Guether, M. F. Manuel, J. E. Robbers, and H. G. Floss, *Lloydia*, 1972, **35**, 272.

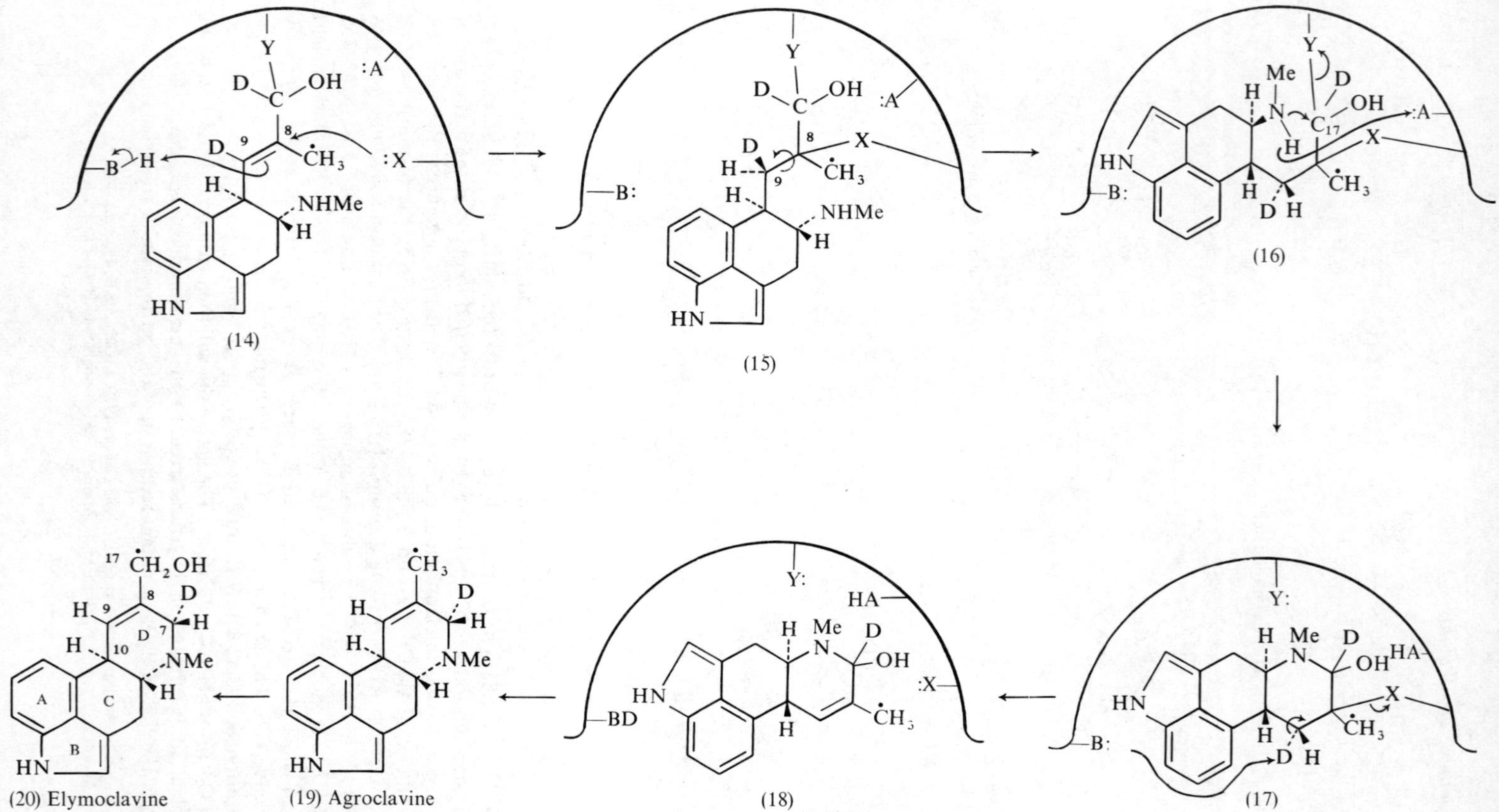

Scheme 4 *Hypothesis for the formation of ring D of the ergot alkaloids*

thence to elymoclavine, the deuterium ends up mainly in the *pro-S* position (as determined by n.m.r.).

Securinine—Unexpectedly Derived from Tyrosine.—An ingenious and plausible biosynthetic route to securinine (23), which is consistent with preliminary tracer work,[33, 34] has been proposed by Parry.[33] This is illustrated with some elaborations in Scheme 5. A Mannich reaction between tyrosine or tyramine and 5-aminopentanal

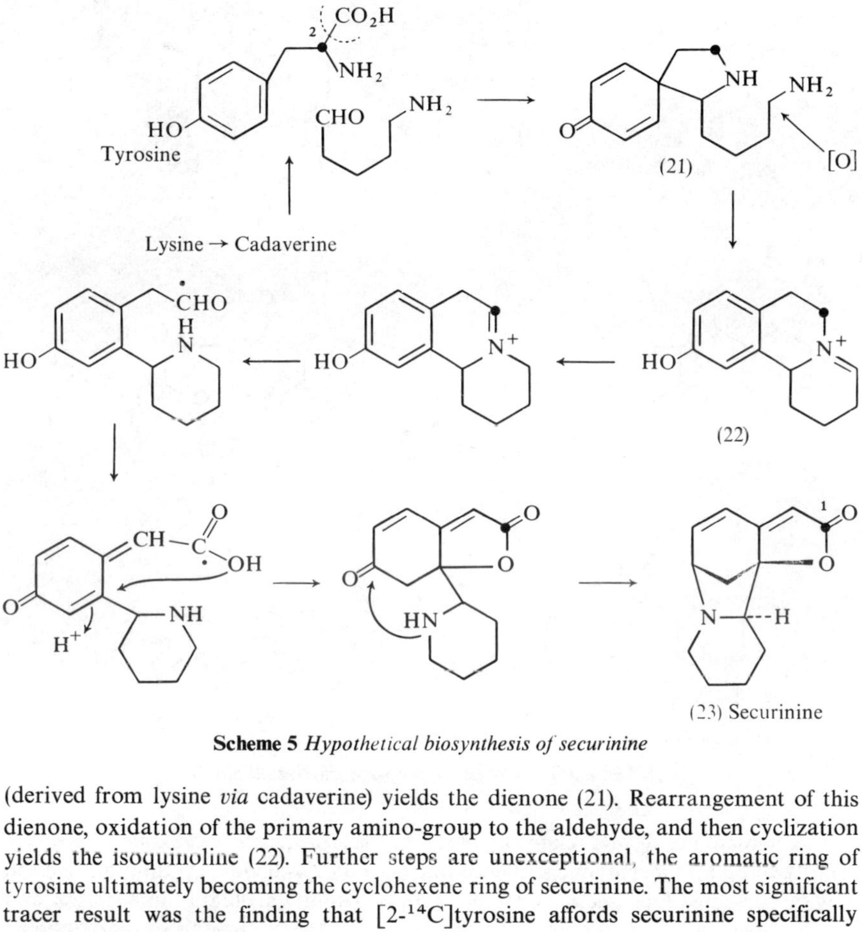

Scheme 5 *Hypothetical biosynthesis of securinine*

(derived from lysine *via* cadaverine) yields the dienone (21). Rearrangement of this dienone, oxidation of the primary amino-group to the aldehyde, and then cyclization yields the isoquinoline (22). Further steps are unexceptional, the aromatic ring of tyrosine ultimately becoming the cyclohexene ring of securinine. The most significant tracer result was the finding that [2-^{14}C]tyrosine affords securinine specifically labelled on its carbonyl group (C-1).

The Furoquinoline Alkaloids: Dictamnine and Skimmianine.—Grundon and co-workers[35] have continued their studies on the origin of the furan ring of dictamnine

[33] R. J. Parry, *Tetrahedron Letters*, 1974, 307.

[34] U. Sankawa, K. Yamasaki, and Y. Ebinzuka, *Tetrahedron Letters*, 1974, 1867.

[35] (*a*) M. F. Grundon, D. M. Harrison, and C. G. Spyropoulos, *J.C.S. Chem. Comm.*, 1974, 51; (*b*) *J.C.S. Perkin I*, 1974, 2181; (*c*) *J.C.S. Perkin I*, 1975, 302; (*d*) J. C. Collins, W. J. Donnelly, M. F. Grundon, and K. J. James, *J.C.S. Perkin I*, 1974, 2177.

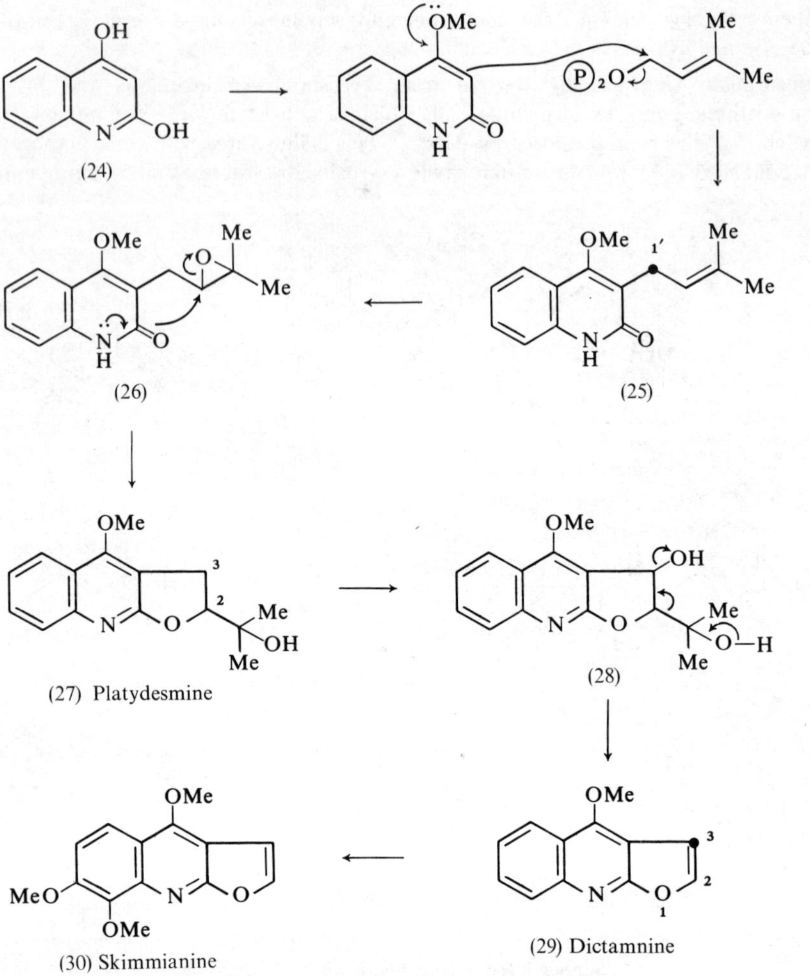

Scheme 6 *Biosynthesis of the furoquinoline alkaloids*

(29), skimmianine (30), and related alkaloids in *Choisya ternata* and *Skimmia japonica*. A fairly complete biosynthetic sequence starting with 2,4-dihydroxyquinoline (24) can now be presented (Scheme 6). 4-Methoxy-3-prenylquinoline (25) is an excellent and specific precursor of dictamnine, skimmianine, and the metho-salt of platydesmine (27). It is suggested that cyclization to the dihydrofuran ring occurs *via* the epoxide (26), so far a hypothetical intermediate. The furan ring is formed with loss of the isopropyl group at C-2, and occurs with retention of one hydrogen at C-3. Therefore, intermediate formation of a 3-oxo-derivative may be eliminated from consideration. The mechanism illustrated involves stereospecific hydroxylation at C-3, followed by an elimination of the isopropyl group as acetone. The metho-salt of platydesmine was not converted into dictamnine, and thus is not part of the main biosynthetic sequence.

The Erythrina Alkaloids.—It had been established previously that (*S*)-*N*-norprotosinomenine (31) is a specific precursor of the *Erythrina* alkaloids,[36] and it has now been shown that isomers of (31), namely *N*-nor-reticuline and *N*-nororientaline are not incorporated into erythraline (38) or β-erythroidine (39).[37] Scheme 7 illustrates the biosynthetic sequence which is considered to proceed *via para–para* oxidative coupling of (31) to yield (32). Ring-opening and reduction of the resultant imine (33) affords the dibenzazonine (34). It was discovered[38] that this dibenzazonine racemized rapidly (half life of 1.2 min at 20 °C) and this indicates probable racemization at this stage *in vivo*. This point may only be of academic interest since it seems likely that the next step is the oxidation to the optically inactive diphenoquinone (35), although its intermediacy in Nature has not been demonstrated. (−)-Erysodienone results from the cyclization of the amino-group onto this quinonoid system. (−)-Erysodienone, but not its (+)-isomer, was incorporated into erythraline and the erythroidines. The precise point of cleavage of ring A to afford the lactone ring of the erythroidines was not determined. A tritium label at C-17 was retained in the ultimate β-erythroidine at the C-17 position. Plausible mechanisms which would lead to this result can be drawn involving cleavage either extradiol (path *a*), or intradiol (path *b*).

3 Enzyme Studies and Non-tracer Experiments Related to Alkaloid Formation

***Ricinus communis* and *Nicotiana rustica*.**—Nicotinic acid and quinolinic acid have been shown to be precursors of the alkaloids nicotine and ricinine (40) and it has been suggested[39] that these compounds are incorporated *via* their *N*-ribosyl derivatives. Some support has been obtained for this hypothesis by measuring the amount of the enzymes which catalyse the formation of the derivatives (41) and (42) (see Scheme 8) from nicotinic and quinolinic acid respectively.[40] The enzymes were assayed by the use of radioactive substrates. In *Ricinus communis* the level of quinolinic acid phosphoribosyl transferase (reaction I) increased 6-fold over a 4-day period, and preceded the onset of ricinine biosynthesis by one day. In *Nicotiana rustica* the level of the enzyme-catalysing reaction II was 38 times higher than the nicotinic acid phosphoribosyl transferase (reaction I).

***Papaver somniferum*.**—Polyphenolase activity in various fractions of the latex of the opium poppy (*Papaver somniferum*) was measured by determining the oxygen uptake manometrically.[41] Tyrosine was a substrate for the enzyme,which is significant since this amino-acid is oxidized to dopa prior to its incorporation into morphine.

***Nicotiana tabacum*.**—The amounts of nicotine, anabasine, anatabine, and nornicotine in *N. tabacum* seedlings were measured daily from germination until the plants were 6 days old.[42] Nicotine was the predominant alkaloid, however all the alkaloids decreased during the first 3 days of growth.

[36] D. H. R. Barton, A. A. L. Gunatilaka, R. Letcher, A. M. F. T. Lobo, and D. A. Widdowson, *J.C.S. Perkin I*, 1973, 874.
[37] D. H. R. Barton, C. J. Potter, and D. A. Widdowson, *J.C.S. Perkin I*, 1974, 346.
[38] D. H. R. Barton, R. D. Brachs, C. J. Potter, and D. A. Widdowson, *J.C.S. Perkin I*, 1974, 2278.
[39] R. D. Johnson and G. R. Waller, *Phytochemistry*, 1974, **13**, 1493, and ref. cited therein.
[40] D. F. Mann and R. U. Byerrum, *Plant Physiol.*, 1974, **53**, 119.
[41] M. F. Roberts, *Phytochemistry*, 1974, **13**, 119.
[42] W. W. Weeks and L. P. Bush, *Plant Physiol.*, 1974, **53**, 73.

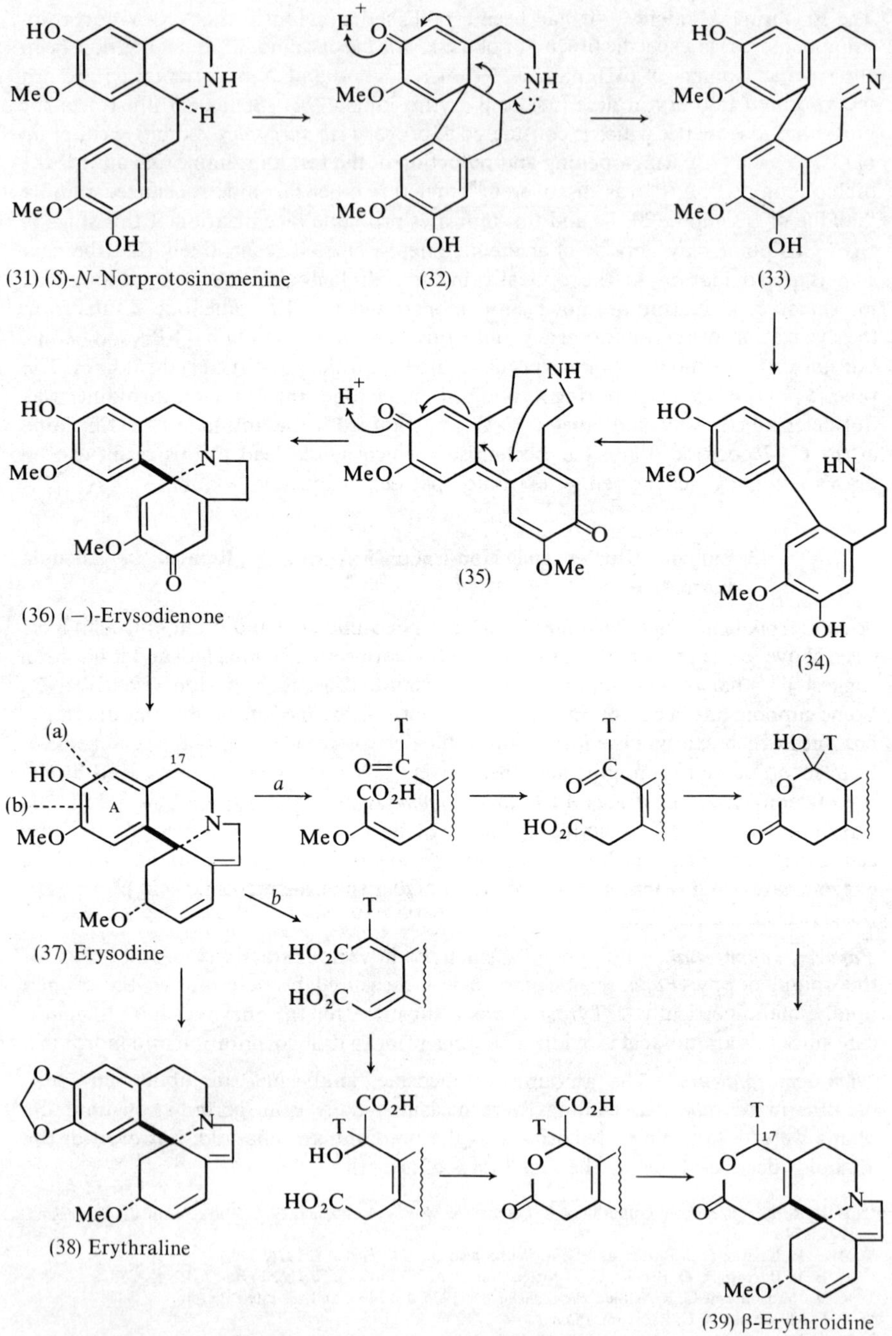

Scheme 7 *Biosynthesis of the Erythrina alkaloids*

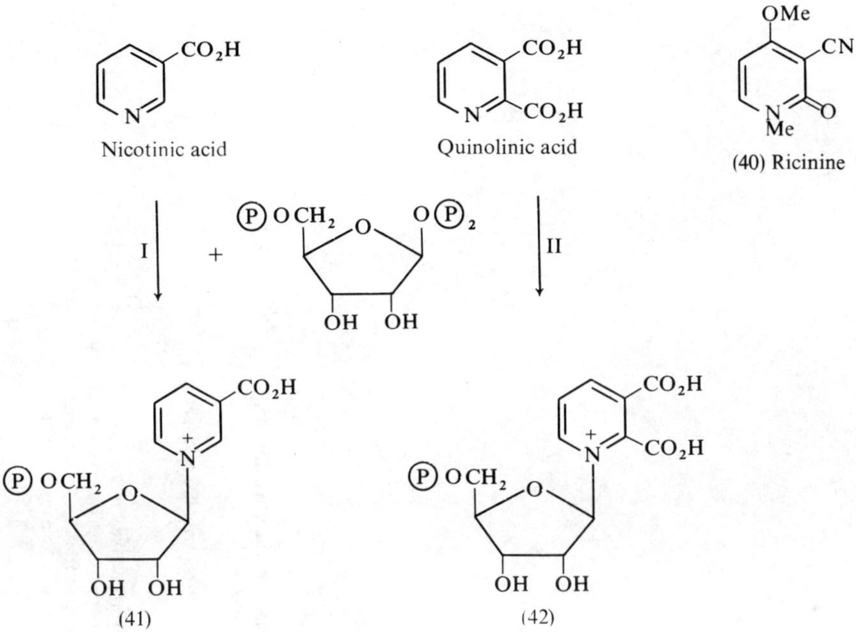

Scheme 8 *Enzymatic reactions studied in* Ricinus communis *and* Nicotiana rustica

Peganum harmala. Increased production of alkaloids (of the β-carboline type: harmine, harmaline) was observed in callus tissues of this plant when 2,4-dichlorophenoxyacetic acid or phosphates were omitted from the growth medium.[43]

4 Table of Tracer Work relating to Alkaloid Biosynthesis (pp. 110—176)

Species are listed in alphabetical order. The positions of isotopes in the precursors fed and the labelled alkaloids are indicated with a heavy dot (if the location is known). Incorporations are reported either as *Absolute incorporations* (the total radioactivity or excess isotope found in the isolated alkaloid, divided by the total radioactivity or excess isotope administered to the plant or micro-organism) or *Specific incorporation* (the specific activity of the isolated alkaloid divided by the specific activity of the administered compound). Specific incorporation, reported in square brackets [ ], is the inverse of *Dilution*, used by some workers to report their biosynthetic experiments. When significant incorporations were obtained, but insufficient data were published to allow calculations of absolute or specific incorporations, this is indicated by the symbol +.

[43] L. Nettleship and M. Slaytor, *J. Exp. Bot.*, 1974, **25**, 1114.

Species / *Compound Administered*	*Alkaloids Labelled*	*Incorporation (%)*	*Ref.*
Adenocarpus foliosus			
DL-[2-^{14}C]Lysine 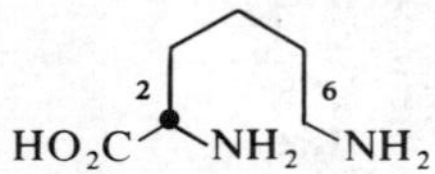 (fed by the wick method for 3 weeks)	Santiaguine 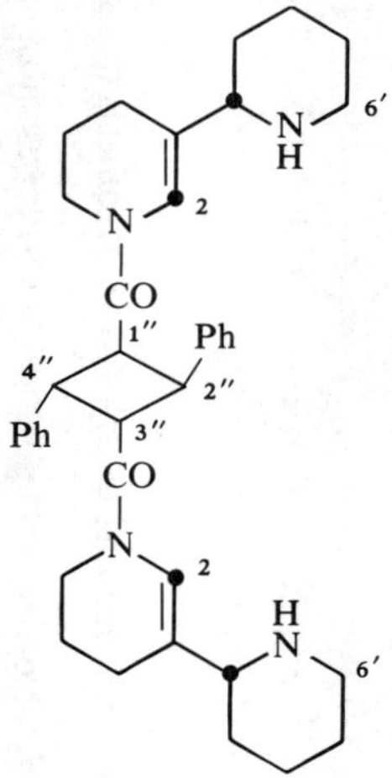 [No activity at the two C-6′ positions or in the α-truxillic acid moiety, activity at the two C-2 positions: 44%]	+	44
DL-[6-^{14}C]Lysine	Santiaguine [No activity at the C-2 positions or in the α-truxillic acid, activity at the C-6′ positions: 46%.]	+	
[2-^{14}C]-Δ^1-Piperideine (A) fed for 24 h, then [3-^{14}C]cinnamic acid (B) was fed (act. A/act. B = 0.2) 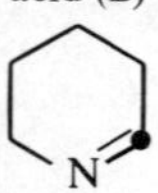 (A) $Ph\dot{C}H{=}CHCO_2H$ (B)	Santiaguine [Activity in tetrahydroanabasine part (I)/activity in α-truxillic acid (II) = 0.7. No activity at the C-6′ positions. 48% of the activity in the tetrahydroanabasine moieties present at the C-2 positions. Cold α-truxillic acid added as a carrier was recovered active (0.02% inc.), labelled at C-2 and C-4]	I: 1.9 II: 0.5[illegible]	

Precursor	Product	Incorporation	Ref.
[2,4-$^{14}C_2$]α-Truxillic acid 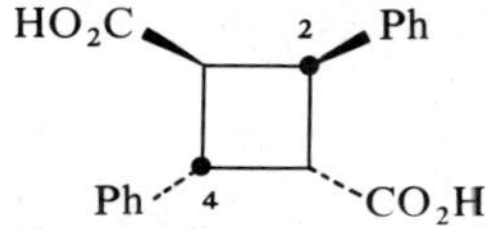	Santiaguine [All activity at C-2″ and C-4″]	0.96	
[9-^{14}C]Adenocarpine 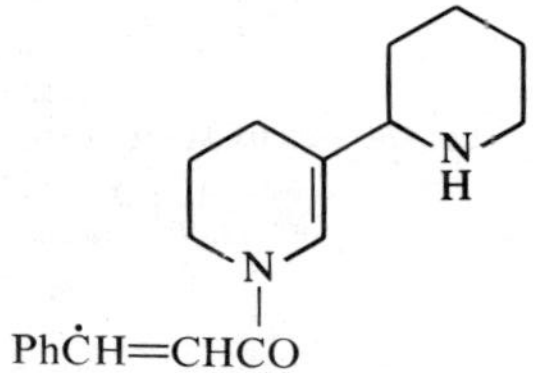	Santiaguine [3 day feeding (*a*) in the dark: (*b*) in the light: These results were interpreted as showing that the cyclobutane is not formed photochemically]	 0.17 0.30	
Anabasis aphylla [2-^{14}C]Lysine	Anabasine 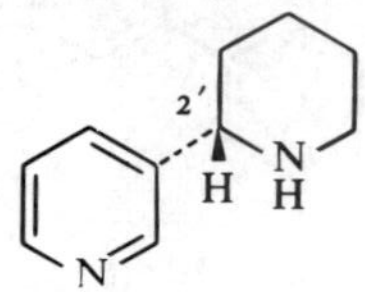 [C-2′: 89 %, pyridine ring: 11 %]	0.14	45
	Lupinine 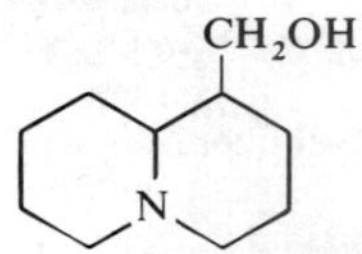 [No degradations]	0.014	

Species *Compound Administered*	*Alkaloids Labelled*	*Incorporation* (%)	*Ref.*
[U-^{14}C]Lysine	Anabasine [C-2′: 15%, piperidine ring: 88%]	0.10	
	Lupinine [No degradations]	0.010	
[U-^{14}C]Aspartic Acid	Anabasine [C-2′: 36%, pyridine ring: 55%]	0.04	46
[1,5-^{14}C]Cadaverine 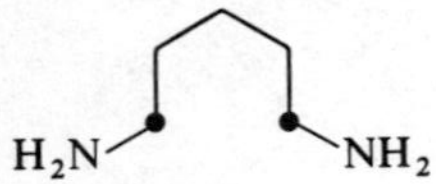 	Anabasine [most of the activity in the piperidine ring]	+	47
	Anabasamine 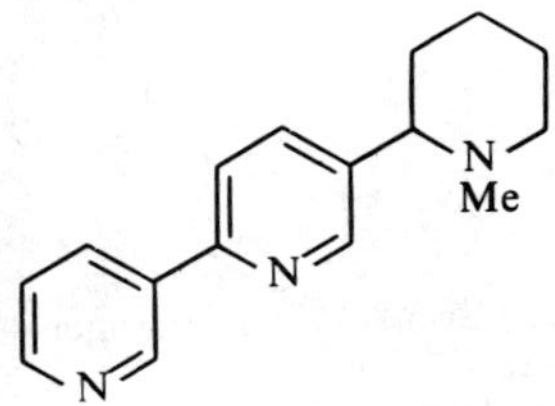 	+	
	Lupinine [The incorporation of activity into anabasine was 6—30 times greater than into the other two alkaloids]	+	
Aspergillus amstelodāmi [2-^{13}C]Glycine (Glycine is an established precursor oı echinulin[49])	Echinulin 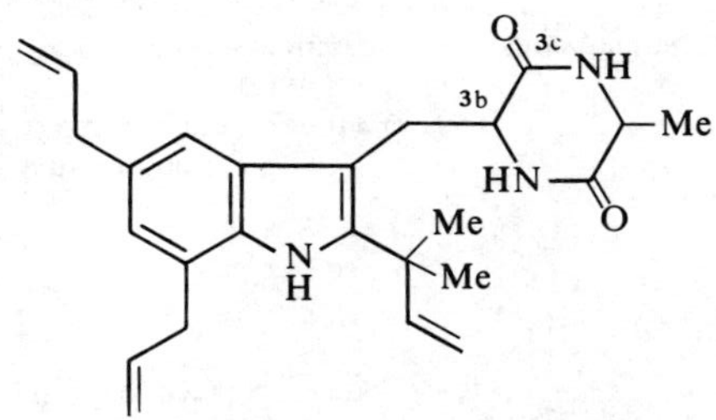 [Enhanced ^{13}C-n.m.r. signal at C-3b]	+	48

Precursor	Alkaloid	Incorporation	Ref.
[1-^{13}C]Glycine	Echinulin [Enhanced ^{13}C-n.m.r. signal at one of the C=O groups, presumably C-3c]	+	
Aspidosperma pyricollum			
Four of the six feeding experiments described have been previously reported,[50] (recorded in Vol. 1, p. 179) [*carbonyl*-^{14}C, 19-^{3}H]Secodine (^{3}H/^{14}C = 3.98)	Apparicine	0.024	51

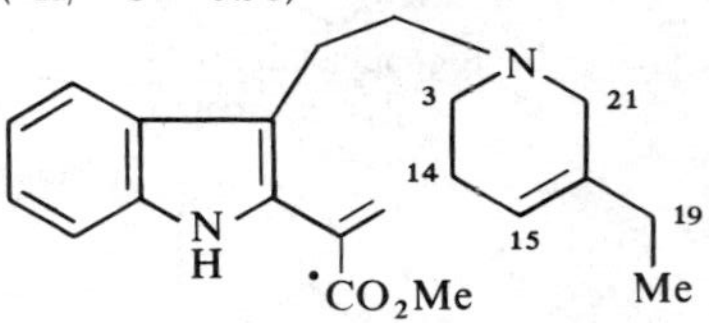

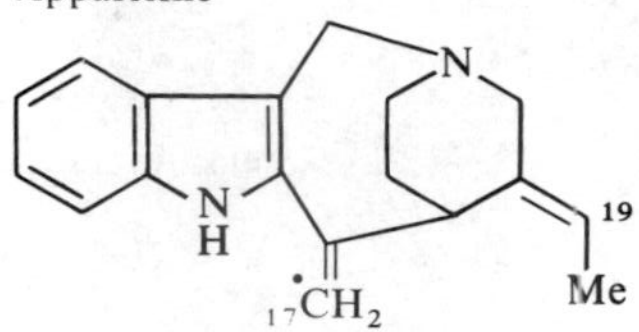

Precursor	Alkaloid	Incorporation	Ref.
	[^{3}H/^{14}C = 2.05, all ^{14}C at C-17, all ^{3}H at C-19]		
[*carbonyl*-^{14}C]Secodine	Apparicine [C-17: 99%]	0.01	
Beauveria bassiana			
L-[*methyl*-^{13}C]Methionine Sodium [1-^{13}C]acetate Sodium [2-^{13}C]acetate Sodium [1,2-$^{13}C_2$]acetate DL-[1-^{13}C]Phenylalanine DL-[2-^{13}C]Phenylalanine	Tenellin. See p. 226 [Preliminary work with ^{14}C labelled precursors indicated that tyrosine was a poorer precursor of tenellin than phenylalanine]	+	52
Brevibacterium iodinum			
D-[6-^{14}C]Shikimic acid	Iodinin	54	53

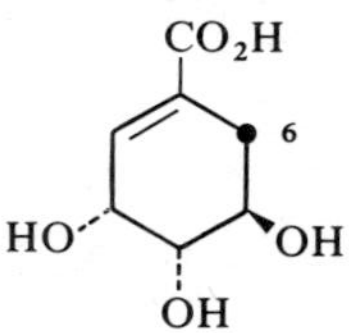

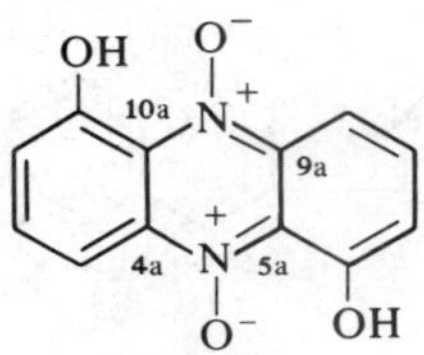

[Degradations indicated that 75% of the activity was located on the central pyrazine carbons (4a, 5a, 9a, 10a), a result different from that obtained by other workers[54] who found 100% of the activity on these carbons]

Species / *Compound Administered*	*Alkaloids Labelled*	*Incorporation* (%)	*Ref.*
Camptotheca acuminata			
[2-^{14}C]Tryptophan	Camptothecin 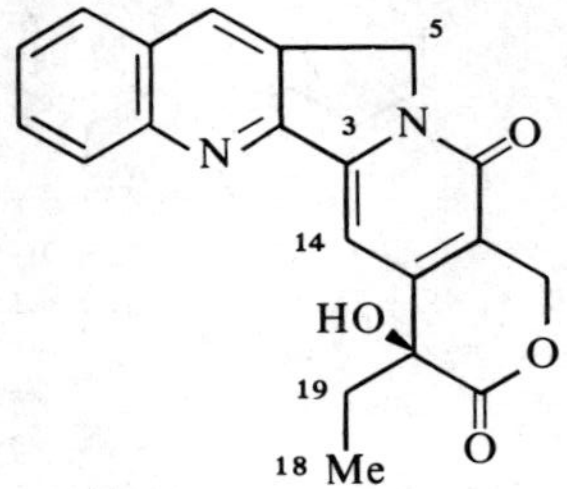 [No degradations]	0.004 [0.009]	2
(3*RS*)-[2-^{14}C]Mevalonic acid	Camptothecin	0.005	
[6-$^{3}H_2$]Secologanin 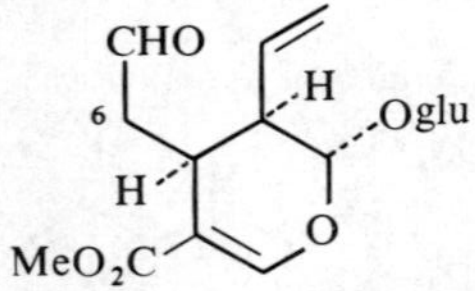	Camptothecin	0.004 [0.02]	
[5-$^{3}H_2$]Vincoside (3*R*) + [5-$^{3}H_2$]Isovincoside (3*S*) 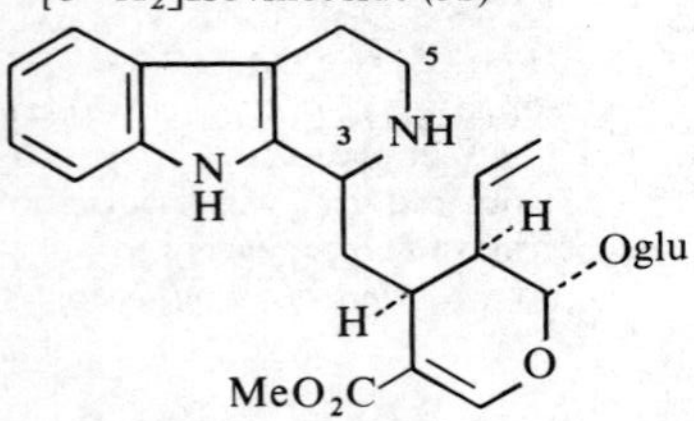	Camptothecin	0.04	

[5-$^3H_2$]Isovincoside lactam 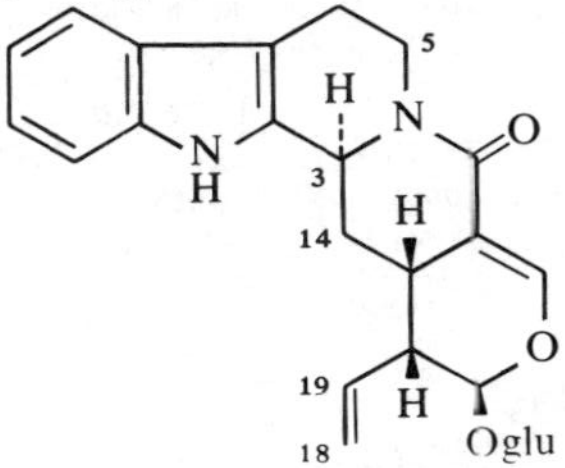	Camptothecin	1.4 [1.8]
[5-^{14}C, 14-$^3H_2$]Isovincoside lactam	Camptothecin [95% retention of ^3H]	2.0 [1.3]
[5-^{13}C]Isovincoside lactam	Camptothecin (see p. 231). [The ^{13}C n.m.r. spectrum indicates enhancement at C-5]	[0.73]
18,19-Dihydro-[5-^{14}C, 14-$^3H_2$]-isovincoside lactam	Camptothecin [93% retention of ^3H]	4.7 [0.8]
[5-$^3H_2$]Vincoside lactam 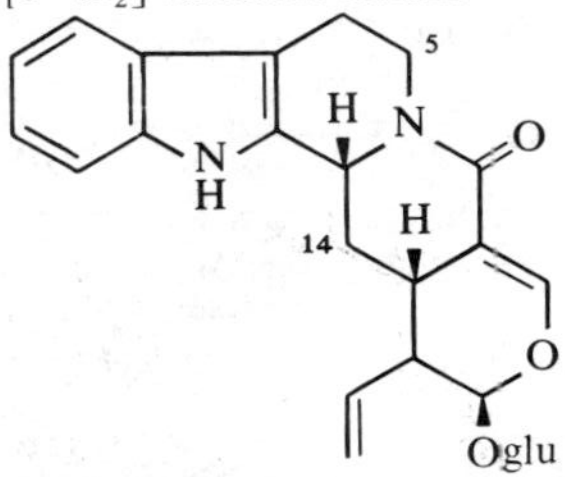	Camptothecin	0.004
[14-$^3H_2$]Vincoside lactam	Camptothecin	0.001
[*aryl*-$^3H_4$]Geissoschizine 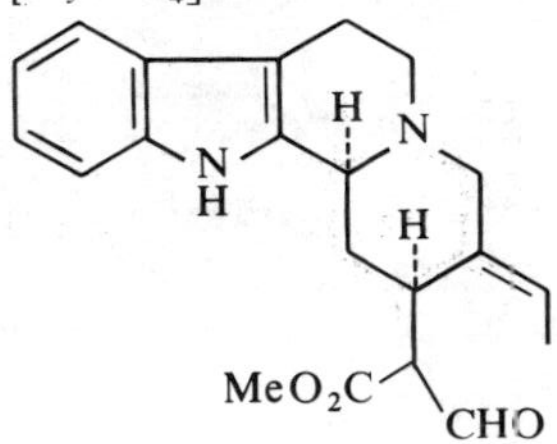	Camptothecin	0.005

Species / *Compound Administered*	*Alkaloids Labelled*	*Incorporation* (%)	*Ref.*
[6,8-$^3H_3$]Loganin 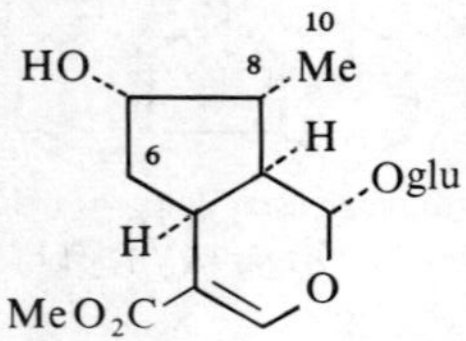	Camptothecin [C-19 : 44% of the ^3H. The loganin had 45% of its activity at C-8] Isovincoside lactam Vincoside lactam [Under the isolation conditions vincoside is converted nonenzymically to the lactam]	0.01 0.01 0.03	
Celosia plumosa (yellow variety : Golden Feather)			
[*carboxy*-^{14}C]-5,6-Dihydroxyindoline-2-carboxylic acid 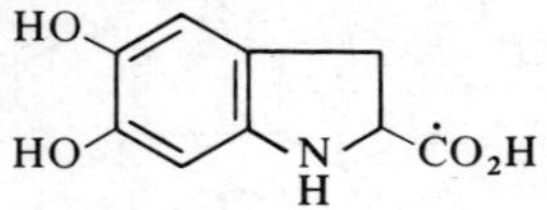	Amaranthine 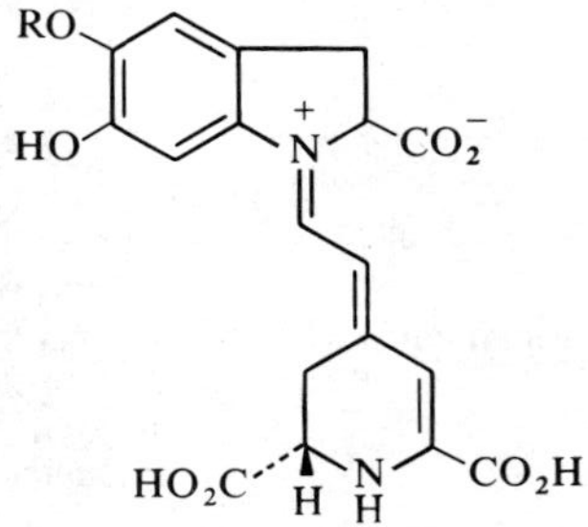 R = 2′-0-(β-D-glucosyluronic acid)-β-D-glucosyl- [This red compound is not a normal component of this variety of *Celosia*. Its production with 100% specific incorporation is thus an example of an aberrant biosynthesis]	[100]	55
Cephalosporium acremonium			
(2*S*,3*S*)-[4-$^2H_3$]Valine (2*S*,3*R*)-[4-$^2H_3$]Valine (2*RS*)-[4-$^2H_6$]Valine Deuterium oxide (80%)	Cephalosporin C, Penicillin N; see p. 248.	+	56 (cf. 57)

Choisya ternata			
[3-^{14}C]Dictamnine 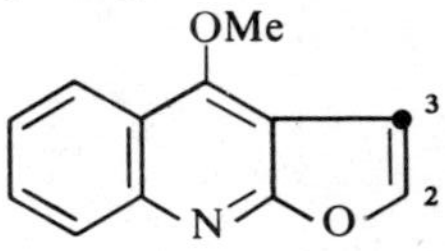	Skimmianine 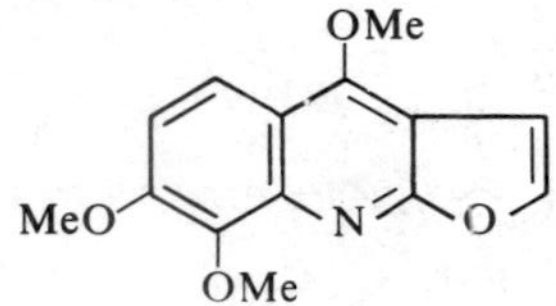	1.3—3.6	35*a*, *b*
	Evoxine 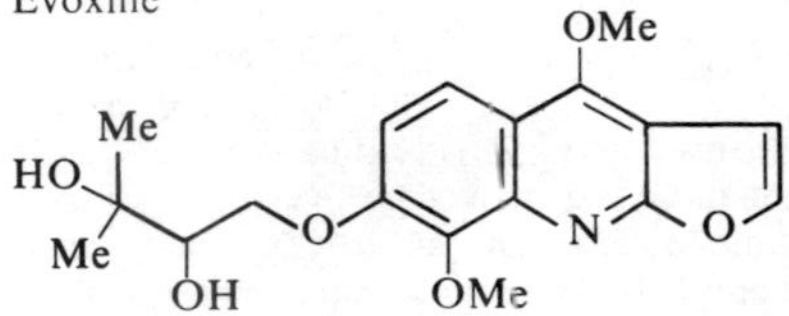	0.1	
	Choisyine 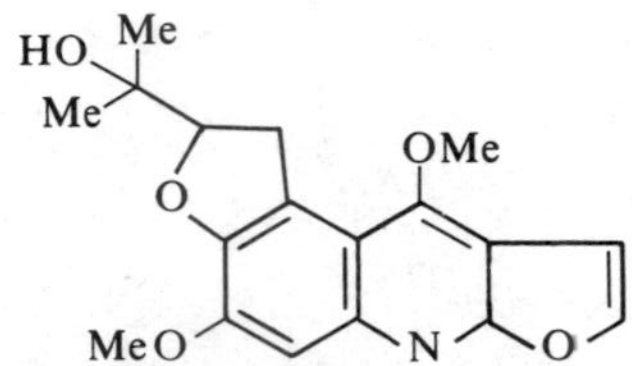	0.1	
[3-^{14}C, 2,3-$^{3}H_2$]Dictamnine (^{3}H/^{14}C = 2.0)	Skimmianine [^{3}H/^{14}C = 2.28]	2.1 (^{14}C)	
	Evoxine [^{3}H/^{14}C = 2.19]	0.06 (^{14}C)	
	Choisyine [^{3}H/^{14}C = 2.3]	0.08 (^{14}C)	
4-Methoxy-3-(3-methyl-[1′-^{14}C, 1′-$^{3}H_2$]but-2-enyl)-2-quinoline 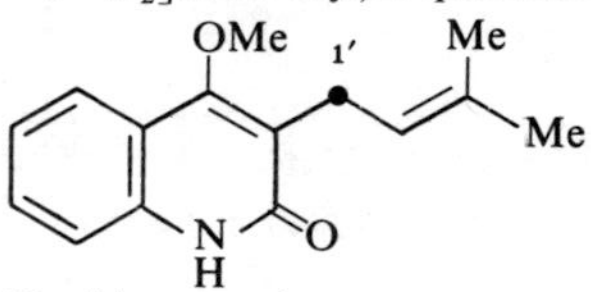 [^{3}H/^{14}C = 2.0]	Skimmianine [^{3}H/^{14}C = 1.07, 1.23. All ^{14}C at C-3]	6.0, 6.4 (^{14}C)	35*a*, c
	Evoxine [^{3}H/^{14}C = 1.16, 1.10]	0.20, 0.34 (^{14}C)	
	Choisyine [^{3}H/^{14}C = 0.97, 0.90]	0.23, 0.27 (^{14}C)	

Species / *Compound Administered*	*Alkaloids Labelled*	*Incorporation (%)*	*Ref.*
4-[*methoxy*-^{3}H]Methoxy-3-(3-methyl-[1′-^{14}C]but-2-enyl)-2-quinolone	Skimmianine [^{3}H/^{14}C = 3.25]	1.1 (^{14}C)	
[^{3}H/^{14}C = 3.0]	Evoxine [^{3}H/^{14}C = 3.31, 2.97]	0.01, 0.005 (^{14}C)	
	Choisyine	inactive	
Claviceps purpurea			
L-[1-^{14}C]Valyl-L-valyl-L-proline	Ergocornine	[0.021]	58

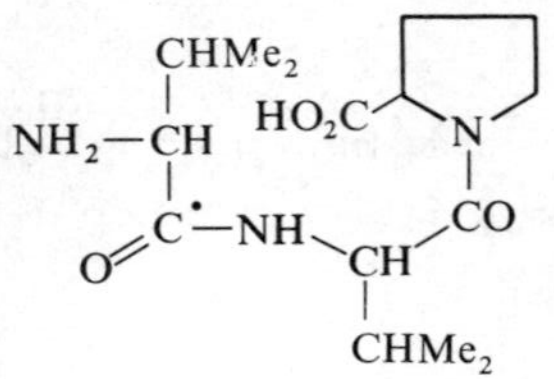

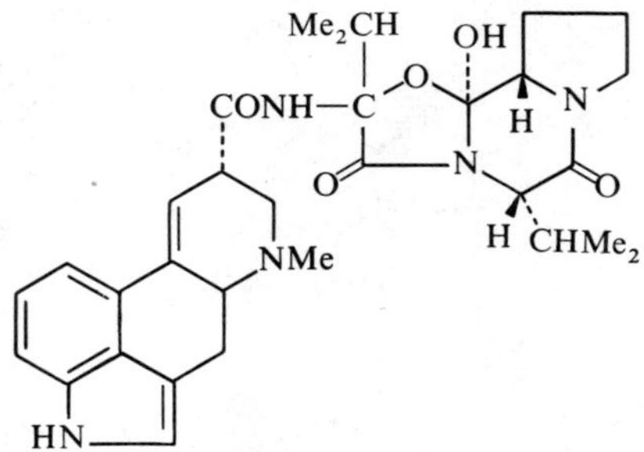

(Degradations indicate that the tripeptide is not incorporated intact. The α-hydroxyvaline moiety contained 46% of the activity and the valine 38%. The distribution of activity in these amino-acids when DL-[1-^{14}C]valine was fed was 42 and 36% respectively.)

	Ergokryptine	[0.015]	

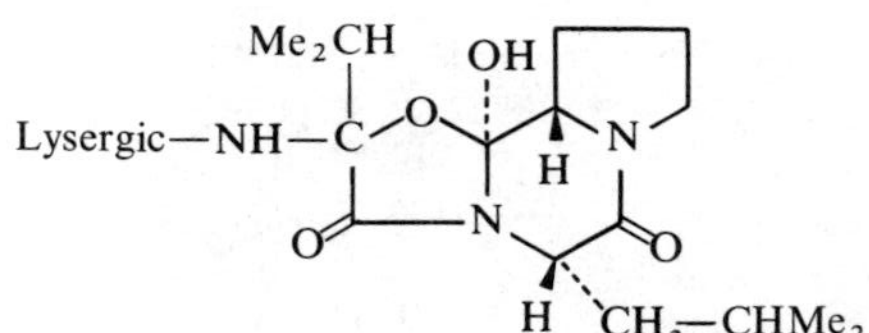

[97% of the activity in the α-hydroxyvaline moiety]

D-Lysergyl-L-[U-^{14}C]alanine methyl ester D-Lysergyl-L-[U-^{14}C]valine methyl ester	Ergotoxine alkaloids (ergocornine, ergokryptine and ergocristine) [These two compounds were cleaved by the culture to lysergic acid and the amino-acids. Neither compound was detected in trapping experiments following the feeding of labelled alanine or valine]		59
Claviceps Strain SD 58			
3R,4R-[2-^{14}C, 4-^{3}H, 3′,4-$^{2}H_5$] Mevalonate (A mixture of this compound with its 3S,4S isomer, which is not biologically active, was fed to replacement cultures, in some experiments together with unlabelled L-tryptophan and mevalonic acid.)	Chanoclavine-I 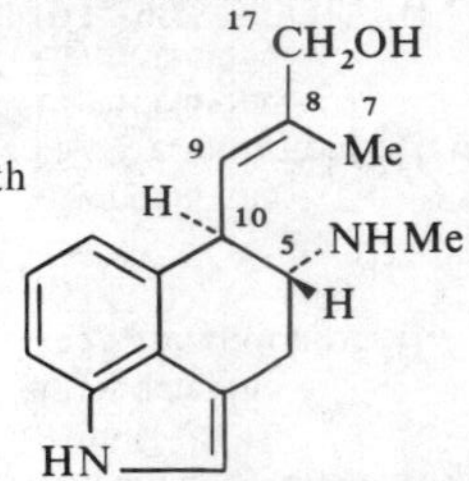 [^{3}H retention relative to ^{14}C was 127—141%; increasing the amounts of unlabelled mevalonate in the culture increased the retention of ^{3}H. Most of deuteriated alkaloid is in the form of trideuteriated molecules, presumably at C-9 and C-17]	+	27
	Elymoclavine 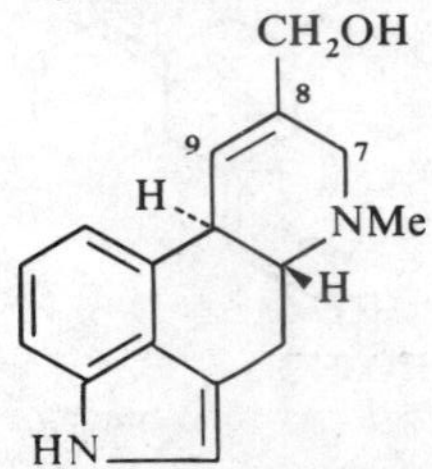 [39—82% retention of ^{2}H relative to ^{14}C]	+	

Species Compound Administered	Alkaloids Labelled	Incorporation (%)	Ref.
	Penniclavine 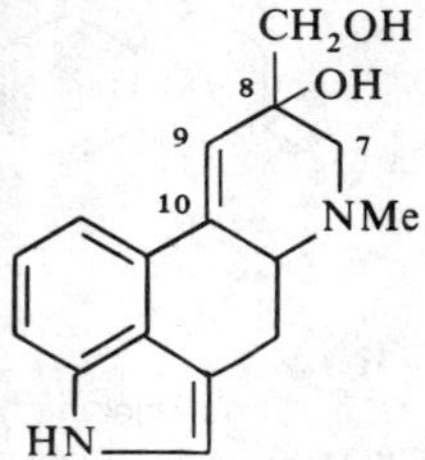 [52—74% retention of ³H relative to ^{14}C. In elymoclavine and penniclavine essentially no trideuteriated molecules were found The distribution of deuterium is consistent with the loss of one D from C-17 of chanoclavine, and an intermolecular loss of some from C-9]	+	
[^{14}C,17-^{3}H]Chanoclavine-I (The ^{14}C was introduced into this compound by feeding [^{14}C]tryptophan to *Claviceps paspali*)	Elymoclavine [53% retention of ^{3}H, 93% of it located at C-7]	17.6 (^{14}C)	
3*R*,4*R*-[2-^{14}C,4-^{3}H]Mevalonolactone	Elymoclavine [57% retention of ^{3}H, negligible ^{3}H located at C-7]	14 (^{14}C)	
[17-^{3}H]Chanoclavine-I	Elymoclavine	5.7, 9.9	
[17-^{3}H]Chanoclavine-I aldehyde 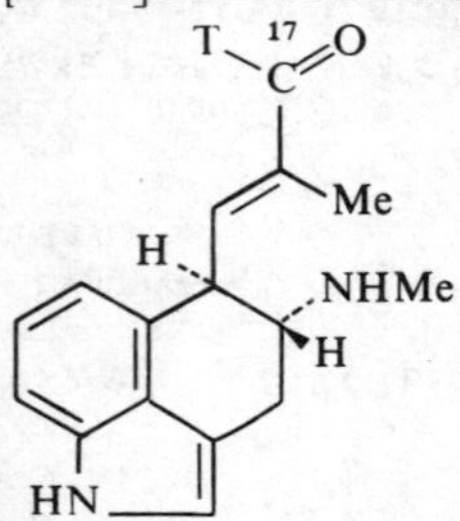	Elymoclavine [94—98.5% of the ^{3}H at C-7]	23.1, 40.1	

[^{14}C,17-^{3}H]Chanoclavine-I aldehyde	Elymoclavine [91% retention of ^{3}H]	34.4 (^{14}C)
[2-^{13}C]Mevalonic acid + [4-$^{2}H_2$]Mevalonic acid (equal amounts of each fed together)	Chanoclavine-I [No molecules containing both ^{13}C and ^{2}H]	+
	Elymoclavine [6.8—8.2% of the molecules contained both ^{13}C and ^{2}H]	+
	Penniclavine [7.1—8.8% of the molecules contain both ^{13}C and ^{2}H]	+
3*R*,4*R*-[2-^{14}C, 4-^{3}H]Mevalonic acid	Agroclavine 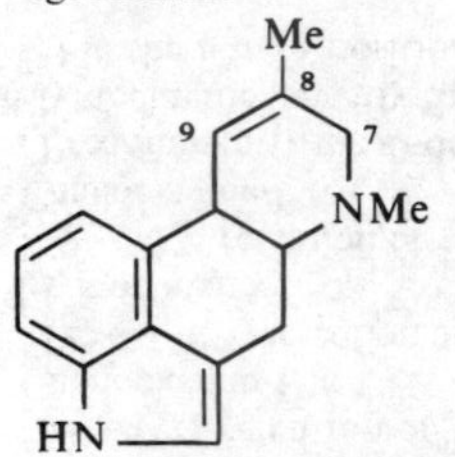 [63% retention of ^{3}H]	+
	Elymoclavine [52% retention of ^{3}H]	+
	Chanoclavine-I [146% retention of ^{3}H]	+
	(−)-Chanoclavine-II 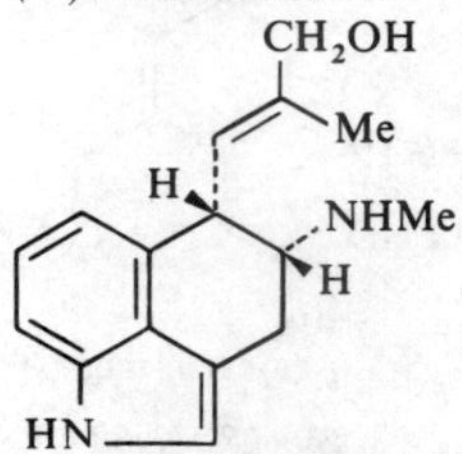 [105% retention of ^{3}H]	+

Species / Compound Administered	Alkaloids Labelled	Incorporation (%)	Ref.
	Isochanoclavine-I 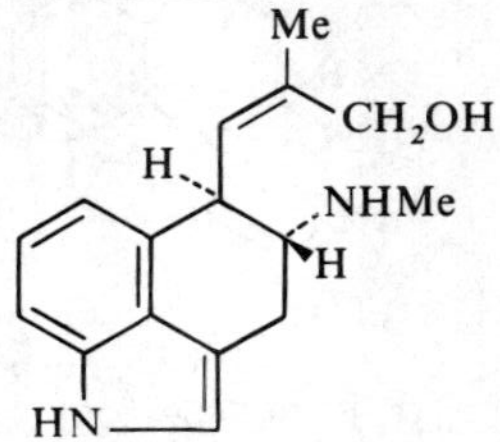 [99% retention of ^{3}H]	+	
3*R*,4*R*-[2-^{14}C, 4-^{3}H]Mevalonate + [4-^{2}H]Mevalonate	Chanoclavine-I [114—117% retention of ^{3}H relative to ^{14}C]	+	
	Elymoclavine [63—67% retention of ^{3}H]	+	
3*R*,4*R*-[2-^{14}C, 4-^{3}H]Mevalonate + [4-^{2}H]Mevalonate + unlabelled mevalonate	Chanoclavine-I [125—130% retention of ^{3}H]	+	
	Elymoclavine [59—66% retention of ^{3}H]	+	
[3′-^{2}H$_3$]Mevalonate	Chanoclavine-I [Predominantly the di-deuteriated species found]	+	
	Elymoclavine [mostly the mono-deuteriated species, with the ^{2}H in the *pro*-7*S* position]	+	
Coffea arabica			
L-[*methyl*-^{14}C]Methionine [administered to leaf discs along with plausible precursors of caffeine]			60
No additions	Caffeine 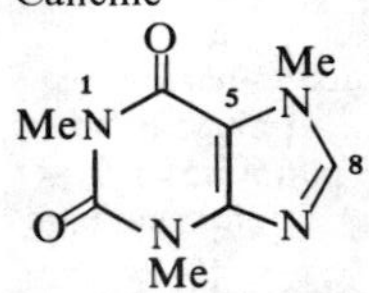	1.0	

Precursor	Alkaloid	Incorporation	Ref.
+ Xanthine 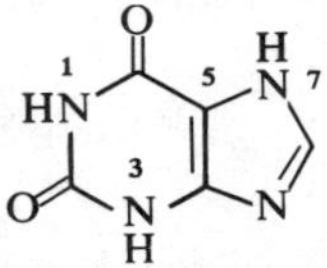	Caffeine	0.85	
+ 1-Methylxanthine	Caffeine	1.4	
+ 7-Methylxanthine	Caffeine	7.3	
+ 1,3-Dimethylxanthine (Theophylline)	Caffeine	1.4	
+ 3,7-Dimethylxanthine (Theobromine)	Caffeine	18.1	
+ Xanthine-9-riboside (Xanthosine)	Caffeine	1.2	
+ 7-Methylxanthosine	Caffeine	4.3—12.3	
+ Guanosine 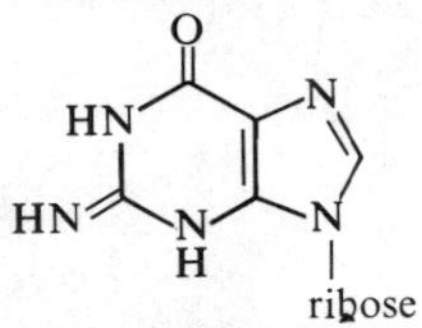	Caffeine	1.4	
+ 7-Methylguanosine	Caffeine	0.63	
+ 7-Methylinosine	Caffeine	0.29	
	[These results are consistent with the following biosynthetic sequence: 7-Methylxanthosine→7-methylxanthine→theobromine→caffeine]		
Colchicum autumnale			
RS-[1-^{13}C]Autumnaline, HCl	Colchicine (see p. 229)	[1.9]	3
Conium maculatum			
L-[*methyl*-^{14}C]Methionine	*N*-Methylconiine 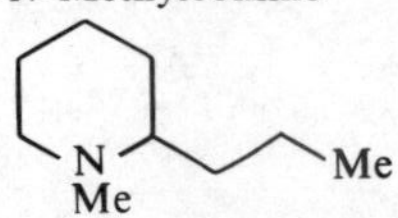	0.05—1.8	61
	[The incorporation depended on the method of feeding and the variety (Messina, Bowles, Chelsea, Minnesota) of the hemlock used. The *N*-methyl group contained essentially all the activity]		

Species / *Compound Administered*	*Alkaloids Labelled*	*Incorporation* (%)	*Ref.*
Corydalis cava			
DL-[2-^{14}C]Dopa 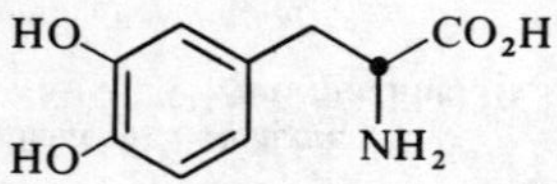	Reticuline (added as a carrier) 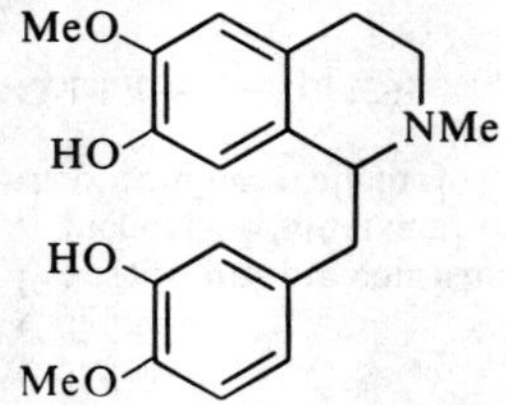	0.014	62
	Protosinomenine (added as a carrier) 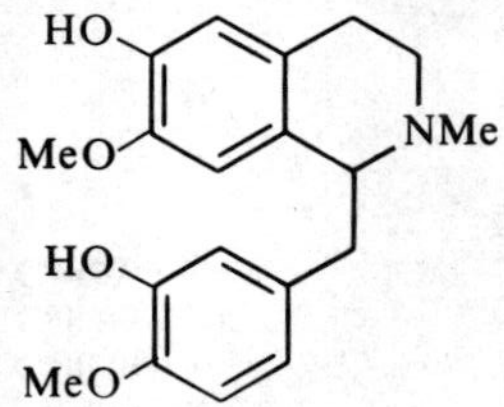	0.0025	
	Bulbocapnine 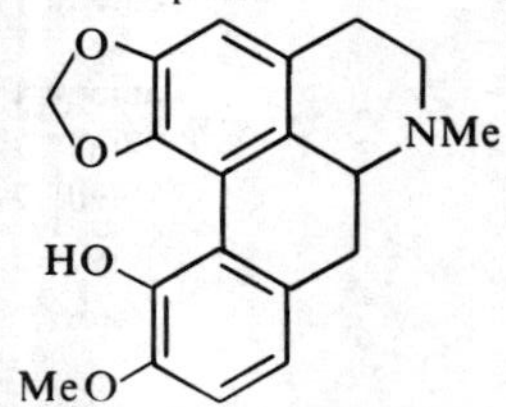	0.65	

	Corydaline 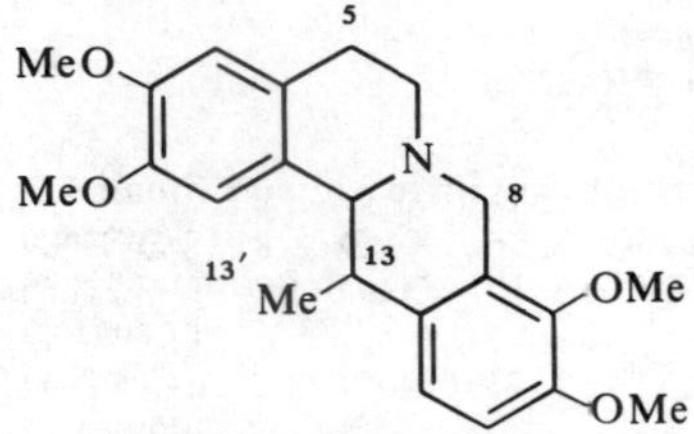 [No degradations. The authors claim that their results favour the formation of sinoacutine from reticuline, however this has been previously established by Barton and co-workers[63]]	0.023	
Corydalis incisa [*N-methyl*-^{14}C]Tetrahydrocoptisine 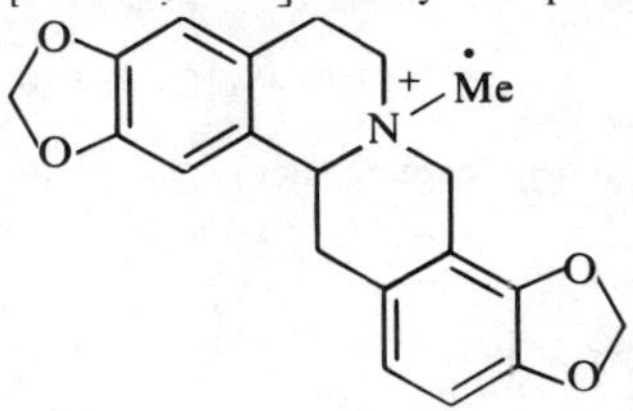	Protopine 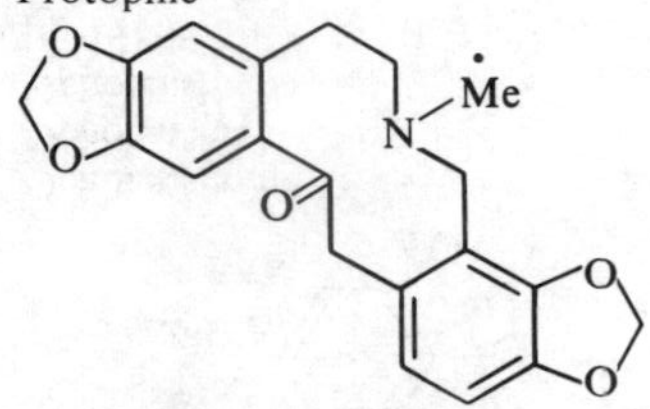 [*N*-Me: 100%]	0.19	64
Corydalis ochotensis L-[*methyl*-^{14}C]Methionine	Ochotensimine 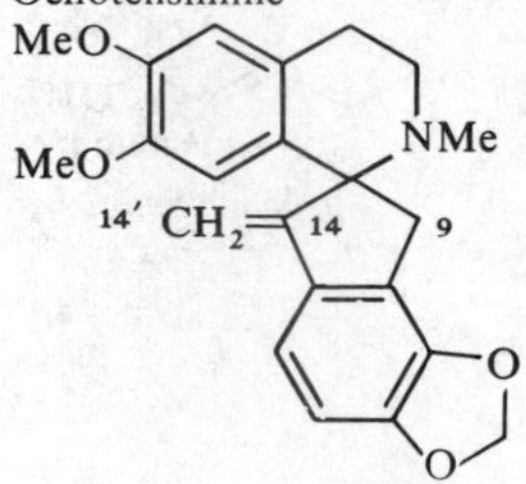 [C-14′: 16%, O—CH_2—O: 16%]	0.89	65

Species / *Compound Administered*	*Alkaloids Labelled*	*Incorporation* (%)	*Ref.*
	Ochotensine 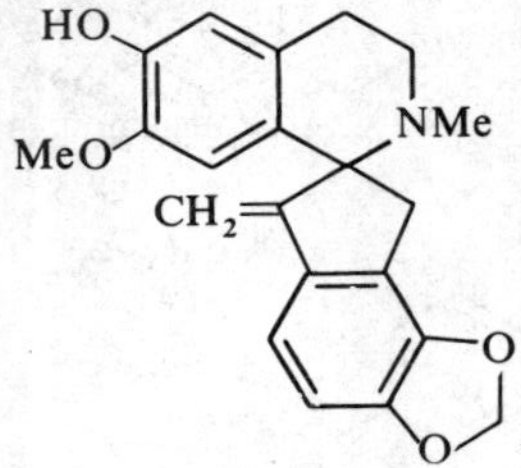	0.0012	
DL-[3-^{14}C]Phenylalanine	Ochotensimine	0.0004	
	Ochotensine	0.00006	
DL-[3-^{14}C]Tyrosine	Ochotensimine [C-14: 49%]	0.078	
	Ochotensine	0.034	
Corydalis solida			
DL-[3-^{14}C]Tyrosine	Corydaline (see *Corydalis cava*) [C-13: 50%, the rest assumed to be at C-5]	0.63	65
L-[*methyl*-^{14}C]Methionine	Corydaline [C-13′: 19%, the four OMe: 64%, C-8: 18%]	0.31	
L-[*methyl*-^{14}C, ^{3}H]Methionine	Corydaline [92% retention of ^{3}H; 94% (17/18) expected if there is no isotope effect. C-8: 18%, C-13′: 19.1%, four OMe: 64%. The tritium is apparently retained at all the carbons labelled with ^{14}C]	0.61 (^{14}C)	

Crotalaria retusa			
[1-^{14}C]Acetate	Monocrotaline	negligible	66
[Fed by various methods: wick, excised stems, injection into stem cavity, hydroponics]	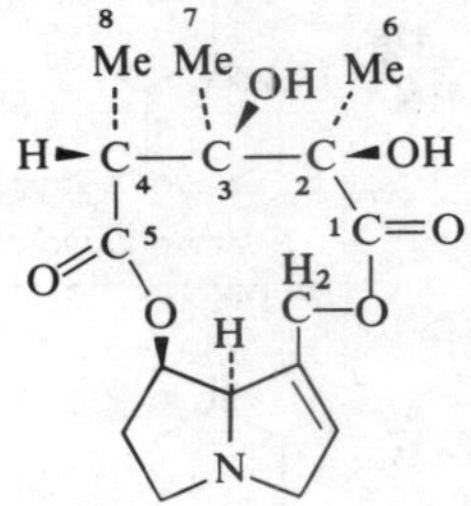		
	[The lack of incorporation of [1-^{14}C]acetate is in conflict with earlier work of Nowacki and Byerrum[67]]		
[2-^{14}C]Acetate	Monocrotaline	0.0004	
[^{3}H]Acetate	Monocrotaline	0.00008	
L-[*U*-^{14}C]Threonine	Monocrotaline [essentially all the activity in the necic acid]	0.014	
L-[*U*-^{14}C]Isoleucine	Monocrotaline [essentially all the activity in the necic acid]	0.0032	
Crotalaria spectabilis			
[2-^{14}C]Acetate	Monocrotaline	<0.00056	66
[^{3}H]Acetate	Monocrotaline	<0.00051	
L-[*U*-^{14}C]Threonine	Monocrotaline	0.0015	
L-[*U*-^{14}C]Isoleucine	Monocrotaline [90% of the activity in the necic acid]	0.0018	
DL-[1-^{14}C]Alanine	Monocrotaline	<0.0001	
L-[*U*-^{14}C]Threonine	Monocrotaline [> 95% in the necic acid, 60% of the activity at C-1,2,3,6,7]	0.05—0.19	
L-[*U*-^{14}C]Isoleucine	Monocrotaline [> 95% in the necic acid, 73% of the activity at C-1,2,3,6,7]	0.11, 0.15	

Species / *Compound Administered*	*Alkaloids Labelled*	*Incorporation* (%)	*Ref.*
Croton sparsiflorus			
DL-[2-^{14}C]Tyrosine	Crotsparine 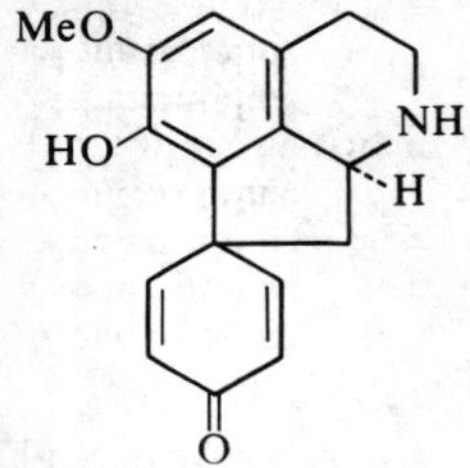 [No degradations]	0.42	68
	Crotsparinine 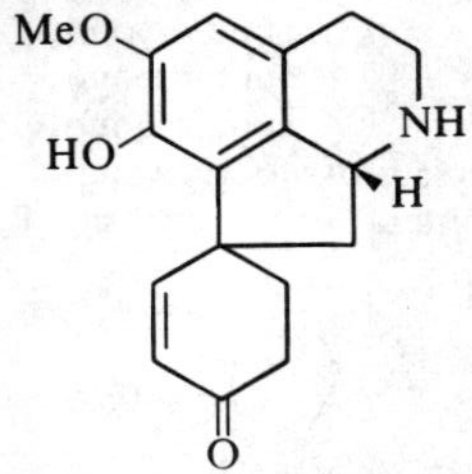 [No degradations]	0.32	
	Sparsiflorine 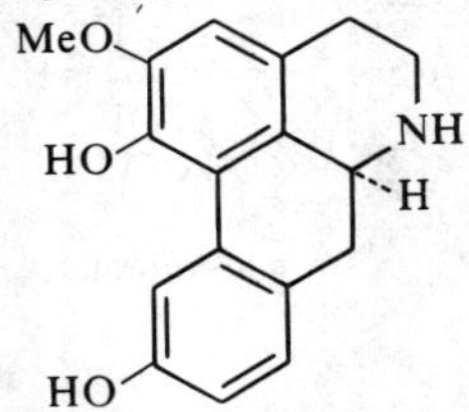 [No degradations]	0.20	

Precursor	Product	Incorporation
RS-[3′,5′,8-$^3H_3$]Coclaurine 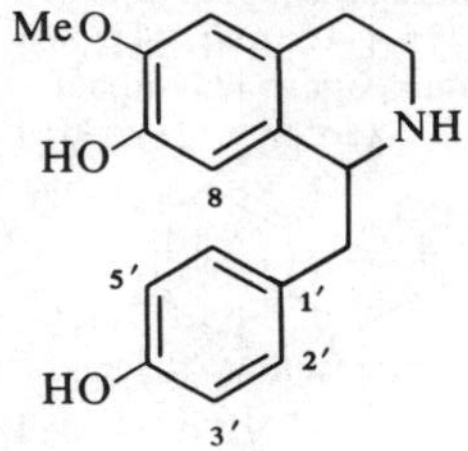	Crotsparine [All ^3H was lost on refluxing the derived *N*-methyltetrahydrocrotsparine with NaOH] Crotsparinine [All ^3H lost on NaOH treatment] Sparsiflorine [All ^3H lost when the derived *N*-methylsparsiflorine was treated with NaOH]	1.90* (*Corrected for loss of ^3H from C-8) 0.62 0.28
RS-[3′,5′, 8-$^3H_3$]Norclaurme 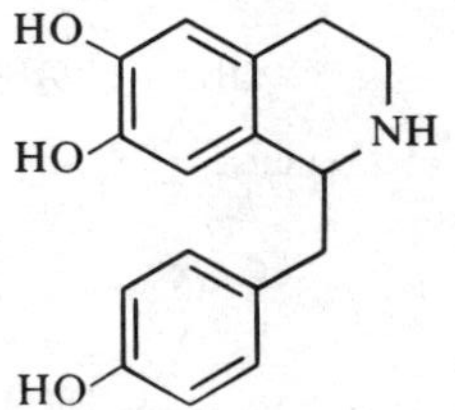	Crotsparine Crotsparinine Sparsiflorine	1.6 0.48 0.24
RS-[3′,5′,5-^3H]Isococlaurine 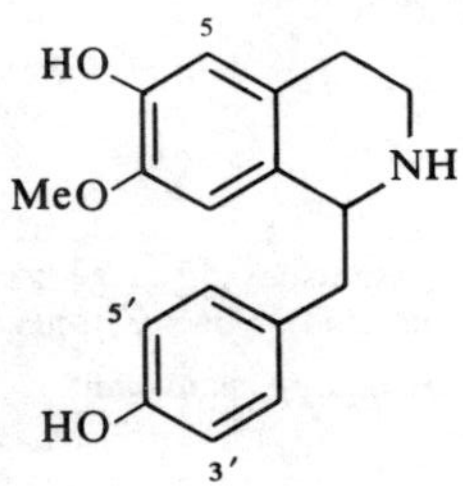	Crotsparine Crotsparinine Sparsiflorine	0.003 inactive 0.002

Species / *Compound Administered*	*Alkaloids Labelled*	*Incorporation (%)*	*Ref.*
Cryptostylis erythroglossa			
DL-[2-^{14}C]Tyrosine	(−)-Cryptostyline I 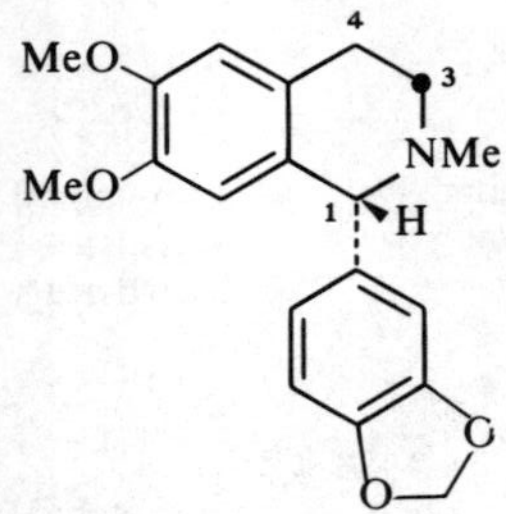 [C-8 : 88 %]	0.019	69
[α-^{14}C]Tyramine	(−)-Cryptostyline I [C-3 : 91 %]	+	
RS-[2-^{14}C]Dopa	(−)-Cryptostyline I [C-3 : 91 %]	+	
[α-^{14}C]Dopamine	(−)-Cryptostyline I [C-3 : 74 %]	+	
[αβ-$^{3}H_2$]3-Hydroxy-4-methoxy-phenethylamine, HCl 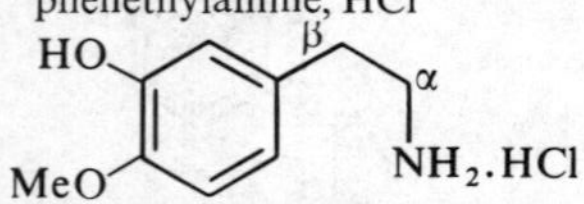	(−)-Cryptostyline I [Activity of degradation products consistent with activity being at C-3 and C-4]	+	
[αβ-$^{3}H_2$]4-Hydroxy-3-methoxyphenethylamine	(−)-Cryptostyline I	very low	
[5-^{3}H]4-Hydroxy-3-methoxyphenethylamine	(−)-Cryptostyline I	very low	
[β-^{14}C]Dopamine	(−)-Cryptostyline I [C-4 : 35 %, the rest assumed to be at C-1]	0.14	70

[α-^{14}C]3,4-Dimethoxyphenethylamine	(−)-Cryptostyline I [C-3 : 91 %. The low incorporation probably indicates that this precursor is demethylated before formation of the alkaloid]	0.0016
N-(3-Hydroxy-4-methoxy-[1′-^{3}H]benzyl)-[αβ-$^3H_2$]3-hydroxy-4-methoxyphenethylamine 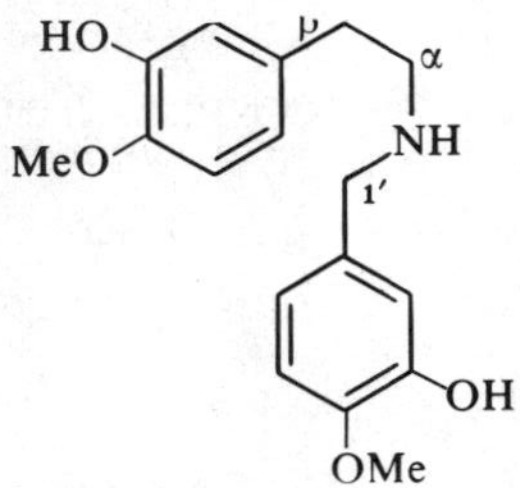 $\frac{^3\text{H at }\alpha,\beta}{^3\text{H at C-1}'} = 0.37$	(−)-Cryptostyline I $\frac{^3\text{H at C-3 + C-4}}{^3\text{H at C-1}} = 3.4$ [This result indicates that this compound is not incorporated intact]	0.0029
[*aryl*-^{3}H]1,2-Dehydrocryptostyline bromide 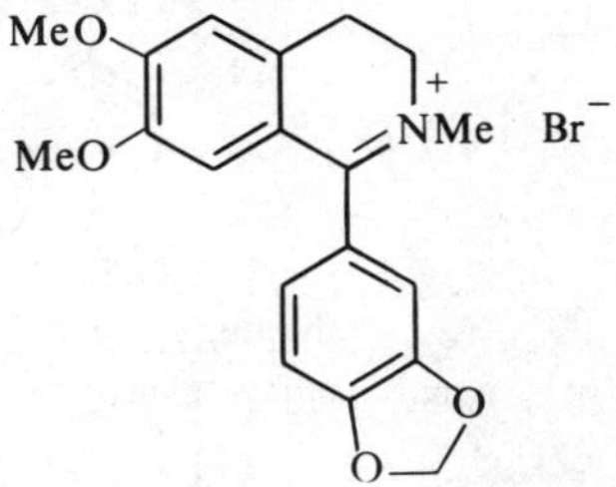	(−)-Cryptostyline I	15

Species / Compound Administered	Alkaloids Labelled	Incorporation (%)	Ref.
[*aryl*-^{3}H]Vanillin 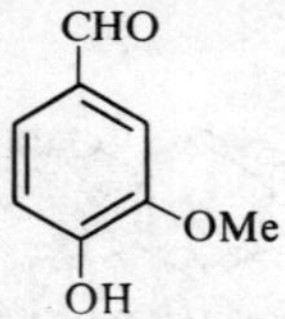 [fed by painting the leaves with a solution in polyethylene glycol]	(−)-Cryptostyline I	0.0023	
[*aryl*-^{3}H]Isovanillin 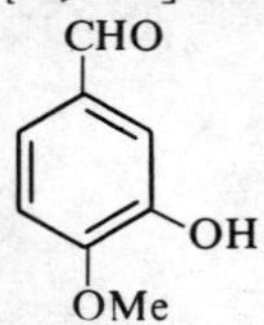	(−)-Cryptostyline I	0.00029	
Datura ceratocaula			
L-[*U*-^{14}C]Isoleucine	6β-[2-(+)-Methylbutanoyl]tropan-3α-ol 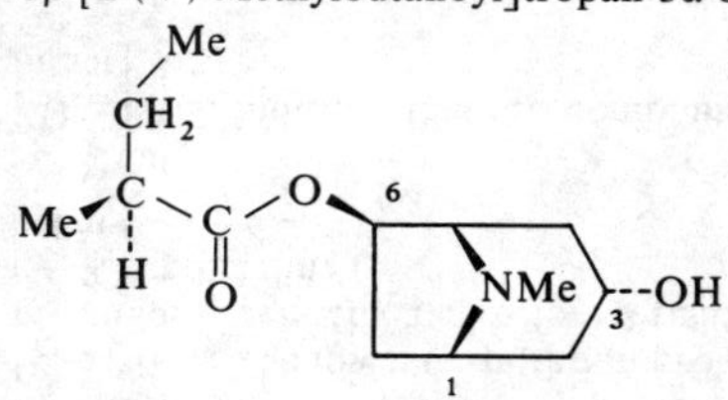 [> 98% in the acid moiety]	0.003	

Datura stramonium

[2-^{14}C]Cinnamoyl-[*N-methyl*-^{14}C]tropine — Atropine — 0.94* (*of the tropine moiety) — 72

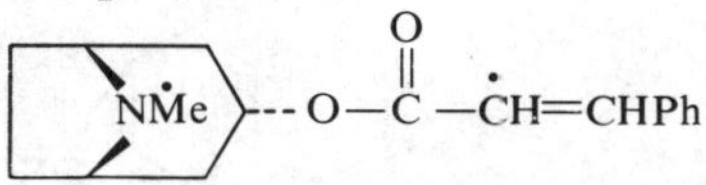

[act. at C-2/ act. at NMe = 0.75]

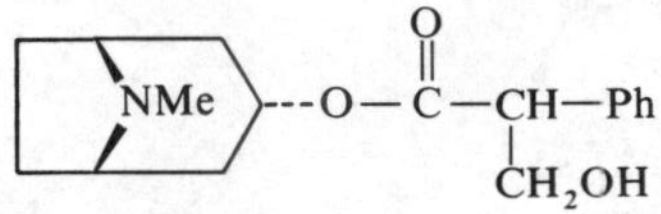

[NMe: 100%, negligible activity in the tropic acid moiety]

Scopolamine — 0.91*

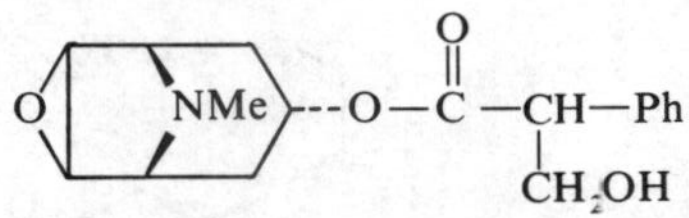

[Negligible activity in the tropic acid moiety]

RS-Littorine {3α-([1-^{14}C]-2-hydroxy-3-phenylpropionyloxy)-[3β-^{3}H]tropane} — Littorine — [100]

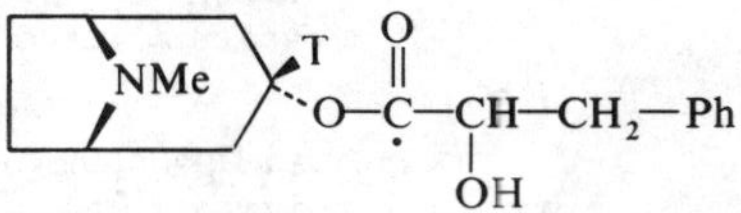

[^{3}H/^{14}C = 6.75]

Littorine: [^{3}H/^{14}C = 6.9]

Atropine — ^{14}C: [0.60] ^{3}H: [2.96]

[^{3}H/^{14}C = 33. All the ^{14}C on the carboxy-group of tropic acid. All the ^{3}H at C-3 of tropine. The drastic change in the ^{3}H/^{14}C ratio indicates that littorine is not converted directly into atropine]

Dolichothele sphaerica

trans-[1-^{14}C]Cinnamic acid — *N*-Cinnamoylhistamine — 0.22 — 73

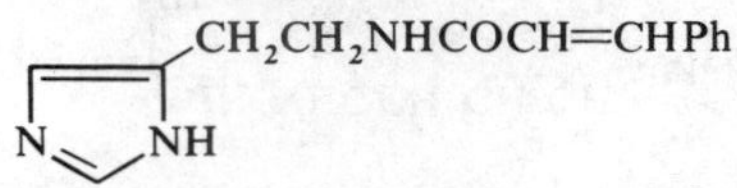

[C=O: 95%. This compound is not found in this species and its formation is an example of an aberrant biosynthesis]

Species / *Compound Administered*	*Alkaloids Labelled*	*Incorporation* (%)	*Ref.*
[2-^{14}C]-4-Aminomethylimidazole + inactive isocaproic acid	4-[*N*-Isocaproylaminomethyl]imidazole	14.1	74
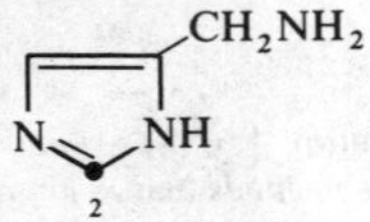	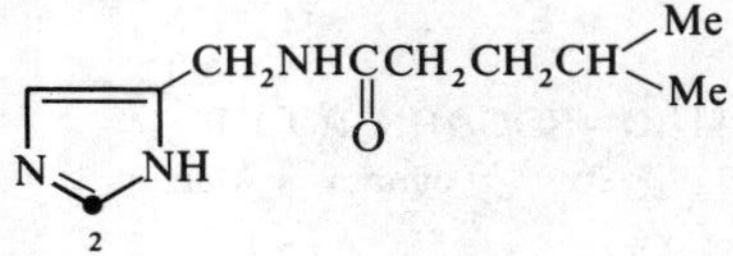 [C-2: 99%]		
[1-^{14}C]Isovaleric acid + inactive 3-aminoethylpyrazole	3-[2-*N*-Isovalerylaminoethyl]pyrazole [C=O: 96%. These two compounds are unnatural alkaloids, the normal component is dolichotheline, *N*-isovalerylhistamine]	1.81	
Erythrina berteroana			
DL-[2-^{14}C]Tyrosine	α- and β-Erythroidine	0.032	37
	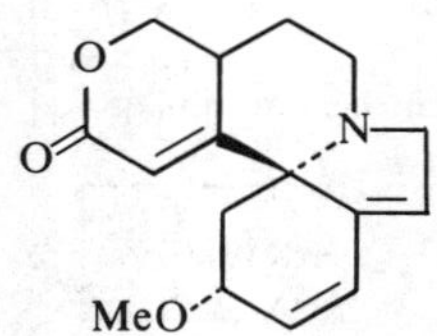 α- 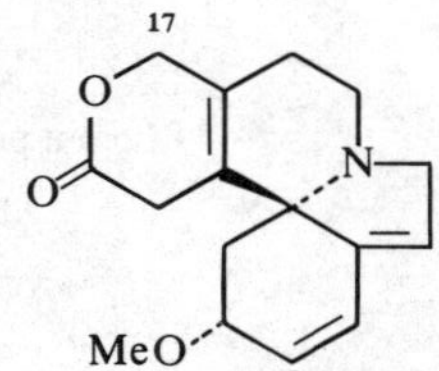 β-		
RS-[5-^{3}H]-*N*-Norprotosinomenine	α-+ β-Erythroidine	0.060	
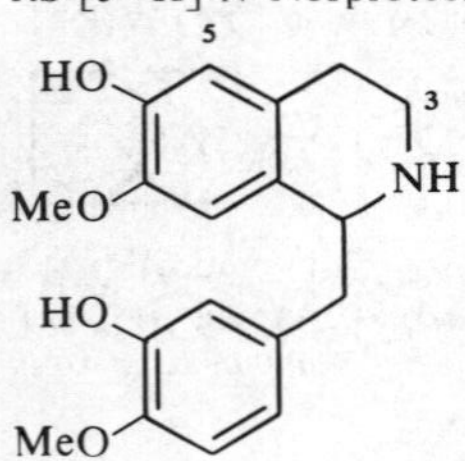			

RS-[2′,6′,8-$^3H_3$]-*N*-Nor-reticuline 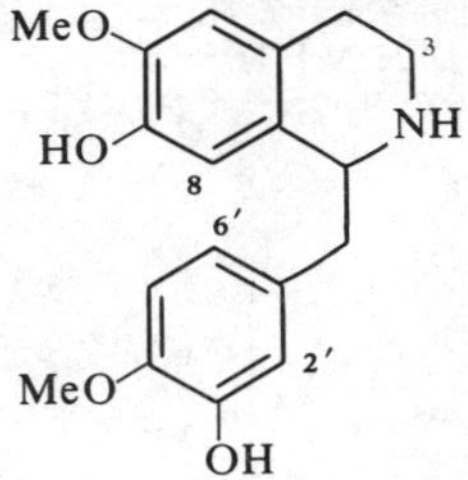	α-+ β-Erythroidine		0.008	
(+)-[5′,8-$^3H_2$]-*N*-Nororientaline* 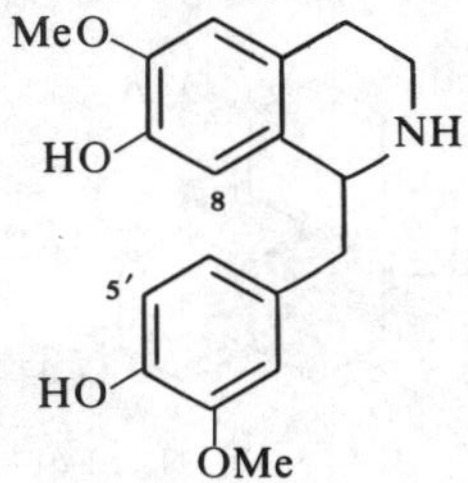 *nomenclature corrected	α-+ β-Erythroidine [These results and those reported later under *E. crista-galli* indicate that only *N*-norprotosinomenine is a specific precursor of the *Erythrina* alkaloids, as previously discovered.[36]]		<0.001	
DL-[2-^{14}C]Tyrosine	α-+ β-Erythroidine		0.32	38
(±)-[5-^{3}H]-*N*-Norprotosinomenine	α-+ β-Erythroidine		0.96	
(±)-[1,17-$^3H_2$]Erysodienone 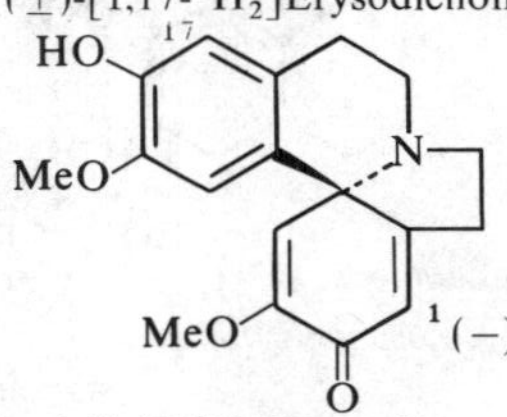 (−)-isomer	α-+ β-Erythroidine	fed 11 d	0.28	
		fed 14 d	0.20	
		fed 33 d	0.078	
(+)-[1,17-$^3H_2$]Erysodienone	α-+ β-Erythroidine		⩽0.045	

Species / *Compound Administered*	*Alkaloids Labelled*		*Incorporation* (%)	*Ref.*
(−)-[1,17-$^{3}H_2$]Erysodienone	α-+ β-Erythroidine		0.33	
[17-^{3}H]Erysodine	α-+ β-Erythroidine	fed 11 d	0.12	
		fed 33 d	0.50	
	[99% of the ^{3}H at C-17]			

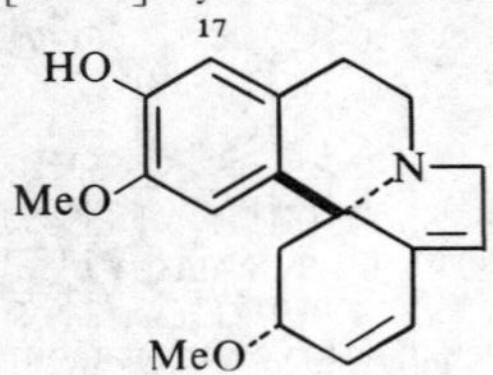

Species / *Compound Administered*	*Alkaloids Labelled*		*Incorporation* (%)	*Ref.*
[14,17-$^{3}H_2$]Erysopine	α-+ β-Erythroidine	fed 11 d	0.15	
		fed 33 d	0.21	

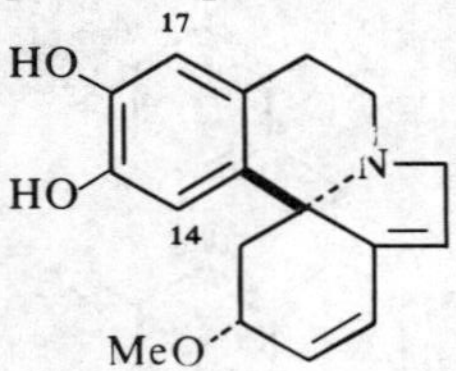

Species / *Compound Administered*	*Alkaloids Labelled*	*Incorporation* (%)	*Ref.*
Erythrina crista-galli			
DL-[2-^{14}C]Tyrosine	Erythraline	0.18	37

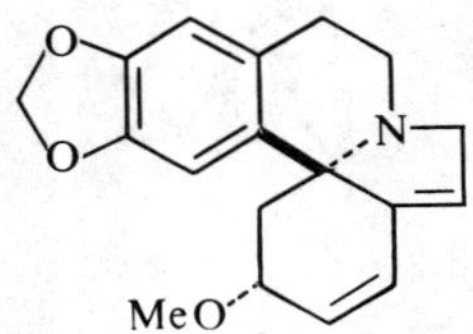

Species / *Compound Administered*	*Alkaloids Labelled*	*Incorporation* (%)	*Ref.*
RS-[5-^{3}H]-*N*-Norprotosinomenine	Erythraline	0.051	
RS-[2′,6′,8-$^{3}H_3$]-*N*-Nor-reticuline	Erythraline	0.001	
(+)-[5′,8-$^{3}H_2$]-*N*-Nororientaline	Erythraline	<0.0001	
(±)-[1,17-$^{3}H_2$]Erysodienone	Erythraline	0.10	38

(+)-[1,17-$^3H_2$]Erysodienone	Erythraline	⩽0.0006	
(−)-[1,17-$^3H_2$]Erysodienone	Erythraline	0.27	
Eupenicillium hirayamae			
Sodium [1-^{14}C]acetate	Desoxyverrucarin E 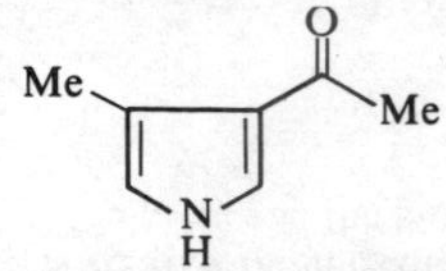 [The authors claimed that the acetic acid obtained by a Kuhn–Roth oxidation had $\frac{1}{4}$ of the specific activity of the pyrrole. However the expected activity is $\frac{1}{3}$: 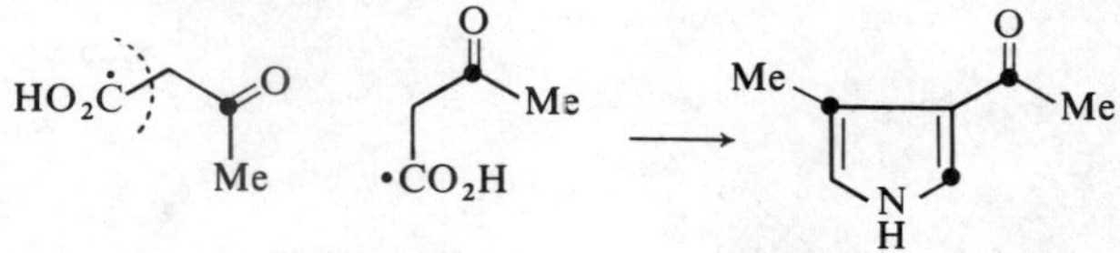	2.0	75
Sodium [2-^{14}C]acetate	Desoxyverrucarin E [It has been previously shown that verrucarin E (3-acetyl-4-hydroxymethylpyrrole) is formed from acetate.[76]]	1.7	
Galega officinalis			
[1-^{14}C]Guanidoacetic acid 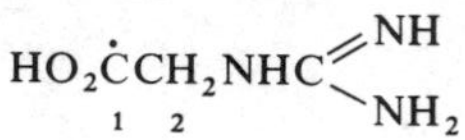	Galegine 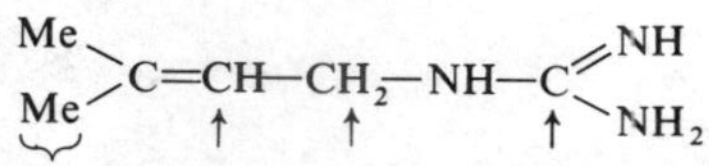 0.4 77.3 8.1 14.2% *Distribution of activity as indicated*	[0.32]	77
[2-^{14}C]Guanidoacetic acid	3.4 6.8 87.6 2.2%	[0.05]	
[*amidine*-^{14}C, 2-^{14}C]Guanidoacetic acid	Galegine [Activity in the CH_2: activity in the amidine group was the same as the ratio of activity at C-2: amidine group in the administered compound]	+	

Species / *Compound Administered*	*Alkaloids Labelled*	*Incorporation* (%)	*Ref.*
Hordeum vulgare			
[2-^{14}C,^{15}N]Glycine	Gramine 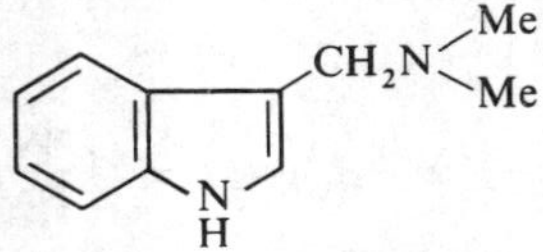 [NMe: 90% of ^{14}C]	^{14}C: [1.44] ^{15}N: [0.27]	78
[α,β-^{14}C, α-^{15}N]Tryptophan 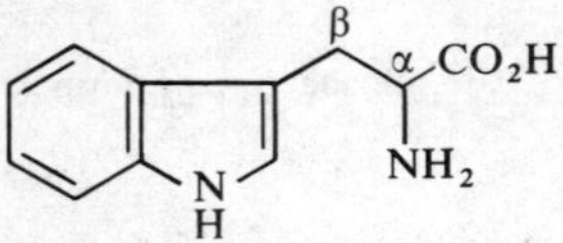 [α:β = 1:1.39]	Gramine [NMe: 5.6% of ^{14}C]	^{14}C: [0.3] ^{15}N: [0.21]	
[α-^{14}C, α-^{15}N]Tryptophan	Gramine	^{14}C: [0.002] ^{15}N: [0.13]	
[β-^{14}C, α-^{15}N]Tryptophan	Gramine	^{14}C: [0.38] ^{15}N: [0.21]	
[β,*N-methyl*-^{14}C$_2$]*N*-Methyl-tryptophan 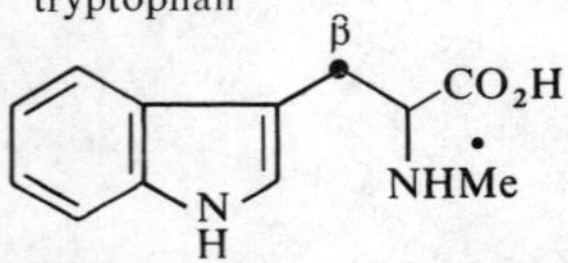 [NMe: β-position = 3.0]	Gramine [NMe: rest of molecule = 2.8. This apparent equal incorporation of the MNe and the β-carbon is probably fortuitous]	[0.06]	
[β-^{14}C]*NN*-Dimethyltryptophan	Gramine [NMe: 25%]	[0.0016]	
[*N-methyl*-^{14}C]*NN*-Dimethyl-tryptophan	Gramine [NMe: 93%]	[0.015]	

[β,N-*methyl*-^{14}C]*NN*-Dimethyl-tryptophan
[NMe : β-position = 0.33]

Gramine
[NMe : rest of molecule = 2.1. The gramine is thus labelled mainly on its NMe groups, presumably from the one-carbon pool, which becomes labelled by demethylation of the administered compound]

[0.0054]

Kreysignia multiflora

RS-[3-^{14}C]Autumnaline

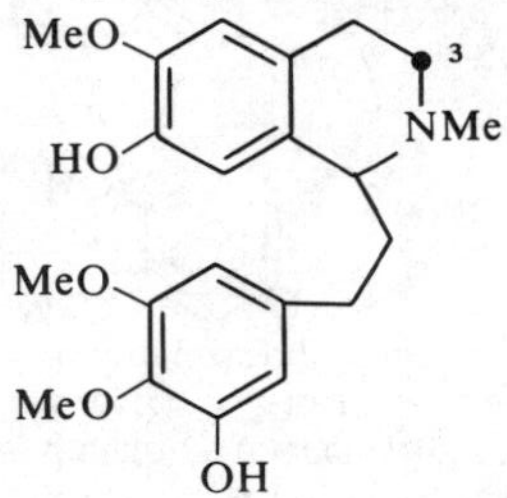

R-Floramultine

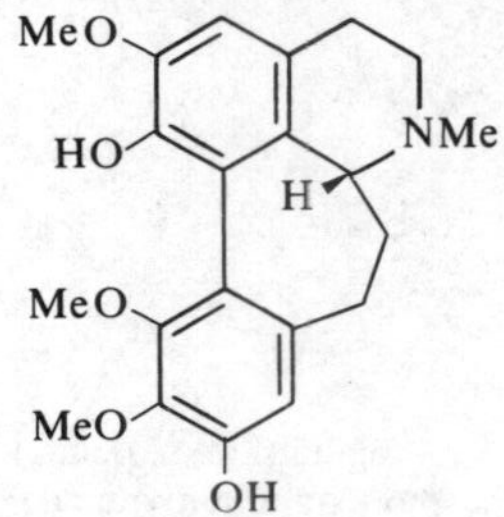

R-Multifloramine

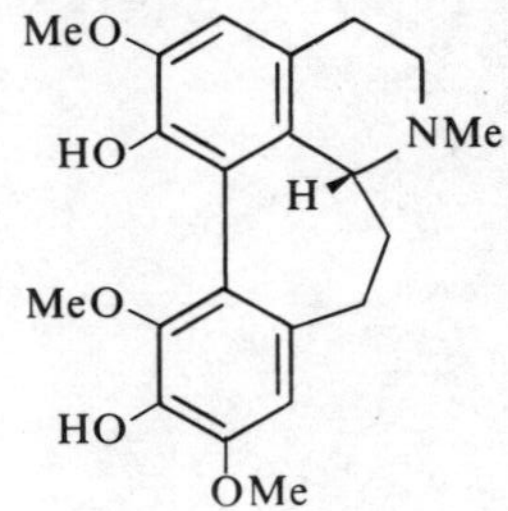

79

RS-Kreysignine

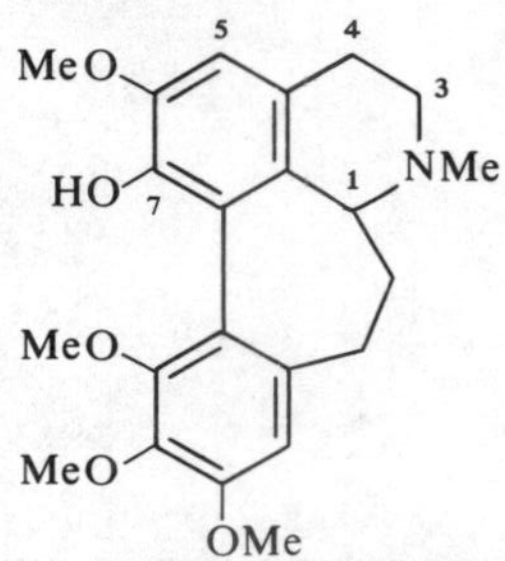

[This mixture of alkaloids was methylated to afford 7-*O*-methylkreysignine (1.8% inc.) C-3 : 86—97%. Thus these homoaporphines are produced by direct *ortho-para* oxidative coupling]

Species / *Compound Administered*	*Alkaloids Labelled*	*Incorporation* (%)	*Ref.*
RS-[3-^{14}C]-1-(4-Hydroxy-3-methoxyphenethyl)-7-hydroxy-6-methyoxy-2-methyl-1,2,3,4-tetrahydroisoquinoline 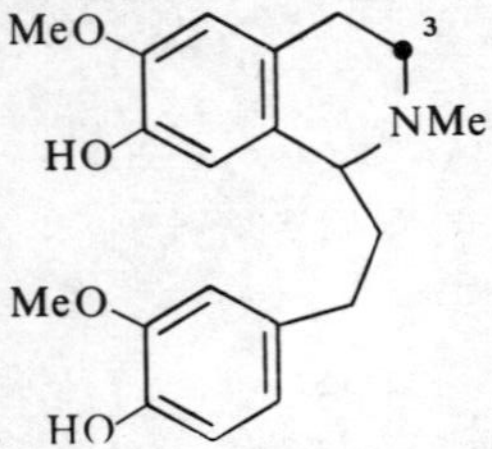	This mixture of alkaloids was converted into *O*-methylkreysignine	0.21	
RS-[3-^{14}C]-1-(4-hydroxy-3,5-dimethoxyphenethyl)-7-hydroxy-6-methoxy-2-methyl-1,2,3,4-tetrahydroisoquinoline 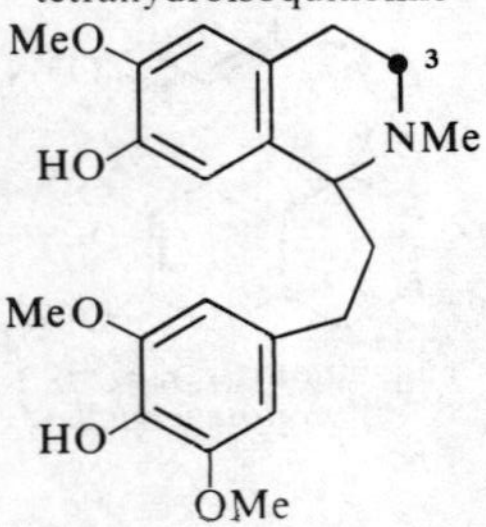	The mixture of alkaloids was converted into *O*-methylkreysignine	<0.014	
Lathyrus sativa			
L-[*U*-^{14}C]Ornithine	Agmatine $H_2N{-}C({=}NH){-}NH(CH_2)_4NH_2$	3.0	80

	Putrescine $H_2N(CH_2)_4NH_2$	3.7
	Spermidine $H_2N(CH_2)_4NH(CH_2)_3NH_2$	0.5
	Spermine $H_2N(CH_2)_3NH(CH_2)_4NH(CH_2)_3NH_2$	0.25
L-[U-^{14}C]Ornithine + inactive Arginine	Agmatine	0.59
	Putrescine	0.96
	Spermidine	0.29
	Spermine	0.14
L-[U-^{14}C]Lysine	Homoagmatine $H_2N{-}C({=}NH){-}NH(CH_2)_5NH_2$	inactive
	Cadaverine $H_2N(CH_2)_5NH_2$	5.2
L-[U-^{14}C]Lysine + inactive Homoarginine	Homoagmatine	inactive
	Cadaverine	4.8
L-[1 to 6-^{14}C]Homoarginine	Homoagmatine	1.3
	Cadaverine	0.47
+ Lysine	Homoagmatine	0.53
	Cadaverine	0.3
+ Homoagmatine	Homoagmatine	0.69
	Cadaverine	0.36
+ N-Carbamylcadaverine	Homoagmatine	0.72
	Cadaverine	0.29
L-[U-^{14}C]Arginine	Agmatine	7.6
	Putrescine	5.2
	Spermidine	3.2
	Spermine	0.4
L-[U-^{14}C]Arginine + inactive Ornithine	Agmatine	6.2
	Putrescine	3.6
	Spermidine	1.8
	Spermine	0.2
+ agmatine	Agmatine	5.1
	Putrescine	3.0
	Spermidine	0.7
	Spermine	0.1

Species / *Compound Administered*	*Alkaloids Labelled*	*Incorporation* (%)	*Ref.*
+ *N*-carbamylputrescine	Agmatine	5.9	
	Putrescine	3.2	
	Spermidine	0.7	
	Spermine	0.2	
+ Putrescine	Agmatine	7.4	
	Putrescine	2.6	
	Spermidine	0.6	
	Spermine	0.1	
Litsea glutinosa			
DL-[2-^{14}C]Tyrosine	Boldine 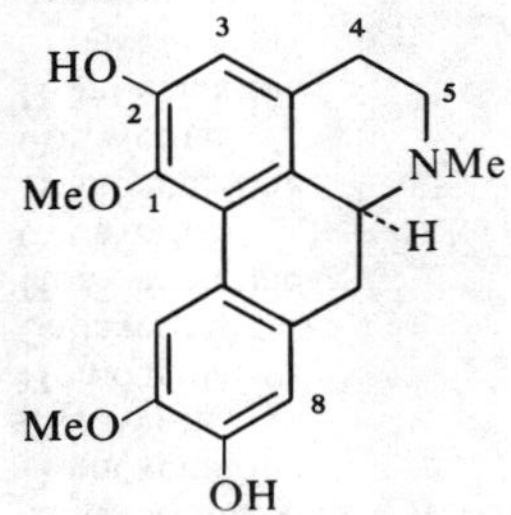	0.16	81
RS-[1-^{3}H]-4′-*O*-Methylnorlaudanosoline 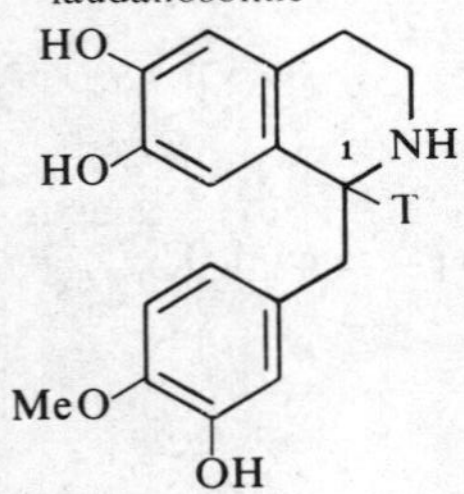	Boldine	0.13	

Precursor	Product	Incorporation (%)
RS-[5′,8-$^3H_2$]-*N*-Nororientaline [see under *Erythrina berteroana*]	Boldine	0.01
RS-[3-^{14}C]-*N*-Norprotosinomenine [see under *Erythrina berteroana*]	Boldine	0.06
RS-[3-^{14}C]-*N*-Nor-reticuline [see under *Erythrina berteroana*]	Boldine [C-5 : 98 %]	0.62
RS-[2′,6′,8-$^3H_3$]Reticuline	Boldine [All ^3H at C-8]	0.44
S-(+)-[2′,6′,8-$^3H_3$]Reticuline	Boldine	0.43

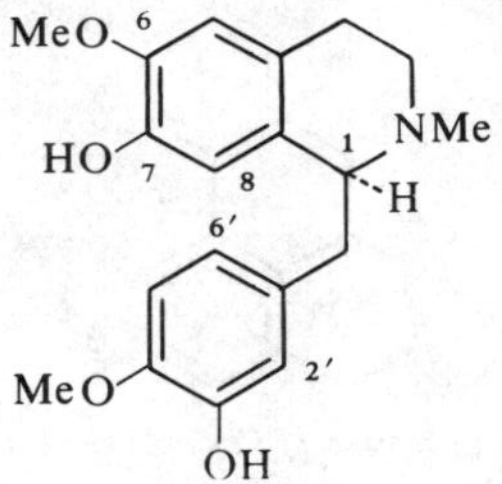

Precursor	Product	Incorporation (%)
R-(−)-[2′,6′,8-$^3H_3$]Reticuline	Boldine	0.01
RS-[6-*O*-*methyl*-^{14}C, 1-^3H] Reticuline	Boldine [There was only a 64 % retention of ^{14}C relative to the ^3H, indicating that the methyl migration from C-6 to C-7 is not intramolecular]	0.90 (^3H)
RS-[2′,6′,8-$^3H_3$]Laudanosine	Boldine	0.006

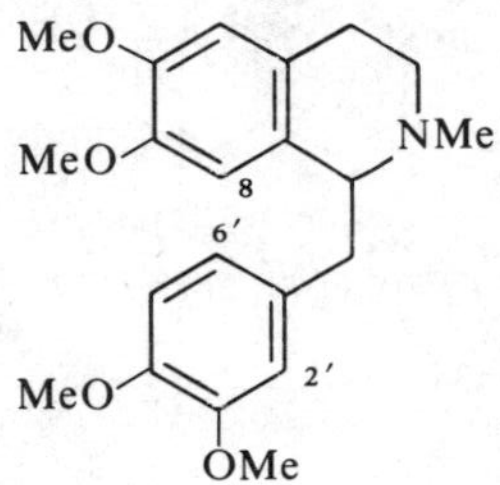

Species / *Compound Administered*	*Alkaloids Labelled*	*Incorporation (%)*	*Ref.*
(+)-[8-^{3}H]Isoboldine 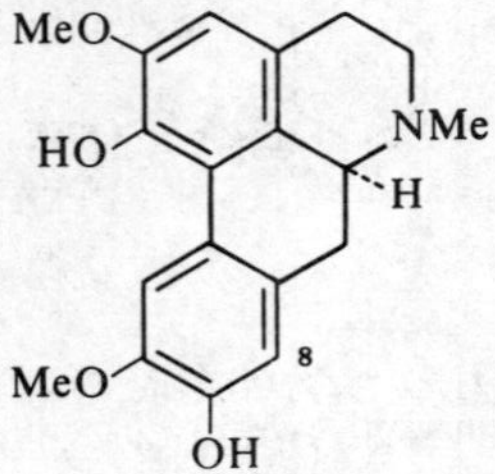	Boldine [All ^{3}H at C-8]	2.0	
(+)-[*aryl*-^{3}H]-*N*-Norboldine 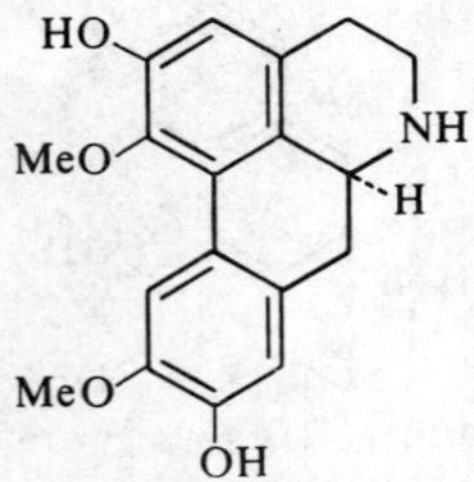	Boldine	1.86	
L-[*U*-^{14}C]Tyrosine	Reticuline (carrier added)	0.16	
	4′-*O*-Methylnorlaudanosoline (carrier added)	0.17	
	These results support the biosynthetic sequence: 4′-O-methyl-norlaudanosoline→nor-reticuline→reticuline→isoboldine→boldine.		

Lophocereus schottii			
DL-[2-^{14}C]Leucine	Lophocerine 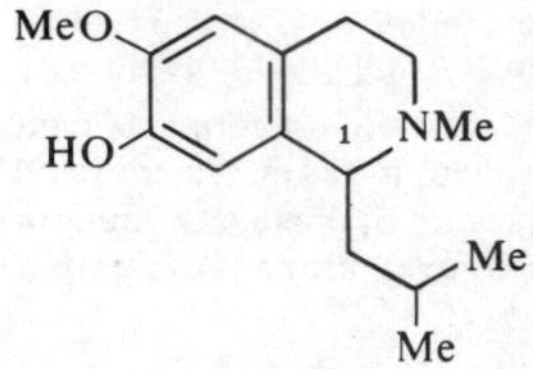	0.023	82
RS-[2-^{14}C]Mevalonic acid	Lophocerine	0.040	
[1-^{14}C]-3-Methylbutanoic acid	Lophocerine	inactive	
[1-^{14}C]-3-Methyl-1-butanol	Lophocerine	0.16	
[1-^{14}C]-3-Methylbutanal	Lophocerine	0.027	
	[Activities of degradation products consistent with all activity being at C-1]		
[1-^{14}C]Isopentenylpyrophosphate	Lophocerine	0.052	
Lophophora williamsii			
L-[*U*-^{14}C]Tyrosine	Total Akaloids	5	83
[plants harvested 4 weeks after injection]	Mescaline 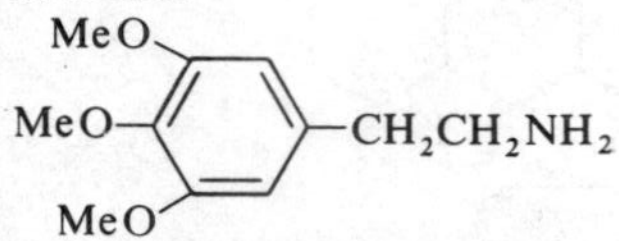	0.59	
	Protein (water soluble)	1.71	
	Tyrosine in the protein	1.51	

Distribution of activity relative to the total alkaloids

D-[*U*-^{14}C]Fructose +	*Total Alkaloids*	*Mescaline*	*Tyrosine in protein*
No additions	100	11.1	65
p-Hydroxyphenylpyruvic acid	89	9.3	34
L-Tyrosine	97	10.0	39

[Apparently alkaloid biosynthesis is not under negative feedback control mechanism]

Species

Compound Administered	*Alkaloids Labelled*	*Incorporation(%)*	*Ref.*
Lycopersicon esculentum (culture of excised tomato root)			
RS-[2-^{14}C]Mevalonic acid	Tomatine 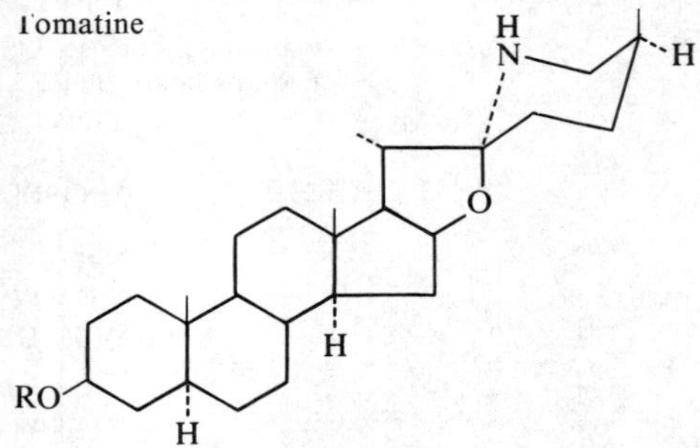 R = D-xylose-D-glucose-D-glucose-D-galactose [Greater incorporations were obtained in the apical region of the root axis]	0.09	84
Macleaya—hybrid of *cordata* and *microcarpa* (Callus tissues)			
[2-^{14}C]Phenylalanine [*U*-^{3}H]Phenylalanine [*methyl*-^{14}C]Methionine [*U*-^{14}C]Tyrosine	Activity in the total alkaloids is reported and apparently no attempt was made to separate them. Surprisingly tyrosine was a poorer precursor of the alkaloids than phenylalanine, although phenylalanine is converted into tyrosine in the callus tissue.	+	85
[^{3}H]Protopine [see under *Corydalis incisa*]	This alkaloid, which is one of the normal components of this species, was recovered (96—97%) essentially unchanged after 24 h incubation with the tissue culture.		

Mitragyna parvifolia 86

[*methoxy*-^{14}C]Mitraphylline

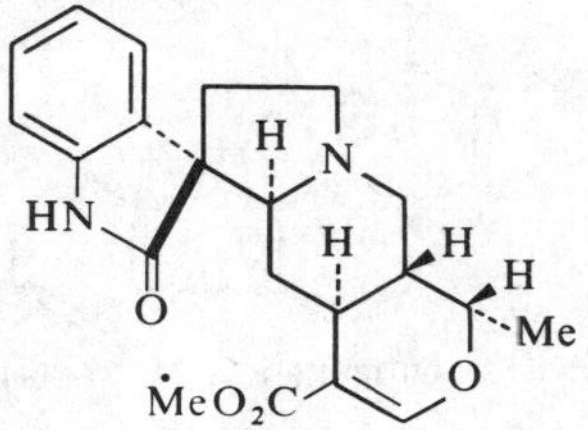

[The plants were harvested various times after feeding the labelled alkaloids. The distribution of activity in these oxindole alkaloids is determined by assay of spots on t.l.c. plates. No attempts were made to crystallize the alkaloids to constant activity]

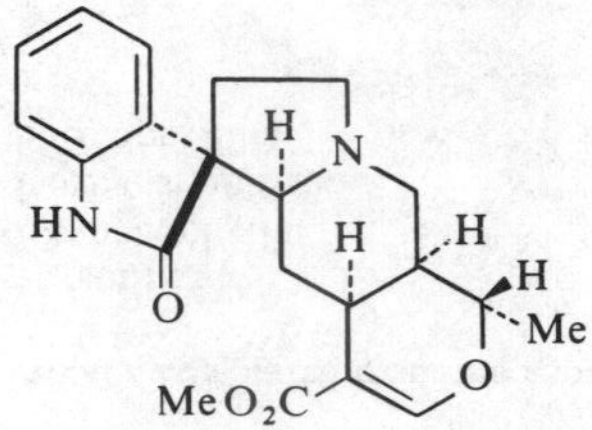

Pteropodine

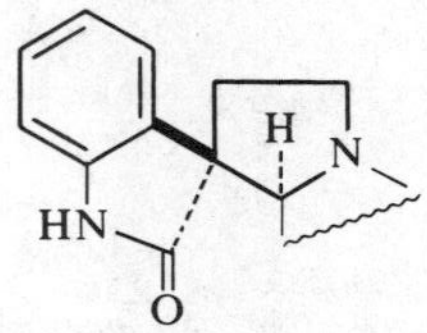

Isopteropodine

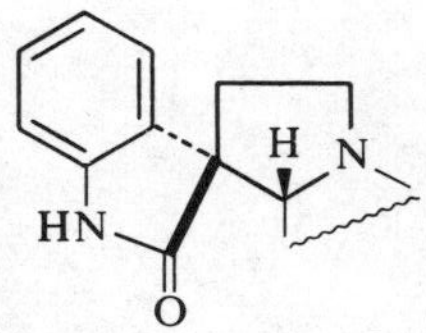

Uncarine F

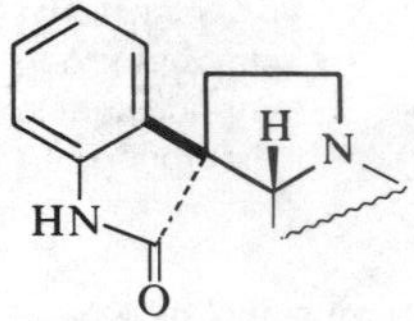

Speciophylline

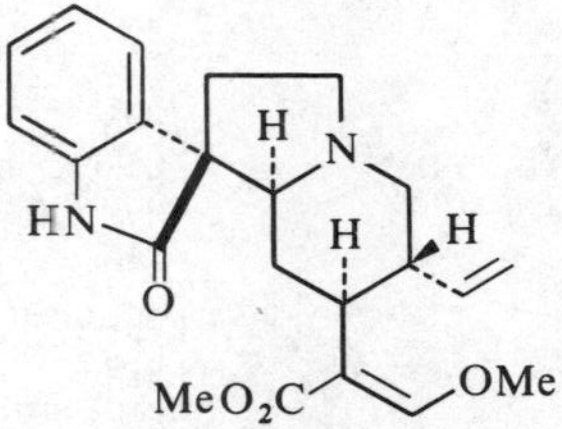

Corynoxeine

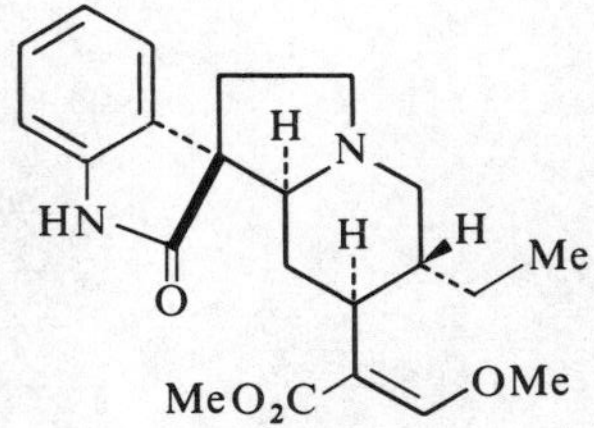

Rhynchophylline

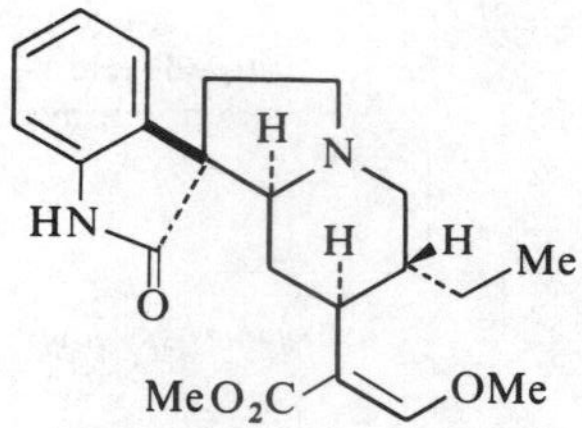

Isorhynchophylline

Species

Compound Administered	*Alkaloids Labelled*					*Incorporation* (%)	*Ref.*
		Distribution of Activity (%) (*Time after feeding*/h)					
		24	30	48	54		
[*methoxy*-^{14}C]Mitraphylline	Pteropodine	3.04	0.97	2.34	2.80		
	Isopteropodine	0.1	0.26	0.19	1.00		
	Uncarine F	6.08	0.28	1.45	1.33		
	Speciophylline	0.1	12.5	15.76	13.7		
	Mitraphylline	90.87	64.87	69.66	41.07		
	Corynoxeine	—	3.81	10.61	6.16		
	Rhynchophylline	—	18.8	—	34.3		
	[These alkaloids were isolated from the leaves]						

Compound Administered	*Alkaloids Labelled*	*Top of bark*	*Lower part of bark*
[*methoxy*-^{14}C]Pteropodine [27 h feeding, activities in the top of the bark and stems, and the lower part of the stems recorded]	Pteropodine	85.3	73.7
	Isopteropodine	3.3	12.8
	Speciophylline	1.1	2.6
	Uncarine F	8.8	7.5
	Mitraphylline	1.1	0.83
	Rhynchophylline	0.41	0.82

[*methoxy*-^{14}C]Hirsutine

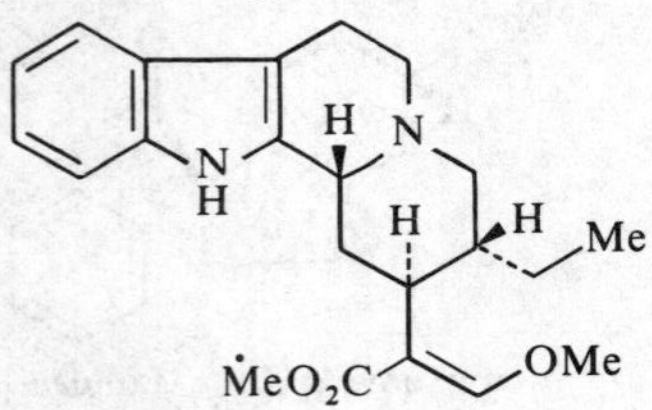

Alkaloids isolated from the root bark 7 d after feeding

	Distribution of Activity (%)
Hirsutine	74.6
Rhynchophylline	3.7
Isorhynchophylline	16.8
Speciophylline	4.9

[*methoxy*-^{14}C]Mitraphylline

Alkaloids isolated from the stem and bark 48 h after feeding

	Distribution of Activity (%)	
	top of bark	*lower part of bark*
Mitraphylline	48.2	19.9
Isopteropodine, Pteropodine	3.8	19.7
Speciophylline	18.1	15.4
Uncarine F	3.9	17.6
Rhynchophylline	18.0	27.2
Corynoxeine	8.0	—

[*methoxy*-^{14}C]Rhynchophylline

Alkaloids isolated from the leaves.

	Distribution of Activity (%) Time after feeding/h		
	6	27	48
Rhynchophylline	63.9	50.4	52.6 (Rhynchophylline + Corynoxeine)
Corynoxeine	15.7	19.2	
Mitraphylline	13.8	17.1	36.6
Isorhynchophylline, Isopteropodine	1.3	6.7	4.2
Pteropodine	1.0	2.8	—
Speciophylline	0.35	1.5	—
Uncarine F	3.2	5.2	—

[*methoxy*-^{14}C]Rhynchophylline

Alkaloids isolated from the root bark

	Distribution of Activity (%)
Rhynchophylline	50.4—51.7
Isorhynchophylline	15.4—15.8
Speciophylline	4.9—23.9
Hirsutine	9.8—14.1

Species / Compound Administered	Alkaloids Labelled	Incorporation (%)	Ref.

Narcissus pseudonarcissus (King Alfred daffodil)
[*O-methyl*-^{14}C, 7-^{3}H]-3-Hydroxy-4-methoxy-*N*-methylbenzylamine

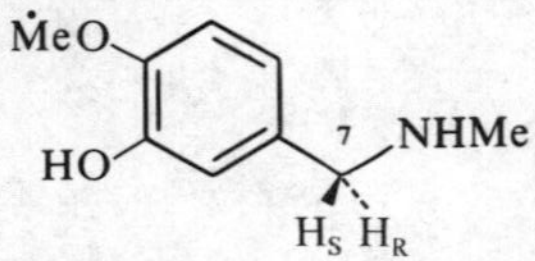

[The enantiomeric 7*S*-^{3}H, 7*R*-^{3}H, and 7*RS*-$^{3}H_2$ labelled compounds were fed in separate experiments.]

Haemanthamine — + — 87

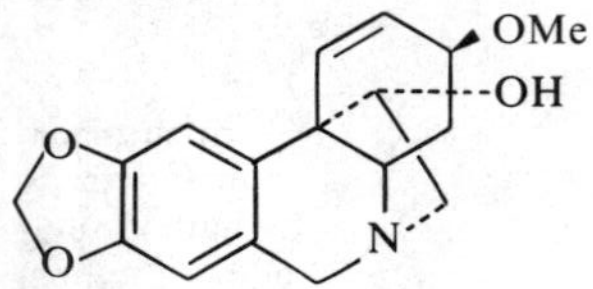

Galanthamine — +

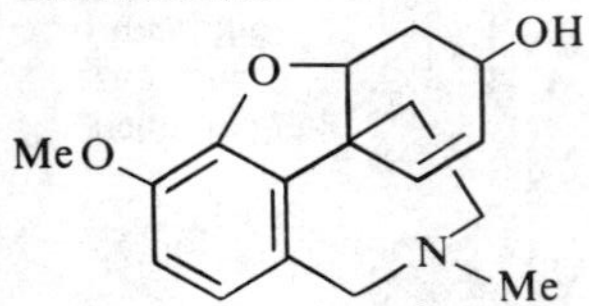

Oduline — +

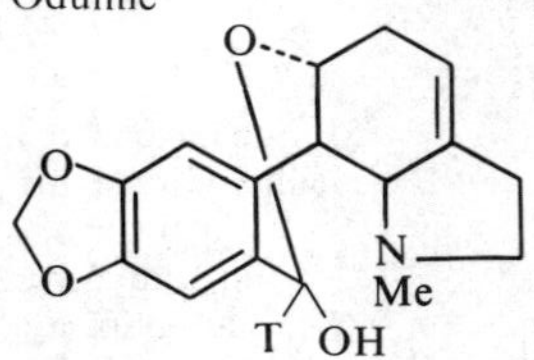

[^{3}H was all at the indicated position]

[Incorporations of 0.03—0.09% were obtained, with 82—85% retention of ^{3}H in all cases. This indicates that oxidation of the precursor to 3-hydroxy-4-methoxybenzaldehyde was not stereospecific, retention of ^{3}H being due to a kinetic isotope effect]

Precursor	Alkaloid	Incorporation	Ref.
(Sempre Avanti variety)			
2*RS*-[1-^{14}C, 2-$^{3}H_2$]-*O*-Methyl-norbelladine	Haemanthamine [48 % ^{3}H retention]	0.9 (^{14}C)	88
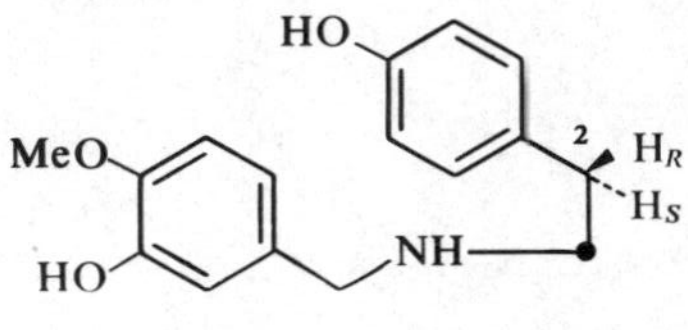	Galanthine 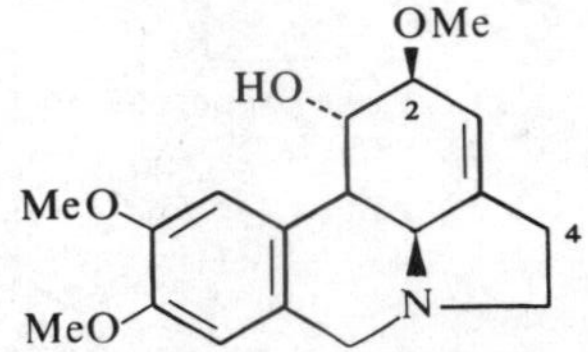 [98 % ^{3}H retention]	0.08 (^{14}C)	
	Narcissidine 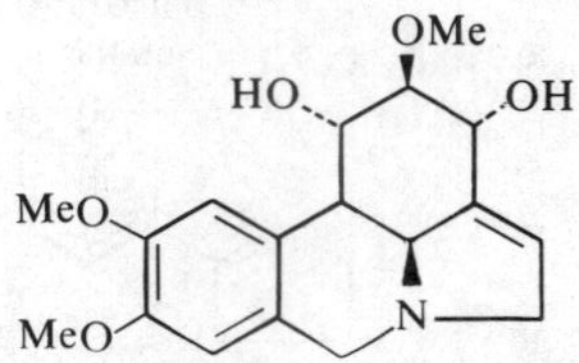 [46 % ^{3}H retention]	0.09 (^{14}C)	
2*R*-[1-^{14}C, 2-^{3}H]-*O*-Methyl-norbelladine	Haemanthamine [5 % ^{3}H retention]	+	
	Galanthine [100 % ^{3}H retention]	+	
	Narcissidine [91 % ^{3}H retention. Thus the *pro-S* hydrogen is lost from C-4 of an intermediate having a lycorane skeleton]	+	
[*aryl*-^{3}H]Galanthine	Narcissidine	5.2	
(*Ismene* hybrid Sulphur Queen)			
2*S*-[1-^{14}C, 2-^{3}H]-*O*-Methyl-norbelladine	Haemanthamine [94 % ^{3}H retention]	0.06 (^{14}C)	
	Narcissidine [7 % ^{3}H retention]	0.008 (^{14}C)	

Species / *Compounds Administered*	*Alkaloids Labelled*	*Incorporation* (%)	*Ref.*
Nicotiana glauca		All ^{14}C inc.	
(−)-[2′-^{3}H]Nicotine + (±)-[2′-^{14}C]Nicotine [3 day wick feeding]	Nicotine	8.0 [19.0]	11
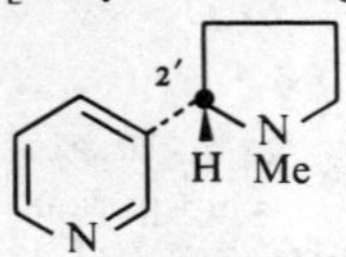 (−)-nicotine	Nornicotine 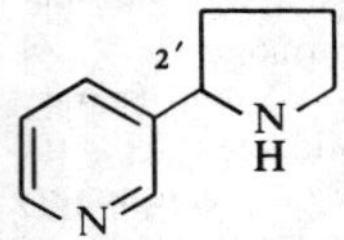	49.5 [26.8]	
	[> 99 % retention of the ^{3}H in (−)-nornicotine, all at C-2′. No ^{3}H in (+)-nornicotine. All ^{14}C at C-2′]		
	Myosmine 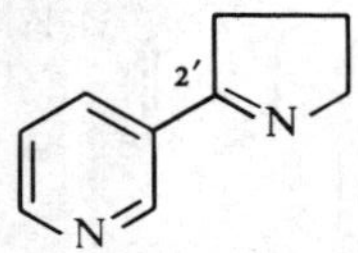	2.05	
	[negligible ^{3}H, C-2′: 100 % of ^{14}C]		
	Anabasine 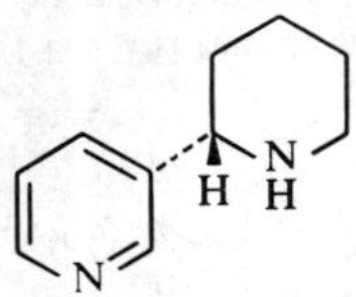	0.01 [0.002]	
	Cotinine (added as a carrier) 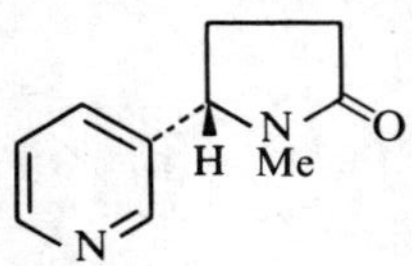	0.01	

	3-Acetylpyridine (added as a carrier) 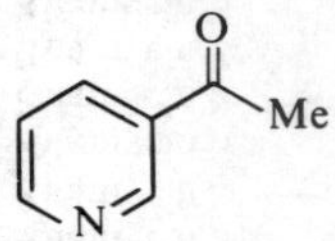	inactive	
	Nicotinic acid (added as a carrier) 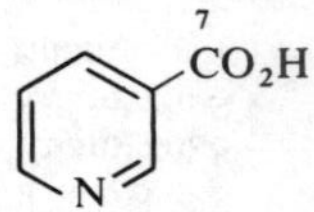	0.02	
$[2'\text{-}^{14}C]$Myosmine	Myosmine	[35]	
	Nicotine	no significant activity	
	Nornicotine	no significant activity	
	Anabasine	no significant activity	
	Nicotinic acid [C-7: 95%]	0.14	
Nicotiana glutinosa			
$[^{14}C]$Carbon Dioxide	Nicotine [76—90% of the activity was located in the pyridine ring. In some feeding experiments unsymmetrical labelling of the pyrrolidine ring was observed. See p. 98 for a discussion of these results]	+	15
Nicotiana tabacum			
$[^{14}C]$-4-Aminobutanoic acid 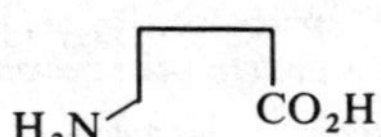	*N*-Nicotinoyl-2-pyrrolidone (added as a carrier) 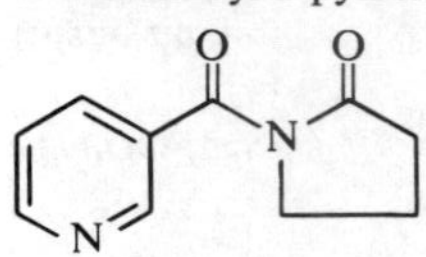	inactive	89

Species / Compounds Administered	Alkaloids Labelled	Incorporation (%)	Ref.
Nocardia mediterranei			
[1-^{14}C]Propionate	Rifamycin S 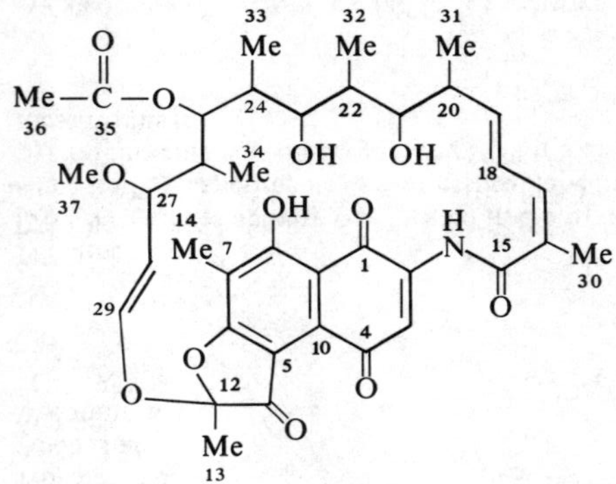	16.8	90
[1-^{14}C]Propionate + Methyl-malonate (Diethyl ester)	Rifamycin S	19.4	
[2-^{14}C]Propionate	Rifamycin S	16.1	
[3-^{14}C]Propionate	Rifamycin S	13.4	
[*methyl*-^{3}H]Methylmalonate	Rifamycin S	5.7	
[*methyl*-^{3}H]Methylmalonate + Propionate	Rifamycin S	0.2	
[1,4-$^{14}C_2$]Succinate	Rifamycin S	4.4	
[2,3-$^{14}C_2$]Succinate	Rifamycin S	5.4	
L-[*U*-^{14}C]Aspartate	Rifamycin S	2.4	
L-[*U*-^{14}C]Glutamate	Rifamycin S	10.9	
[1-^{14}C]Acetate	Rifamycin S	1.4	
[2-^{14}C]Acetate	Rifamycin S	7.3	
[2-^{14}C]Malonate (diethyl ester)	Rifamycin S [Significantly lower activity at the acetoxy-group C-35, C-36]	5.6	

Precursor	Product	Ratio of Activity in chromophore C-1 to C-14/*ansa* chain		Ref.
[2-^{13}C]Malonate (diethyl ester)	Rifamycin S (see p. 204) [^{13}C n.m.r. spectra indicates enhancement at C-5 and C-18, little enrichment at C-36]			
D-[1-^{14}C]Glucose	Rifamycin S	0.52	2.0	
D-[2-^{14}C]Glucose	Rifamycin S	0.79	1.2	
D-[3,4-$^{14}C_2$]Glucose	Rifamycin S	2.56	1.1	
D-[6-^{14}C]Glucose	Rifamycin S	0.54	3.5	
D-[*U*-^{14}C]Glucose	Rifamycin S	0.71	8.9	
D-[1-^{14}C]Ribose	Rifamycin S	0.73	4.4	
D-[*U*-^{14}C]Ribose	Rifamycin S	0.97	3.1	
D-[1-^{14}C]Glucosamine	Rifamycin S	0.56	2.5	
D-[*U*-^{14}C]Glucosamine	Rifamycin S	0.67	2.8	
DL-[1-^{14}C]Glycerate	Rifamycin S	4.78	1.9	
[*U*-^{14}C]Myoinosate	Rifamycin S	0.44	4.2	
[*U*-^{14}C]Benzoate	Rifamycin S	0.45	5.8	
[*U*-^{14}C]Shikimate	Rifamycin S	0.30	0.2	
DL-[*U*-^{14}C]Tryptophan	Rifamycin S	0.36	0.3	
L-[*U*-^{14}C]Tyrosine	Rifamycin S	0.55	0.4	
	[Any ratio > 0.485 indicates that the administered compound is a significant precursor of the naphthoquinone moiety]			
[1-^{13}C]Glucose	Rifamycin S (see p. 204) [^{13}C n.m.r. indicated enrichment of C-1, C-10, and C-36]		+	
[1-^{13}C]Propionate, [2-^{13}C]Propionate, [3-^{13}C] Propionate, [1-^{13}C] Acetate	Rifamycin W (see p. 204)		+	91
[^{14}C]Rifamycin W [obtained biosynthetically by feeding [1-^{14}C]propionate to *N. mediterranei*]	Rifamycin B (see p. 204)		+	

Species				
Compounds Administered	*Alkaloids Labelled*		*Incorporation* (%)	*Ref.*
Pancratium maritimum				
[1,4-$^3H_2$]-11-Hydroxyvittatine 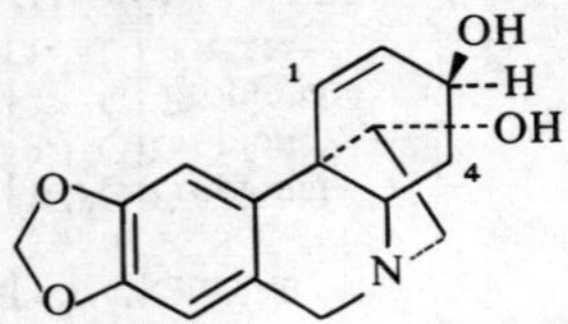	Narciclasine 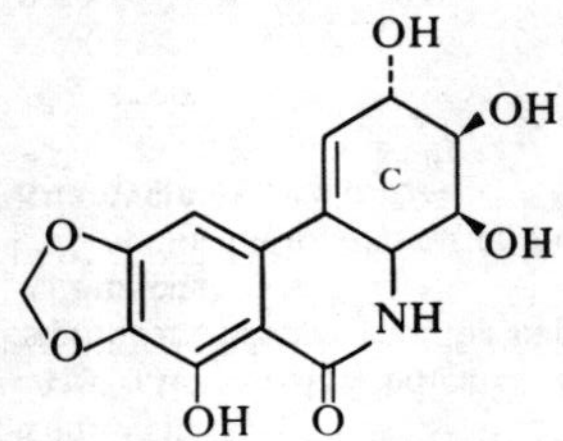 [70% of the ^{3}H is retained on aromatization of ring C]		5.5	92
	Haemanthidine 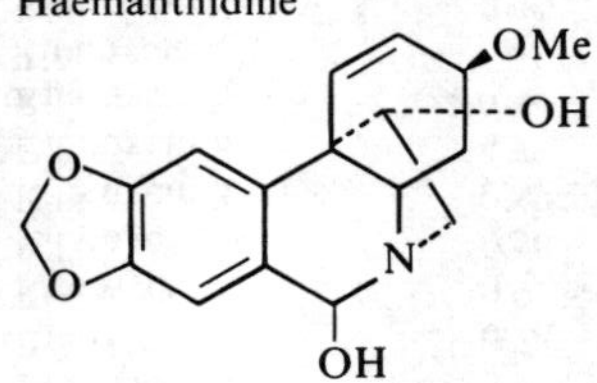		3	
Papaver somniferum				
DL-[2-^{14}C]Dopa	Morphine 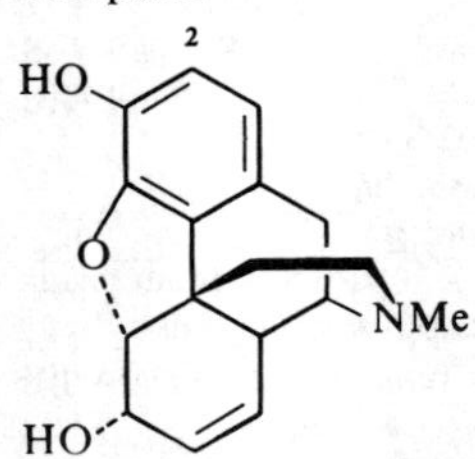 [> 97% of the activity in the ethanamine bridge]		0.78	93

Precursor	Product		Ref.
DL-[2-^{14}C]Dopa	Morphine [Conversion was achieved in centrifuged latex obtained from the stem and capsules of the plant. Efficient conversion was obtained in the pellet obtained from the capsule, but was negligible in the stem]	+	94
[2-^{3}H]Morphine	Metabolism occurred to non-alkaloid materials		
Passiflora edulis			
L-[β-^{14}C]Tryptophan	Harman 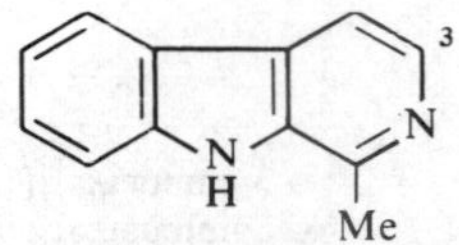 [The radioactive harman reaches a maximum after about 5 d and then decreases in activity. This decrease is possibly due to excretion of the alkaloid through the roots. The feeding of tryptophan to the plant increased the amount of harman 10-fold]	+	95
[3-^{14}C]Harman	The alkaloid was excreted from the plant, possibly *via* the roots.		
Peganum harmala (callus tissue)			
L-[β-^{14}C]Tryptophan fed along with potential intermediates of the β-carboline alkaloids: 5-Hydroxytryptophan 6-Hydroxytryptophan 5-Hydroxytryptamine 6-Hydroxytryptamine 6-Methoxytryptamine Harmalol, Harmaline 6-Methoxyharmalan, Harmol	Harmalol 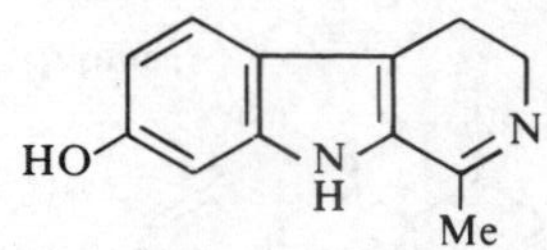 Harmine 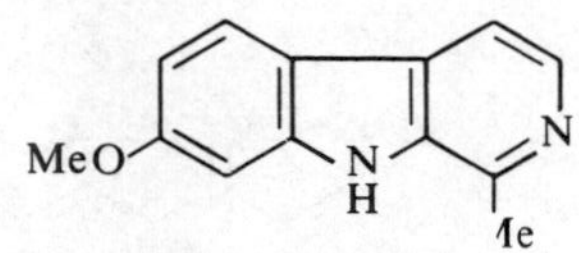 [None of these compounds caused significant inhibition of the incorporation of activity into the above alkaloids. Appreciable (10%) activity was trapped in 5-hydroxytryptamine, but surprisingly little in 6-hydroxytryptamine.]		96
[α-^{14}C]Tryptamine	5-Hydroxytryptamine [Little activity detected in harmine]	+	

Species / Compounds Administered	Alkaloids Labelled	Incorporation (%)	Ref.
[10-^{14}C]Harmaline	Dihydroruine + Ruine	3	
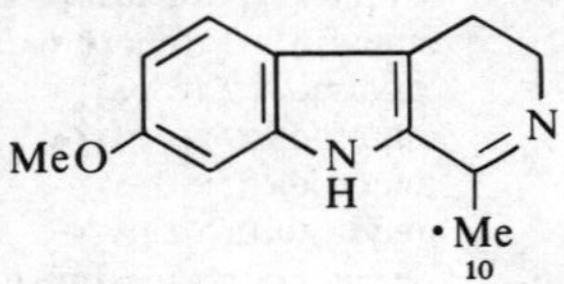	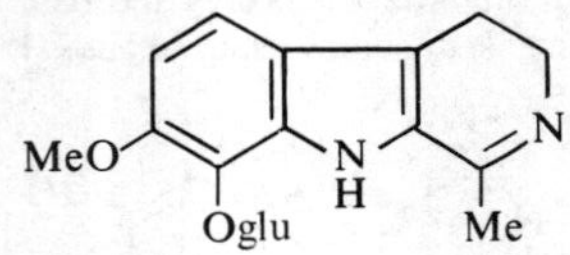 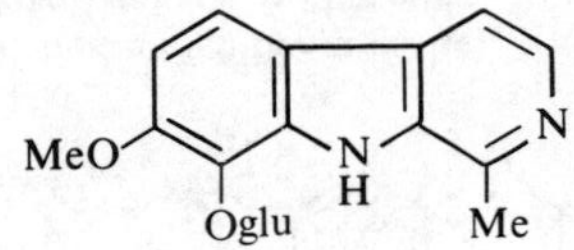		
	Harmine	+	
Penicillium brevicompactum			
DL-[β-^{14}C]Tryptophan	Brevianamide A	5.1* [17.6] (*based on utilization of L-isomer only)	97
	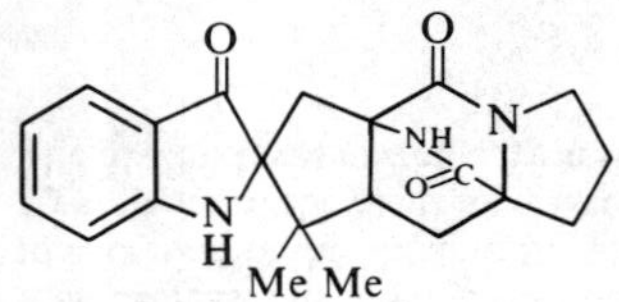		
L-[5-^{3}H]Proline	Brevianamide A	0.4	
RS-[2-^{14}C]Mevalonolactone	Brevianamide A	0.16	
cyclo-L-[β-^{14}C]Tryptophyl-L-proline	Brevianamide A	1.8 [14.0]	
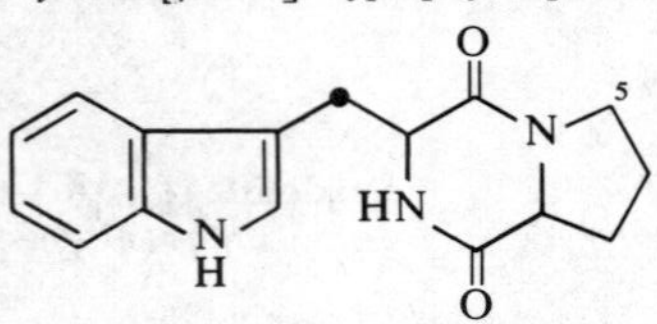			
cyclo-L-[β-^{14}C]Tryptophyl-L-[5-^{3}H]proline	Brevianamide A [100% retention of ^{3}H]	3.2, 3.6	
Penicillium chrysogenum			
(2*RS*, 3*S*)-[4-^{13}C]Valine	Penicillin V (see p. 248)	+	98
(2*RS*)-[*methyl*-$^{2}H_6$]Valine	Penicillin V (see p. 248)	+	99

Phoma species S 298

[2-^{13}C]Acetate — Cytochalasin B (Phomin) — 0.11 — 100

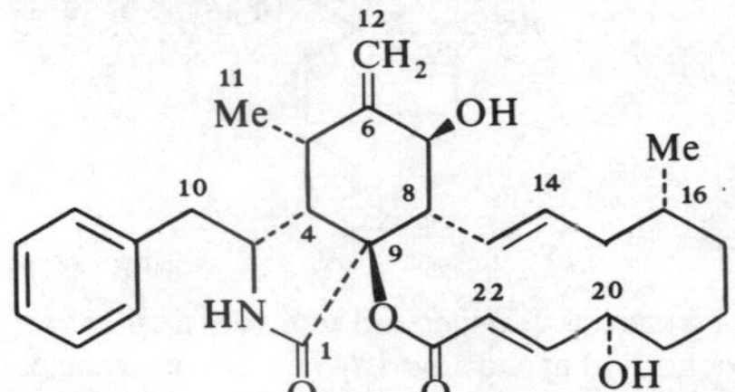

[^{13}C n.m.r. indicated enrichment at C-11,6,8,9,14,18,20,22. Enrichment was also present at another unassigned peak which was assumed to be C-16 on biogenetic grounds]

Physalis peruviana

L-[U-^{14}C]Isoleucine (fed by hydroponic method to the roots of intact plants) — Tigloidine — from leaves: [0.6]; from roots: [4.6] — 101

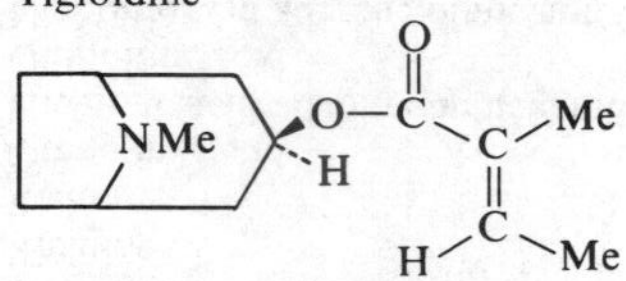

3-α-Tigloyltropane — from leaves: [4]; from roots: [14]

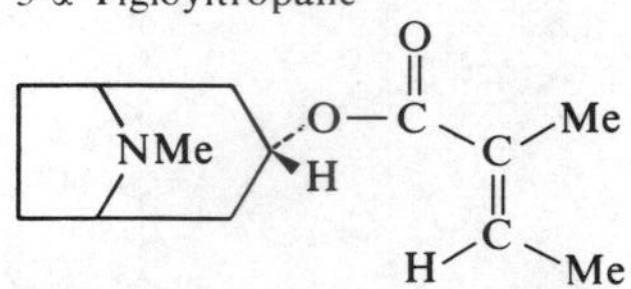

[> 98 % in the tigloyl moieties]

Ricinus communis

The effect of various compounds (some inhibitors of the pyridine nucleotide cycle) on ricinine biosynthesis from quinolinic acid. — 39

Species			
Compound Administered	*Alkaloids Labelled*	*Incorporation(%)*	*Ref.*
[6-^{14}C]Quinolinic acid 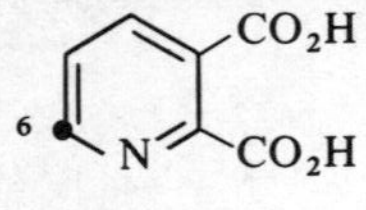	Ricinine 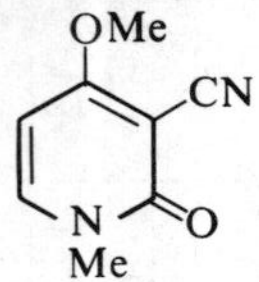		
no additions	Ricinine	21—36	
+ azaserine	Ricinine	12—20	
+ azaleucine	Ricinine	3—6	
+ ricinine	Ricinine	5—7	
+ ethionine	Ricinine	8—6	
no additions	Quinolinic acid	19.8	
	Nicotinic acid adenine dinucleotide	2.5	
+ azaserine	Quinolinic acid	38.3	
	Nicotinic acid adenine dinucleotide	6.2	
+ azaleucine	Quinolinic acid	41.2	
	Nicotinic acid adenine dinucleotide	1.5	
	[similar results obtained with increased amounts of inhibitors]		

Sceletium strictum

Most of the data presented in this reference has been previously published in preliminary form[102] and has been reviewed (see Volume 1, pp. 223, 224). Precursors fed, not previously described, as listed below.

[1-^{14}C]-4′-*O*-Methylnorbelladine 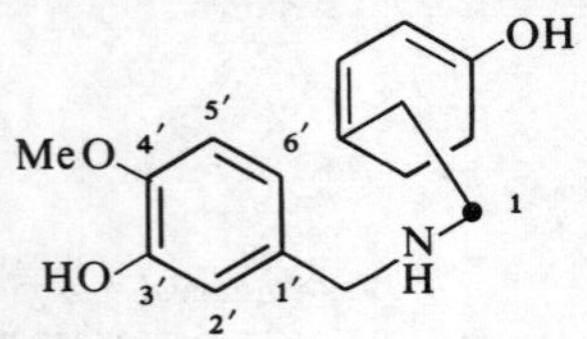	Mesembrenol 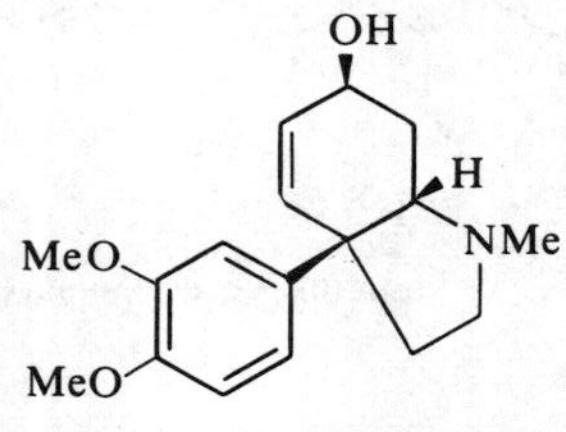	[0.0043]	103

Precursor	Alkaloid	Incorporation	Ref.
[1-^{14}C]-3′-*O*-Methylnorbelladine	Mesembrenol	[0.010]	
[1-^{14}C.2′,6′-^{3}H$_2$]*N*-Methyl-norbelladine	Mesembrine 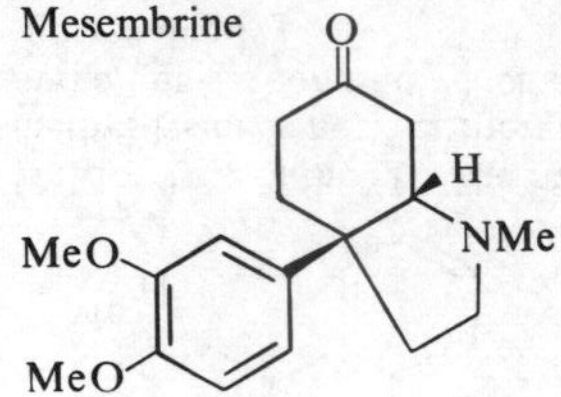	inactive	
[1-^{14}C,3′,5′-^{3}H$_2$]-3′-Deoxy-norbelladine 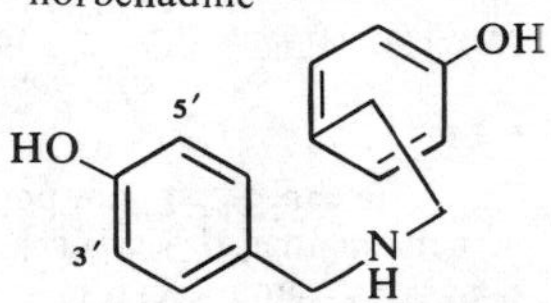	Mesembrine	inactive	
[1-^{14}C,3′,5′-^{3}H$_2$]-3′Deoxy-*N*-methylnorbelladine	Mesembrine	inactive	
Securinega suffruticosa			
DL-[2-^{14}C]Tyrosine [fed by cotton wick for 14 d]	Securinine 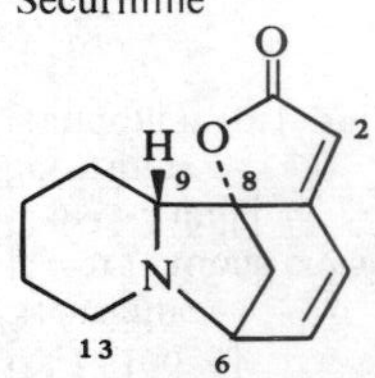 [C-1 : 92%]	0.0035	33
L-[*G*-^{3}H]Phenylalanine	Securinine	inactive	34
DL-[2-^{14}C]Lysine	Securinine	0.11	
L-[*U*-^{14}C]Lysine	Securinine	0.02	
[1,5-^{14}C$_2$]Cadaverine	Securinine + Allosecurinine [Degradations consistent with the activity being at C-9 and C-13]	0.02	

Species *Compounds Administered*	*Alkaloids Labelled*	*Incorporation* (%)	*Ref.*
DL-[2-^{14}C]Tyrosine	Securinine [C-1 : 106%]	0.028	
L-[*U*-^{14}C]Tyrosine	Securinine [Degradations consistent with the intact incorporation of a C-6–C-2 unit]	0.038	
[1-^{14}C]Homogentisic acid 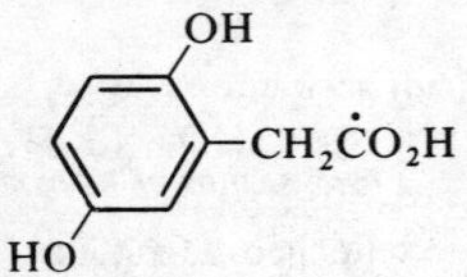	Securinine [random incorporation]	0.0063	
DL-[2-^{14}C]Dopa	Securinine	0.0011	
[1-^{14}C]Tyramine	Securinine	0.0047	
L-[*aryl*-^{14}C]Phenylalanine	Securinine	0.00027	
Sodium [1-^{14}C]acetate	Securinine	0.0053	
Sedum acre			
DL-[2-^{14}C,*N-methyl*-^{3}H]-ε-*N*-Methyl-lysine 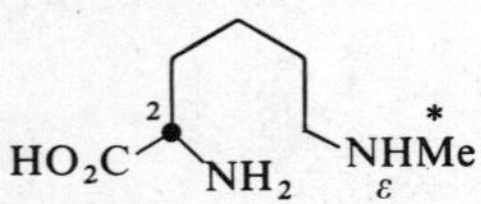 [administered to excised shoots; however, the absorption of tracer was poor]	Sedamine 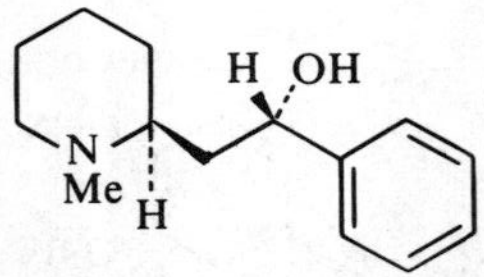 [NMe: 91% of the ^{3}H. The results indicate that the ε-*N*-methyl-lysine is not incorporated intact, but is degraded to lysine and a one-carbon unit which are then incorporated independently]	^{14}C: 0.009 ^{3}H: 0.006	104

Precursor	Product / comments		Ref.
Senecio magnificus			
L-[G-^{3}H]Isoleucine (58%) + L-[G-^{3}H]Alloisoleucine (42%) + L-[U-^{14}C]Isoleucine	Senecionine 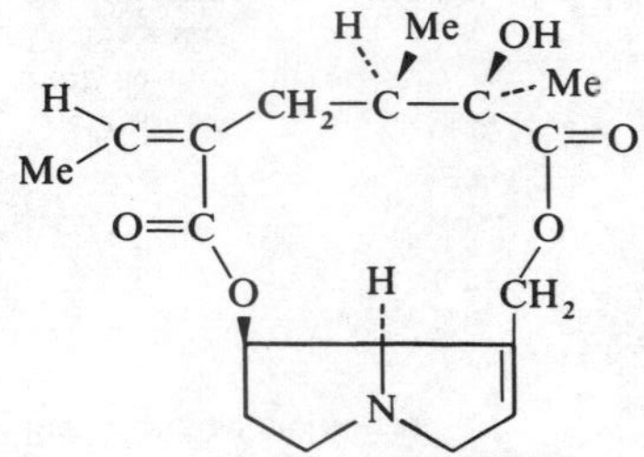 [65% retention of ^{3}H. Previous studies[106] have shown that isoleucine is incorporated exclusively into the senecic acid moiety]	+	105
DL-[G-^{3}H]Alloisoleucine + L-[U-^{14}C]Isoleucine	Senecionine [5.3, 8.6% retention of ^{3}H in two different feeding experiments]	+	
D-[G-^{3}H]Isoleucine (58%) + D-[G-^{3}H]Alloisoleucine (42%) + L-[U-^{14}C]Isoleucine	Senecionine [2% retention of ^{3}H] These results indicated that the senecic acid is formed from L-isoleucine and none of its isomers.	+	
Serratia marcescens			
Most of the L-[U-^{14}C]-amino-acids found in casein hydrolysate were fed.	Prodigiosin 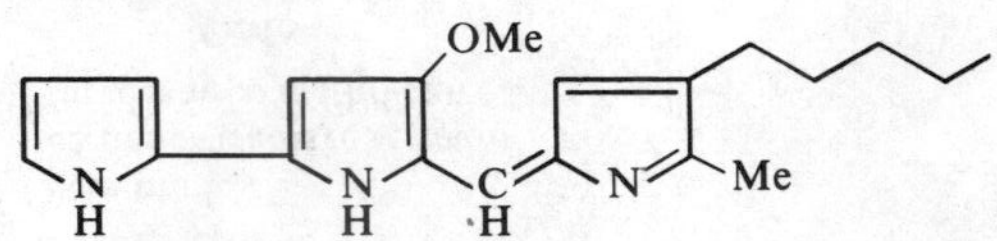 Proline (1.63% inc.) and alanine (0.53% inc.) were significant precursors. For some reason these authors ignored completely the elegant and definitive work of Wasserman and co-workers[108] on the biosynthesis of prodigiosin.		107

Species / *Compounds Administered*	*Alkaloids Labelled*	*Incorporation* (%)	*Ref.*
Skimmia japonica			
[2,4-$^{14}C_2$]-4-Hydroxy-2-quinolone	Dictamnine (see under *Choisya ternata*)	1.3	35*d*
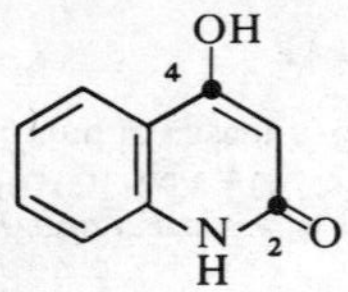	Platydesmine metho-salt 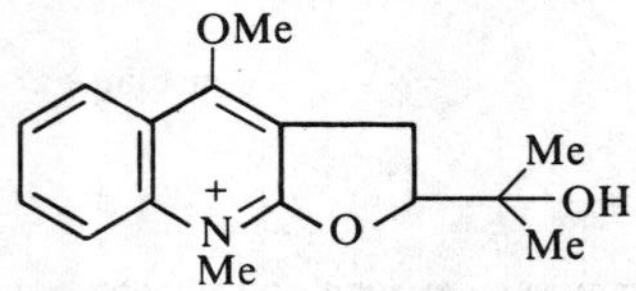	2.0	
[fed as a solution in Na_2CO_3 for 3 days to excised shoots]	Skimmianine (see under *C. ternata*)	0.6	
[2,4-$^{14}C_2$]-4-Methoxy-2-quinolone	Dictamnine	1.5	
	Platydesmine metho-salt	2.1	
[fed as a dispersion in $CHCl_3$–H_2O to excised shoots]			
[2,4-$^{14}C_2$]-2,4-Dimethoxyquinoline	Dictamnine	inactive	
4-Hydroxy-3-(3-methyl-[1′-^{14}C] but-2-enyl)-2-quinoline	Dictamnine [> 91% at C-3]	3.8	
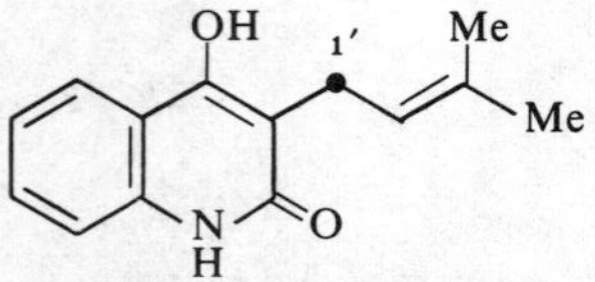	Platydesmine metho-salt	4.7	
4-Methoxy-3-(3-methyl-[1′-^{14}C] but-2-enyl)-2-quinolone	Dictamnine	3.6	
	Platydesmine metho-salt	4.8	
4-Methoxy-3-(3-methyl-[1′-^{14}C] but-2-enyl)-2-quinolone + *cold* platydesmine	Dictamnine	0.82	
	Platydesmine (not a normal alkaloid of this species)	0.86	

RS-[3-^{14}C]Platydesmine 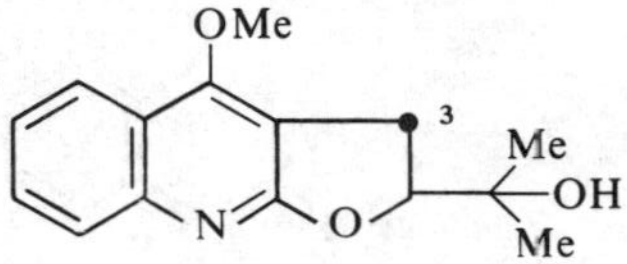 [fed in DMSO + H_2O]	Dictamnine [C-3: 100%. The reported inc. is probably a minimum value since it seems likely that only one isomer is utilized]	18.8	
	Platydesmine metho-salt	4.3	
	Skimmianine	0.1	
RS-[3-^{14}C]Platydesmine methiodide	Dictamnine [The metho-salt is thus not a major intermediate in the biosynthesis]	0.03	
[3-^{14}C]-2,3-Dihydrodictamnine	Dictamnine	0.1	
Skimmia species (unspecified) This species contained skimmianine and dictamnine in the ratio of 100:9.7			
[3-^{14}C]Dictamnine	Skimmianine	1.8—2.4	35*b*
Solanum laciniatum			
[2-^{14}C]Acetate [fed to detached leaves]	Solanine 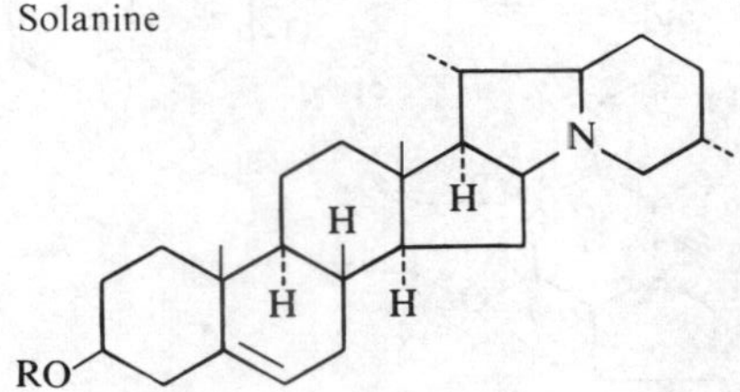 R = D-galactose (D-glucose, L-rhamnose)	0.085	109
RS-[2-^{14}C]Mevalonic acid	Solanine	5.33	

Species *Compounds Administered*	*Alkaloids Labelled*	*Incorporation(%)*	*Ref.*
Sphacelia sorghi			
RS-[2-^{14}C,5-$^{3}H_2$]Mevalonic acid	Dihydroergosine	1.1 (^{14}C)	110
	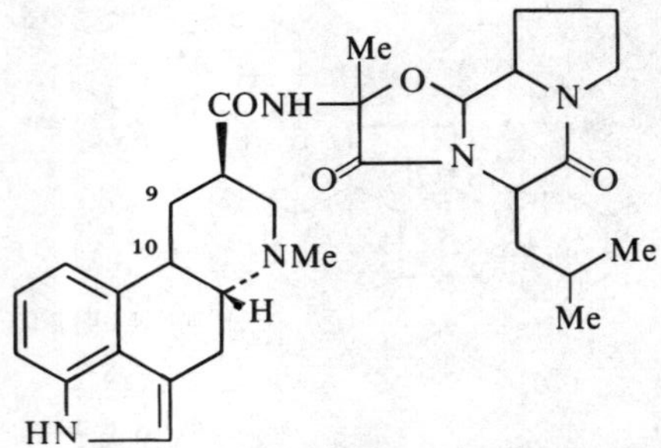 [45% retention of ^{3}H. Since C-5 of mevalonic acid becomes C-10 of the ergoline skeleton, the C-9–C-10 unsaturated alkaloids cannot be intermediates in this biosynthesis]		
[labelled ?]Agroclavine (see under *Claviceps* Strain SD 58)	Dihydroergosine	1.8	
[?]Tryptophan	Dihydroergosine	11	
[?]Festuclavine 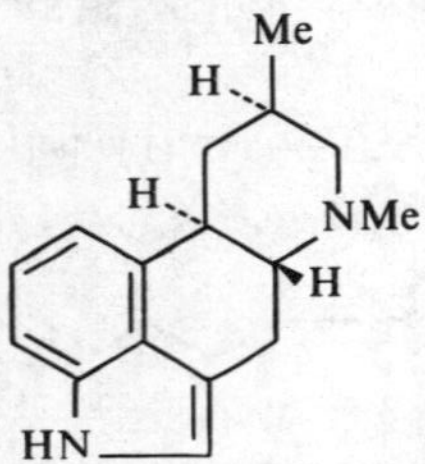	Dihydroergosine	31	

[?]Dihydroelymoclavine 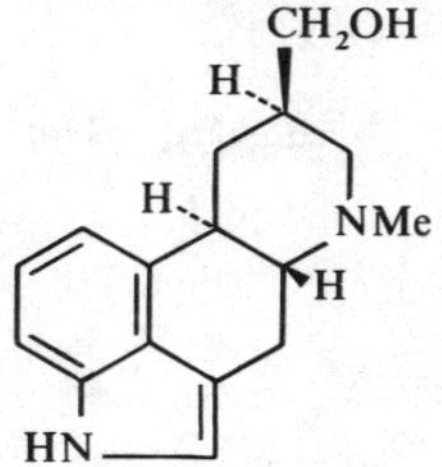	Dihydroergosine	33	
[?]Dihydrolysergic acid 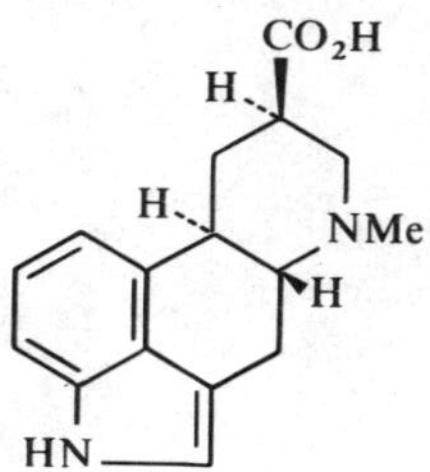	Dihydroergosine	32	
Sphaerophysa salsula			
DL-[2-^{14}C]Arginine + L-[*amidine*-^{14}C]Arginine [56.6% on the amidine carbon]	Spherophysine 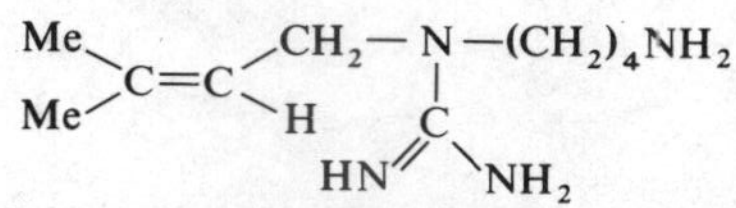 [56.4% on the amidine carbon]	0.21	111

Species / *Compound Administered*	*Alkaloids Labelled*	*Incorporation(%)*	*Ref.*
Stephania japonica			
DL-[2-^{14}C]Tyrosine	Hasubanonine 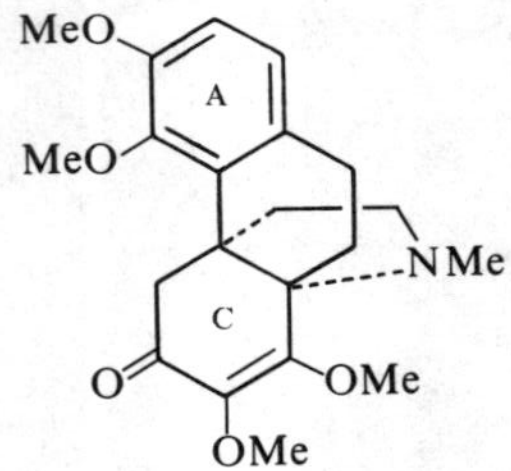 [Activity in ring C ethanamine bridge: 53%]	0.64—3.1	112
	Protostephanine 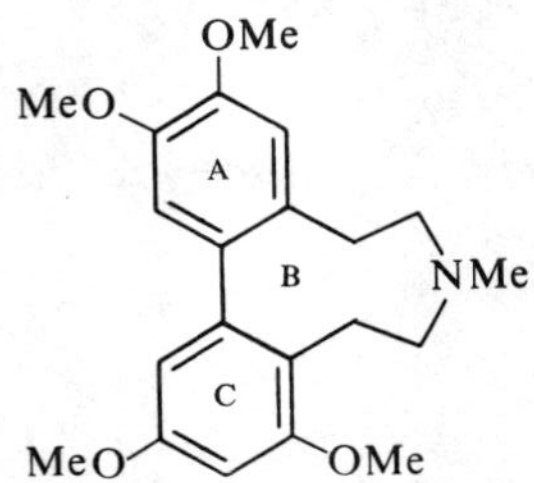 [Activity in ring C ethanamine bridge: 57%]	0.028—0.18	
DL-[2-^{14}C]Dopa	Hasubanonine [98% of activity on ring C ethanamine bridge]	0.02	
	Protostephanine [92% on ring C ethanamine bridge]	0.008	
[2-^{14}C]Tyramine	Hasubanonine [99% on ring C ethanamine bridge]	1.2	

	Protostephanine [93% on ring C ethanamine bridge]	0.005
[2-^{14}C]Dopamine	Hasubanonine [99% on ring C ethanamine bridge]	0.17
	Protostephanine [94% on ring C ethanamine bridge]	0.005
[1-^{14}C]-3′,4′,5′-Trihydroxyphenethylamine	Hasubanonine [99% on ring C ethanamine bridge]	1.0
	Protostephanine	0.017
[1-^{14}C]-3′,4′-Dihydroxy-5-methoxyphenethylamine	Hasubanonine [99% on ring C ethanamine bridge]	1.1
	Protostephanine	0.011
[1-^{14}C]-4′-Hydroxy-3′,5′-dimethoxyphenethylamine	Hasubanonine	<0.002
	Protostephanine	<0.0018
[1-^{14}C]-3′-Hydroxy-4′,5′-dimethoxyphenethylamine	Hasubanonine	<0.002
	Protostephanine	<0.0018
[3-^{14}C]Norlaudanosoline	Hasubanonine + protostephanine	neg. inc.
[3-^{14}C]Norlaudanosoline-1-carboxylic acid	Hasubanonine + protostephanine	neg. inc.
[3-^{14}C]-1,2-Dehydronorlaudanosoline	Hasubanonine + protostephanine	neg. inc.
[1-^{14}C]-Bis-(3′,4′-dihydroxyphenethyl)amine	Hasubanonine + protostephanine	neg. inc.

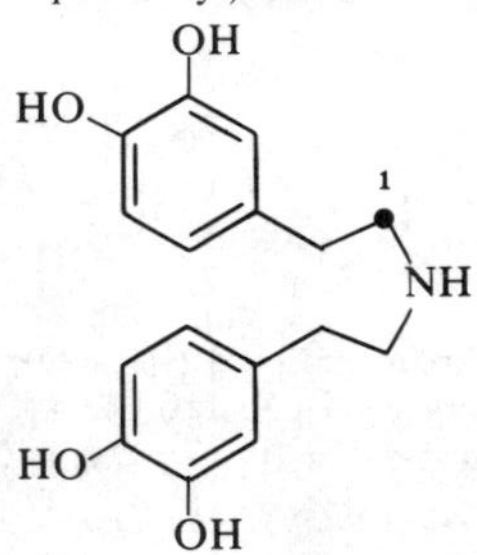

Species / Compound Administered	Alkaloids Labelled	Incorporation (%)	Ref.
$[3\text{-}^{14}C]$-1-(3-Hydroxy-4-methoxy-benzyl-6-hydroxy-7,8-dimethyl 2-methyl-1,2,3,4-tetrahydro-isoquinoline 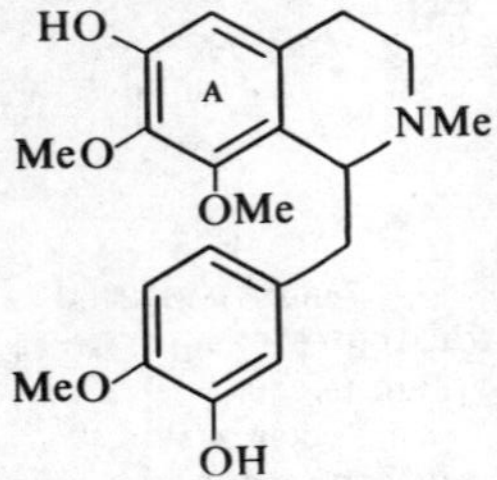	Hasubanonine + protostephanine [several isomeric methyl ethers with a different pattern of methylation in ring A also failed to afford significantly labelled alkaloids]	neg. inc.	
Streptomyces hygroscopicus var. *geldanus* var. *nova*			
[*methyl*-^{14}C]Methionine	Geldanamycin (see p. 208)	[22.7]	113
[*methyl*-^{13}C]Methionine	Geldanamycin	1.16 [11.1]	
Sodium $[1\text{-}^{14}C]$propionate	Geldanamycin	3.7	
Sodium $[1\text{-}^{13}C]$propionate	Geldanamycin	0.6 [13.3]	
Sodium $[1\text{-}^{14}C]$acetate	Geldanamycin	0.90	
Sodium $[1,3\text{-}^{14}C_2]$malonate	Geldanamycin	0.27	
Sodium $[^{14}C]$formate	Geldanamycin	0.04	
Streptomyces refuineus var. *thermotolerans*			
L-$[7a\text{-}^{14}C]$Tryptophan 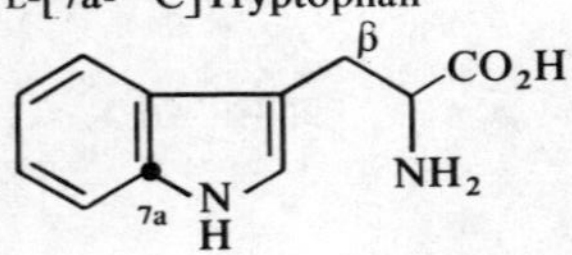	Anthramycin 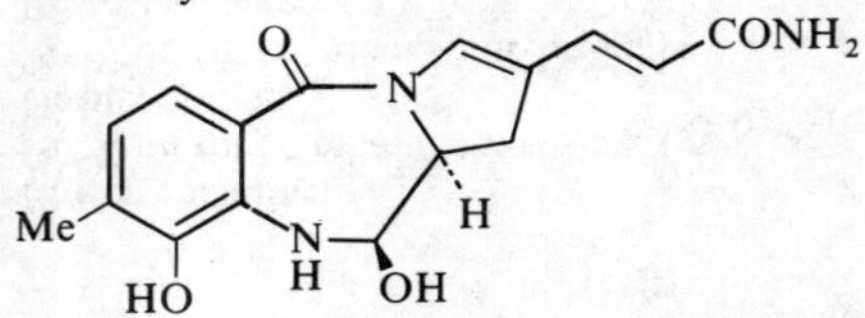	13.4	114

DL-[β-^{14}C]Tryptophan	Anthramycin	<0.1
L-[*methyl*-^{14}C]Methionine	Anthramycin	20.6
	[C-Me: 42%, the rest of the activity is postulated to be on the carboxamide group]	
L-[*methyl*-^{14}C,^{3}H$_3$]Methionine	Anthramycin	+
	[C-Me: 44%, the essentially all the ^{3}H retained]	
L-[*U*-^{14}C]Tyrosine	Anthramycin	9.9
	[No activity in the 4-methyl-3-hydroxy-anthranilic acid part]	
L-[1-^{14}C,3′,5′-^{3}H$_2$]Tyrosine	Anthramycin	+
	[50% retention of ^{3}H]	
L-[1-^{14}C]Dopa	Anthramycin	14.8
L-[*U*-^{14}C]Phenylalanine	Anthramycin	<0.1
D-[1-^{14}C]Glucose	Anthramycin	<0.1
D-[6-^{14}C]Glucose	Anthramycin	<0.1
Sodium [1-^{14}C]acetate	Anthramycin	<0.1
L-[*U*-^{14}C]Proline	Anthramycin	<0.1
[4-^{14}C]-δ-Aminolevulinic acid	Anthramycin	<0.1

Streptomyces verticillatus

Mitomycin A

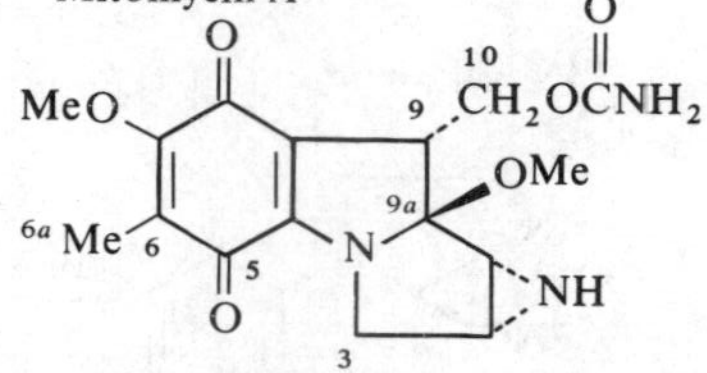

Mitomycin B

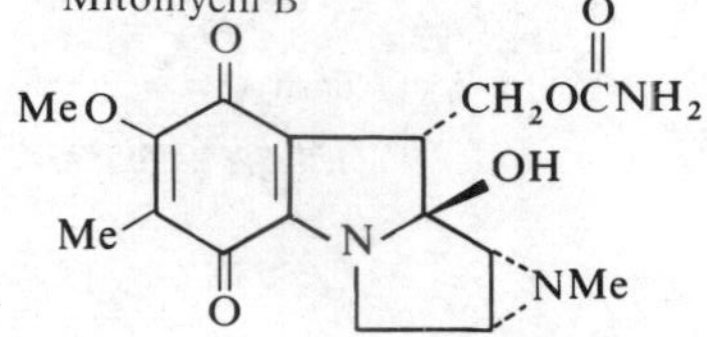

Species Compound Administered	Alkaloids Labelled	Incorporation(%)	Ref.
	Mitomycin C 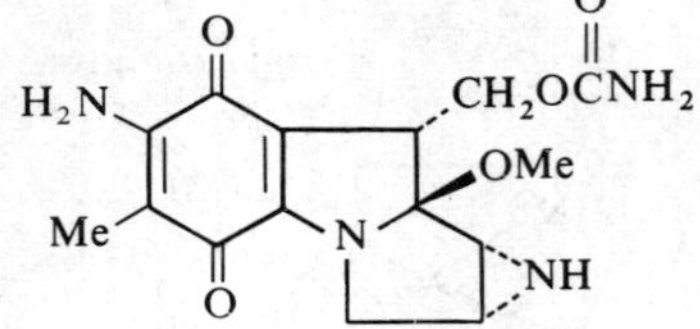 Porfiromycin 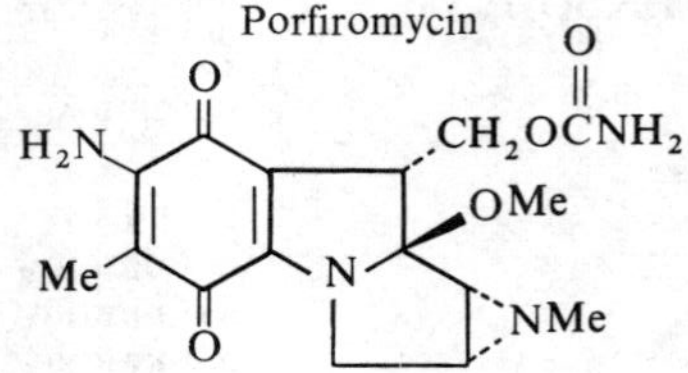		
D-[1-^{14}C, 1-^{13}C, ^{15}N]Glucosamine 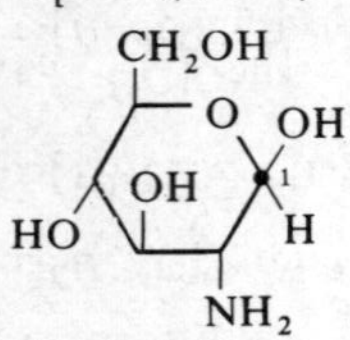	Mixture of Mitomycins Mitomycin B [Mass spectrometry indicated that the ^{13}C was at C-3. It has previously been established that the bulk of the nitrogen from glucosamine enters the aziridine ring.[116]]	1.8 (^{14}C) ^{14}C: [41.2] ^{13}C: [36.9]	115
D-[6-^{14}C] Glucosamine	Mixture of Mitomycins Mitomycin C [C-10: 72%]	5.1	

L-[*ureido*-^{14}C]Citrulline	Mitomycins [most of the activity in the carbamoyl group]	8.2	
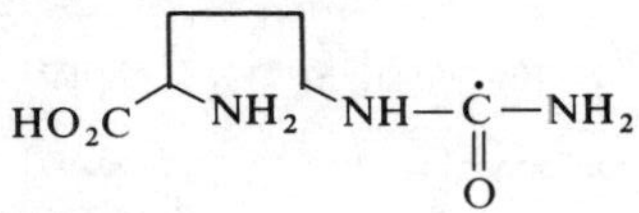			
L-[*ureido*-^{14}C]Citrulline + non-labelled arginine	Mitomycins	10.4	
L-[*guanido*-^{14}C]Arginine	Mitomycins	5.7	
L-[*guanido*-^{14}C]Arginine + non-labelled citrulline	Mitomycins [Most of the activity in the carbamoyl group]	5.8	
The following compounds were fed along with unlabelled D-glucosamine to suppress their incorporation into the C_6N unit (C-3,2,1,9a,9,10)			
D-[1-^{14}C]Glucose	Mitomycins Mitomycin C [C-6 + C-6a: 4.6%]	0.7	117
D-[2-^{14}C]Glucose	Mitomycins Mitomycin C [C-6: 21.2, C-6a: 5.9%]	0.8	
D-[3,4-$^{14}C_2$]Glucose	Mitomycins Mitomycin C [C-6: 22.4, C-6a: 13.8%]	0.3	
D-[6-^{14}C]Glucose	Mitomycins Mitomycin C [C-6 + C-6a: 0.4%]	0.8	
[1-^{14}C]Pyruvate	Mitomycins Mitomycin C [C-6: 0, C-6a: 28.6%]	0.2	
[2-^{14}C]Pyruvate	Mitomycins Mitomycin C [C-6: 30.2, C-6a: 2.6%]	0.2	
[3-^{14}C]Pyruvate	Mitomycins Mitomycin C [C-6 + C-6a: 2.0%]	0.5	

Species / *Compound Administered*	*Alkaloids Labelled*	*Incorporation*(%)	*Ref.*
D-[1-^{14}C,4-^{3}H]Glucosamine	Mitomycins [35% retention of ^{3}H]	4.0 (^{14}C)	118
DL-[*guanido*-^{14}C]Arginine	Mitomycins	8.0	
L-[*ureido*-^{14}C]Citrulline	Mitomycins	11.2	
Potassium [^{14}C]cyanate	Mitomycins	0.1	
D-[5-^{3}H]Ribose	Mitomycins	0.3	
L-[*U*-^{14}C]Arabinose	Mitomycins	0.01	
L-[1-^{14}C]Arabitol	Mitomycins	0.01	
D-[1-^{14}C]Altroheptulose	Mitomycins	0.05	
Sodium [1-^{14}C]acetate	Mitomycins	0.1	
Sodium [2-^{14}C]propionate	Mitomycins	0.3	
Sodium L-[4-^{14}C]malate	Mitomycins	0.2	
Sodium [2,3-$^{14}C_2$]succinate	Mitomycins	0.3	
Vinca rosea			
[*aryl*-^{3}H]Secodine [see under *Aspidosperma pyricollum*]	Catharanthine 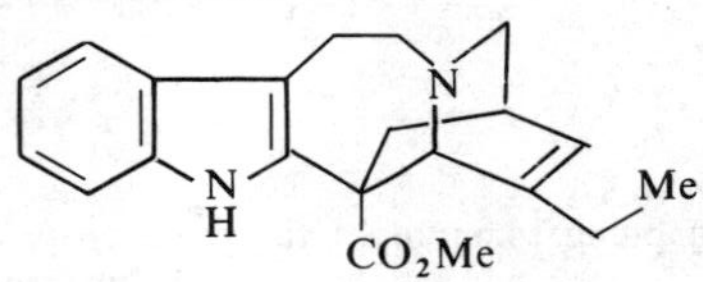	0.002—0.005	119
[*carbonyl*-^{14}C]Secodine	Catharanthine [all activity on the carbonyl group]	0.006	
[*carbonyl*-^{14}C, 3,14,15,21-$^{3}H_4$]-Secodine	Catharanthine [38% retention of ^{3}H]	+	
[*carbonyl*-^{14}C, *aryl*-^{3}H]Secodine	Catharanthine [86% retention of ^{3}H]	0.0008 (^{14}C)	
[*carbonyl*-^{14}C, 19-^{3}H]Secodine	Catharanthine [102% retention of ^{3}H, all ^{14}C on the carbonyl group]	0.001	

Precursor	Product	Incorporation	Ref.
[*methoxy*-^{14}C,6,6,8-$^3H_3$]Loganin 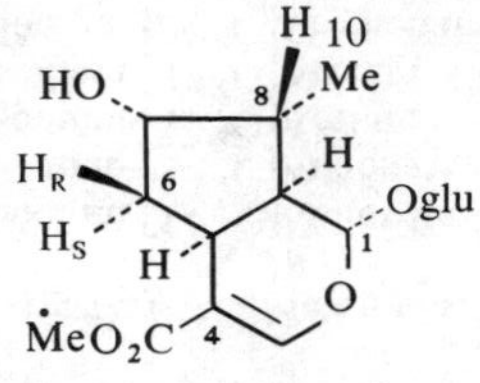 [^{3}H distribution: C-8 44%, C-6 *pro-R* 44%, C-6 *pro-S* 11%]	Vincaleukoblastine (VLB) 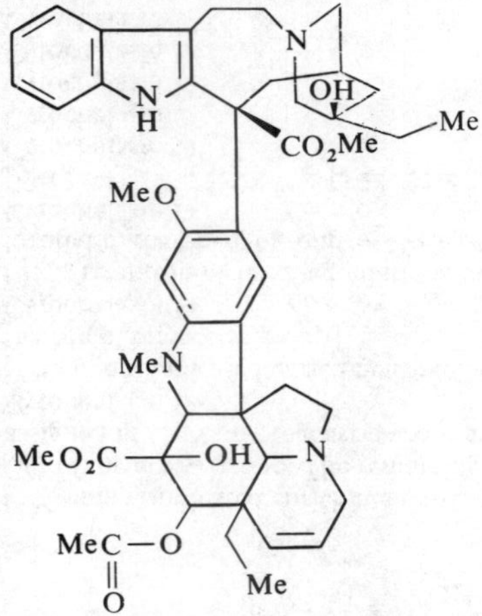 [91.5% retention of ^{3}H, all ^{14}C on the carbomethoxy-groups]	[0.069]	120
	Catharanthine [106% retention of ^{3}H]	[0.39]	
	Vindoline 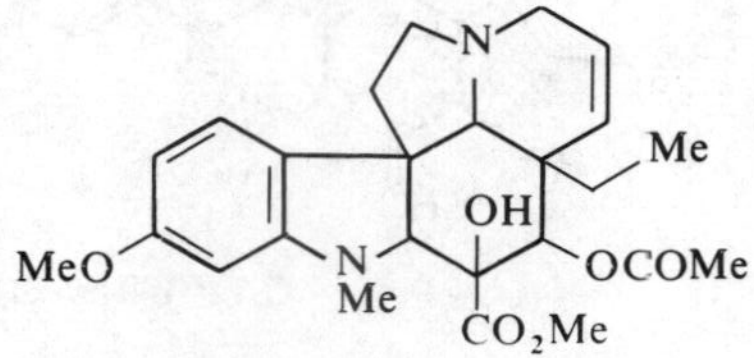 [73% retention of ^{3}H]	[0.26]	

Species / *Compounds Administered*	*Alkaloids Labelled*	*Incorporation(%)*	*Ref.*
Zygosporium masonii			
[1-^{13}C]Acetate	Cytochalasin D	0.86	100

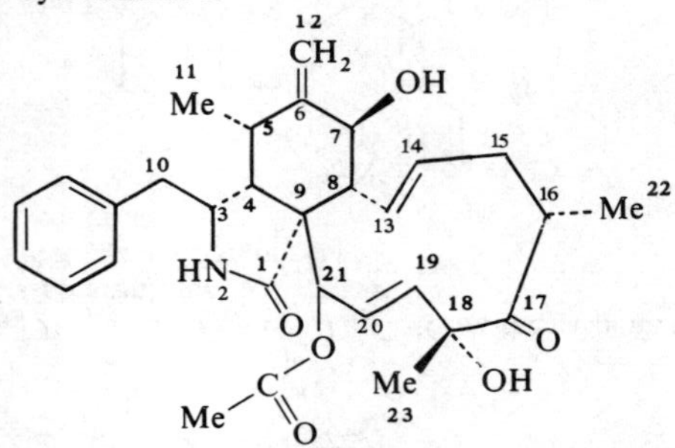

	[^{13}C-n.m.r. indicated enrichment at C-1,5,7,13,15,17,21, and the CO of the acetyl group. The remaining enriched peak was assigned to C-19 on biogenetic grounds]		
[2-^{14}C]Acetate	Cytochalasin D [^{13}C-n.m.r. indicated enrichment at C-6,8,9,11,14,16,18,20, and the Me of the acetyl group]	1.5	
L-[*U*-^{14}C,4′-^{3}H]Phenylalanine	Cytochalasin D [95% retention of ^{3}H. Degradations consistent with the ^{14}C being located in the phenyl group + C-10,3,4]	14.1 (^{14}C)	121
L-[*methyl*-^{14}C]Methionine	Cytochalasin D [C-12:34%, C-22: 28.7%, C-23: 31.2%]	16.2	
Sodium [1-^{14}C]Malonate	Cytochalasin D	0.125	
Sodium [2-^{14}C]Malonate	Cytochalasin D	0.244	
Sodium [1-^{14}C]Acetate	Cytochalasin D	2.58	
Sodium [2-^{14}C]Acetate	Cytochalasin D	1.37	
Sodium [2-^{14}C]Propionate	Cytochalasin D [It is assumed that this precursor is converted into [1-^{14}C]acetate prior to incorporation]	1.80	

[44] D. G. O'Donovan and P. B. Creedon, *J.C.S. Perkin I*, 1974, 2524.
[45] M. Y. Lovkova, E. Nurimov, and G. S. Il'in, *Biokhimiya*, 1974, **39**, 388.
[46] M. Y. Lovkova, E. Nurimov, and G. S. Il'in, *Biokhimiya*, 1974, **39**, 523.
[47] E. Nurimov, M. Y. Lovkova, and B. A. Abdusalamov, *Priklad. Biokhim. Mikrobiol.*, 1974, **10**, 785.
[48] R. Cardillo, C. Fuganti, G. Gatti, D. Ghiringhelli, and P. Grasselli, *Tetrahedron Letters*, 1974, 3163.
[49] A. J. Birch, C. E. Blance, S. David, and H. Smith, *J. Chem. Soc.*, 1961, 3138.
[50] J. P. Kutney, J. F. Beck, C. Ehert, G. Poulton, R. S. Sood, and N. D. Westcott, *Bioorganic Chem.*, 1971, **1**, 194.
[51] J. P. Kutney, J. F. Beck, and G. B. Fuller, *Heterocycles*, 1973, **1**, 5.
[52] A. G. McInnes, D. G. Smith, J. A. Walter, L. C. Vining, and J. L. C. Wright, *J.C.S. Chem. Comm.*, 1974, 282.
[53] R. B. Herbert, F. G. Holliman, and P. N. Ibberson, *Tetrahedron Letters*, 1974, 151.
[54] U. Hollstein and D. A. McCarrey, *J. Org. Chem.*, 1973, **38**, 3415.
[55] S. Sciuto, G. Oriente, M. Piattelli, G. Ompellizzeri, and V. Amico, *Phytochemistry*, 1974, **13**, 947.
[56] H. Kluender, F. C. Huang, A. Fritzberg, H. Schnoes, C. J. Sih, P. Fawcett, and E. P. Abraham, *J. Amer. Chem. Soc.*, 1974, **96**, 4054.
[57] (*a*) N. Neuss, C. H. Nash, J. E. Baldwin, P. A. Lemke, and J. B. Grutzner, *J. Amer. Chem. Soc.*, 1973, **95**, 3797, 6511; (*b*) H. Kluender, C. H. Bradley, C. J. Sih, P. Fawcett, and E. P. Abraham, *ibid.*, p. 6150.
[58] H. G. Floss, M. Lin-Tcheng, H. Kobel, and P. Stadler, *Experientia*, 1974, **40**, 1369.
[59] W. Maier, D. Erge, and D. Gröger, *Biochem. Physiol. Pflanz.*, 1974, **165**, 479.
[60] E. Looser, T. W. Bauman, and H. Wanner, *Phytochemistry*, 1974, **13**, 2515.
[61] M. F. Roberts, *Phytochemistry*, 1974, **13**, 1841.
[62] G. Blaschke, G. Waldheim, M. von Schantz, and P. Peura, *Arch. Pharm.*, 1974, **307**, 122.
[63] D. H. R. Barton, A. J. Kirby, and G. W. Kirby, *Chem. Comm.*, 1965, 52; *J. Chem. Soc.*, (*C*), 1968, 929.
[64] C. Tani and K. Tagahara, *Chem. and Pharm. Bull.*, 1974, **22**, 2457.
[65] H. L. Holland, M. Castillo, D. B. MacLean, and I. D. Spenser, *Canad. J. Chem.*, 1974, **52**, 2818.
[66] D. J. Robins, N. M. Bale, and D. H. G. Crout, *J.C.S. Perkin I*, 1974, 2082.
[67] E. Nowaki and R. U. Byerrum, *Life Sci.*, 1962, **5**, 157.
[68] D. S. Bhakuni, S. Satish, H. Uprety, and R. S. Kapil, *Phytochemistry*, 1974, **13**, 2767.
[69] S. Agurell, I. Granelli, K. Leander, B. Luning, and J. Rosenblom, *Acta Chem. Scand.* (*B*), 1974, **28**, 239.
[70] S. Agurell, I. Granelli, K. Leander, and J. Rosenblom, *Acta Chem. Scand.* (*B*), 1974, **28**, 1175.
[71] P. J. Beresford and J. G. Woolley, *Phytochemistry*, 1974, **13**, 2511.
[72] E. Leete and E. P. Kirven, *Phytochemistry*, 1974, **13**, 1501.
[73] H. Rosenberg and S. J. Stohs, *Lloydia*, 1974, **37**, 313.
[74] H. Rosenberg, S. J. Stohs, and A. G. Paul, *Phytochemistry*, 1974, **13**, 823.
[75] R. R. Arndt, C. W. Holzapfel, N. P. Ferreira, and J. J. March, *Phytochemistry*, 1974, **13**, 1865.
[76] P. Pfäffli and Ch. Tamm, *Helv. Chim. Acta*, 1969, **52**, 1921.
[77] J. Steiniger and G. Reuter, *Biochem. Physiol. Pflanz.*, 1974, **166**, 275.
[78] D. Gross, H. Lehmann, and H. R. Schütte, *Biochem. Physiol. Pflanz.*, 1974, **166**, 281.
[79] A. R. Battersby, P. Böhler, M. H. G. Munro, and R. Ramage, *J.C.S. Perkin I*, 1974, 1399.
[80] S. Ramakrishna and A. P. Radhakantha, *Phytochemistry*, 1974, **13**, 2161.
[81] S. Tewari, D. S. Bhakuni, and R. S. Kapil, *J.C.S. Chem. Comm.*, 1974, 940.
[82] D. G. O'Donovan and E. Barry, *J.C.S. Perkin I*, 1974, 2528.
[83] H. Rosenberg and S. J. Stohs, *Phytochemistry*, 1974, **13**, 1861.
[84] J. G. Roddick, *Phytochemistry*, 1974, **13**, 1459.
[85] D. Neumann and E. Müller, *Biochem. Physiol. Pflanz.*, 1974, **165**, 271.
[86] E. J. Shellard and P. J. Houghton, *Planta Medica*, 1974, **25**, 80.
[87] C. Fuganti, D. Ghiringhelli, P. Grasselli, and P. Mazza, *Tetrahedron Letters*, 1974, 2261.
[88] C. Fuganti, D. Ghiringhelli, and P. Grasselli, *J.C.S. Chem. Comm.*, 1974, 350.
[89] B. P. Mundy, K. B. Lipkowitz, M. Lee, and B. R. Larsen, *J. Org. Chem.*, 1974, **39**, 1963.
[90] A. Karlsson, G. Satori, and R. J. White, *Eur. J. Biochem.*, 1974, **47**, 251.
[91] R. J. White, E. Martinelli, and G. Lancini, *Proc. Nat. Acad. Sci. U.S.A.*, 1974, **71**, 3260.
[92] C. Fuganti, *Gazzetta*, 1973, **103**, 1255.
[93] A. R. Battersby and R. Kazlauskas, unpublished work referred to in ref. 112.
[94] J. W. Fairbairn, F. Hakim, and Y. El Kheir, *Phytochemistry*, 1974, **13**, 1133.
[95] L. Copeland and M. Slaytor, *Physiol. Plantarum*, 1974, **31**, 327.
[96] L. Nettleship and M. Slaytor, *Phytochemistry*, 1974, **13**, 735.
[97] J. Baldas, A. J. Birch, and R. A. Russell, *J.C.S. Perkin I*, 1974, 50.
[98] D. J. Aberhart and L. J. Lin, *J.C.S. Perkin I*, 1974, 2320.
[99] D. J. Aberhart, J. Y.-R. Chu, N. Neuss, C. H. Nash, J. Occolowitz, L. L. Huckstep, and N. De La Higuera, *J.C.S. Chem. Comm.*, 1974, 564.
[100] W. Graf, J. L. Robert, J. C. Vederas, Ch. Tamm, P. H. Solomon, I. Miura, and K. Nakanishi, *Helv. Chim. Acta*, 1974, **57**, 1801.
[101] P. J. Beresford and J. G. Woolley, *Phytochemistry*, 1974, **13**, 2143.

[102] P. W. Jeffs, H. F. Campbell, D. S. Farrier, and G. Molina, *Chem. Comm.*, 1971, 228.
[103] P. W. Jeffs, H. F. Campbell, D. S. Farrier, G. Ganguli, N. H. Martin, and G. Molina, *Phytochemistry*, 1974, **13**, 933.
[104] P. Korzan and T. J. Gilbertson, *Phytochemistry*, 1974, **13**, 435.
[105] N. M. Davis and D. H. G. Grout, *J.C.S. Perkin I*, 1975, 2079.
[106] D. H. G. Crout, N. M. Davis, E. H. Smith, and D. Whitehouse, *J.C.S. Perkin I*, 1972, 671.
[107] S. M. H. Quadri and R. P. Williams, *Canad. J. Microbiol.*, 1974, **20**, 461.
[108] H. H. Wasserman, R. J. Sykes, P. Peverada, C. K. Shaw, R. J. Cushley, and S. R. Lipsky, *J. Amer. Chem. Soc.*, 1973, **95**, 6874.
[109] A. R. Guseva, V. A. Paseshnichenko, and M. G. Borikhina, *Priklad. Biokhim. Mikrobiol.*, 1973, **9**, 764.
[110] K. D. Barrow, P. G. Mantle, and F. R. Quigley, *Tetrahedron Letters*, 1974, 1557.
[111] J. Steiniger and G. Reuter, *Pharmazie*, 1974, **29**, 422.
[112] A. R. Battersby, R. C. F. Jones, R. Kazlauskas, C. Poupat, C. W. Thornber, S. Ruchirawat, and J. Staunton, *J.C.S. Chem. Comm.*, 1974, 773.
[113] R. D. Johnson, A. Haber, and K. L. Rinehart, *J. Amer. Chem. Soc.*, 1974, **96**, 3316.
[114] L. H. Hurley and M. Zmijewski, *J.C.S. Chem. Comm.*, 1974, 337.
[115] U. Hornemann, J. P. Kehrer, C. S. Nunez, and R. L. Ranieri, *J. Amer. Chem. Soc.*, 1974, **96**, 320.
[116] (*a*) U. Hornemann and J. C. Cloyd, *J.C.S. Chem. Comm.*, 1971, 301; (*b*) U. Hornemann and M. J. Aikman, *J.C.S. Chem. Comm.*, 1973, 88.
[117] U. Hornemann, J. P. Kehrer, and J. E. Eggert, *J.C.S. Chem. Comm.*, 1974, 1045.
[118] U. Hornemann, J. P. Kehrer, C. S. Nunez, R. L. Ranieri, and Y. K. Ho, *Develop. Ind. Microbiol.*, 1974, **15**, 82.
[119] J. P. Kutney and N. J. Eggers, *Heterocycles*, 1973, **1**, 11.
[120] P. E. Daddona and C. R. Hutchinson, *J. Amer. Chem. Soc.*, 1974, **96**, 6806.
[121] C. R. Lebet and Ch. Tamm, *Helv. Chim Acta*, 1974, **57**, 1785.

6
Non-protein Amino-acids, Cyanogenic Glycosides, and Glucosinolates

BY A. KJÆR AND P. OLESEN LARSEN

1 Introduction

The reasons for treating non-protein amino-acids, cyanogenic glycosides, and glucosinolates under one heading were given in a previous Report in this series[1] and are not repeated here. As before, main emphasis is placed on constituents of higher plants, but products of microbial origin are commented upon wherever the context is appropriate. Biosynthetic pathways to protein amino-acids are included only insofar as they are obviously related to those traversed in the *in vivo* synthesis of the non-protein amino-acids. The delineation from alkaloids (Chapter 5) is, as always, difficult and somewhat artificial.

This Report covers contributions which appeared during 1973 and 1974; some background material is presented and a few references to papers published in early 1975 are included.

2 Non-protein Amino-acids

General.—The field of non-protein amino-acids is characterized today by an increasing interest in the enzymological aspects, including those concerning enzyme specificity. In addition to the interest in the anabolic relationship to protein amino-acids, much attention seems centred on the catabolism of amino-acids and on the ecological importance of non-protein amino-acids. Reviews have appeared which cover the general biosynthesis of protein amino-acids in plants[2] and the control of the biosynthesis of non-protein amino-acids, especially those with branched carbon chains[3] (see p. 190).

Serine and Cysteine Derivatives.—In the previous Report[1] the general importance of the pathway from serine to *O*-acetylserine and thence to a number of amino-acids including cysteine was stressed. This has been further emphasized by subsequent work. L-Serine acetyltransferase has been isolated from germinating seeds of *Brassica napus* L. The enzyme is different from the enzyme previously described from kidney-bean seedlings. *O*-Acetyl-L-serine sulphydrylase (cysteine synthase) has been isolated from germinating seeds of a number of plants including several *Brassica* species.[4] The latter enzyme has also been partly purified from *Allium cepa*.[5] The enzyme from *Allium*,

[1] A. Kjær and P. Olesen Larsen, in 'Biosynthesis', ed. T. A. Geissman (Specialist Periodical Reports), The Chemical Society, London, Vol. 2, 1973, p. 171.

[2] B. J. Miflin, in 'Biosynthesis and its Control in Plants', ed. B. V. Milborrow, Academic Press, London and New York, 1973, p. 49.

[3] L. Fowden, in ref. 2, p. 323.

[4] T. T. Ngo and P. D. Shargool, *Canad. J. Biochem.*, 1974, **52**, 435.

[5] B. Granroth, *Acta Chem. Scand.* (*B*), 1974, **28**, 813.

like the enzyme from *Escherichia coli*, is unspecific with regard to thiol, catalysing reactions not only with hydrogen sulphide, but also with MeSH, EtSH, $CH_2:CH\text{-}CH_2\cdot SH$, BuSH, and $PhCH_2\cdot SH$. Thus, lack of specificity may explain the previously observed recycling of side-chains in substituted cysteines.[6]

It may be of interest in this connection that in yeast *O*-acetylserine sulphydrylase is identical with *O*-acetylhomoserine sulphydrylase,[7] whereas *O*-acetylhomoserine is a poor substrate for the plant enzyme.[4]

The action of cysteine desulphydrase from *Aerobacter aerogenes* has been shown to be reversible, so that L-cysteine can be formed from hydrogen sulphide, pyruvic acid, and ammonia; MeSH, EtSH, or PrSH may be substituted for hydrogen sulphide.[8]

γ-L-Glutamyl-*S*-(*trans*-1-propenyl)-L-cysteine sulphoxide has been isolated from *Santalum album* L. The configuration at the sulphur atom is opposite to that of the same compound isolated from *Allium*, lending further support to the operation of stereospecific sulphide oxidation in plants.[9]

The formation of 3-cyanoalanine from cysteine and cyanide has been demonstrated with an enzyme preparation from *Acia georginae*. When cyanide was not supplied, pyruvate was produced. 3-Cyanoalanine is not a normal constituent in *A. georginae* whereas other serine derivatives are present (albizziine, 3-acetyldiaminopropionic acid, djenkolic acid), suggesting low specificity of the enzymatic system in this plant. The enzyme was not able, however, to produce 3-fluoroalanine with fluoride ions, although this had previously been proposed as a possible first step in the biosynthesis of fluoroacetic acid. It should be noted, however, that the variety of *A. georginae* used contained no fluoroacetic acid whereas other varieties of the same species do.[10]

The derivation of diaminopropionic acid from *O*-acetylserine or cysteine and ammonia, though likely, has never been established in plants. It has, however, been demonstrated that diaminopropionic acid in the peptide antibiotic viomycin is derived from serine, although the individual steps are unknown.[11]

2-Alanyl-3-isoxazolin-5-one (1) is present in *Pisum sativum* seedlings,[12] together with β-(2-β-D-glucopyranosylisoxazolin-5-one-4-yl)alanine (2).[13] Recently, these two amino-acids have been found also in *Lathyrus odoratus*, together with 2-(2-cyanoethyl)-3-isoxazolin-5-one (3), 2-(β-D-glucopyranosyl)-3-isoxazolin-5-one (4), and 2-carboxymethyl-3-isoxazolin-5-one (5),[14] as well as 2-amino-4-(isoxazolin-5-one-2-yl)butyric acid (6), 2-aminoethylisoxazoline-5-one (7), and the γ-glutamyl derivative of the latter.[15] The formation of 2-alanyl-3-isoxazolin-5-one from *O*-acetylserine and isoxazolin-5-one is catalysed by extracts from seedlings of *Pisum sativum* and *P.*

[6] B. Granroth and A. Sarnesto, *Acta Chem. Scand.* (*B*), 1974, **28**, 814.
[7] S. Yamagata, K. Takeshima, and N. Naiki, *J. Biochem.* (*Japan*), 1974, **75**, 1221.
[8] H. Kumagai, Y. Choi, S. Sejima, and H. Yamada, *Biochem. Biophys. Res. Comm.*, 1974, **59**, 789.
[9] R. Kuttan, N. G. Nair, A. N. Radhakrishnan, T. F. Spande, H. J. Yeh, and B. Witkop, *Biochemistry*, 1974, **13**, 4394.
[10] R. J. Mead and W. Segal, *Phytochemistry*, 1973, **12**, 1977.
[11] J. H. Carter, R. H. Du Bus, J. R. Dyer, J. C. Floyd, K. C. Rice, and P. D. Shaw, *Biochemistry*, 1974, **13**, 1221.
[12] F. Lambein, N. Schamp, L. Vandendriessche, and R. Van Parijs, *Biochem. Biophys. Res. Comm.*, 1969, **37**, 375.
[13] F. Lambein and R. Van Parijs, *Biochem. Biophys. Res. Comm.*, 1970, **40**, 557.
[14] V. Van Rompuy, F. Lambein, R. De Gussem, and R. Van Parijs, *Biochem. Biophys. Res. Comm.*, 1974, **56**, 199.
[15] F. Lambein and R. Van Parijs, *Biochem. Biophys. Res. Comm.*, 1974, **61**, 155.

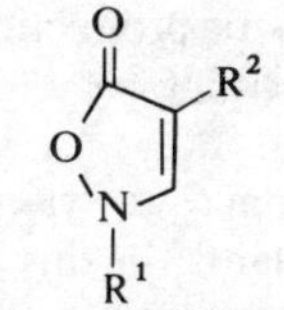

(1) $R^1 = CH_2CHNH_3CO_2^-$; $R^2 = H$
(2) R^1 = β-D-Glucopyranosyl; $R^2 = CH_2CH\overset{+}{N}H_3CO_2^-$
(3) $R^1 = (CH_2)_2CN$; $R^2 = H$
(4) R^1 = β-D-Glucopyranosyl; $R^2 = H$
(5) $R^1 = CH_2CO_2H$; $R^2 = H$
(6) $R^1 = (CH_2)_2CH\overset{+}{N}H_3CO_2^-$; $R^2 = H$
(7) $R^1 = (CH_2)_2NH_2$; $R^2 = H$

arvense.[16] It is proposed that 2-(β-D-glucopyranosyl)-3-isoxazolin-5-one is the precursor of β-(2-β-D-glucopyranosyl)isoxazolin-5-one-4-yl)alanine.[15] Nothing definite can be said about the origin of the side-chains in the other members of this group of compounds. It has been proposed that the compound with a C_4-side-chain is derived from the heterocycle and *O*-acetylhomoserine,[15] but chain elongation from the alanine derivative or derivation from azetidine-2-carboxylic acid are other possibilities (see below). No information is available on the origin of the heterocyclic system.

β-(3,5-Dioxo-1,2,4-oxazolidin-2-yl)-L-alanine (quisqualic acid) (8) has been isolated from seeds of *Quisqualis indica* L. and *Q. indica* L. var. *villosa* (Clarke) (Combretaceae). The formation of this amino-acid from *O*-acetylserine and 3,5-dioxo-1,2,4-oxazolidine is catalysed by extracts from *Q. indica* and from *Pisum sativum*, but not by those from *Leucaena* or *Citrullus* although these extracts catalyse the formation of mimosine (9) and β-pyrazol-l-ylalanine, respectively.[17] Information on the origin of the ring is not available. Experiments with addition of unlabelled β-pyrazol-l-ylalanine to root tips of *Cucumis melo*, synthesizing this amino-acid, indicated that the amino-acid did not act as a regulator of its own biosynthesis.[3]

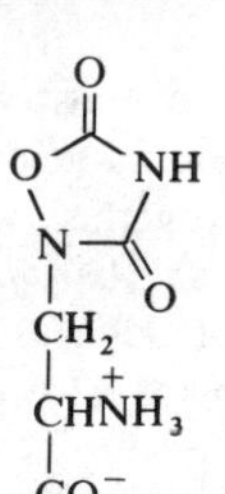

(8)

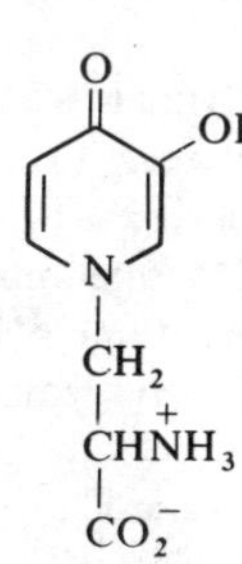

(9)

The production of 3-amino-1,2,4-triazol-1-ylalanine when the herbicide 3-amino-1,2,4-triazole is administered to plants is well known.[18] This reaction can take place

[16] I. Murakoshi, F. Kato, J. Haginiwa, and L. Fowden, *Chem. and Pharm. Bull.* (*Japan*), 1973, **21**, 918.
[17] I. Murakoshi, F. Kato, J. Haginiwa, and T. Takemoto, *Chem. and Pharm. Bull.* (*Japan*), 1974, **22**, 473.
[18] D. M. Frisch, P. M. Dunnill, A. Smith, and L. Fowden, *Phytochemistry*, 1967, **6**, 921.

also in *Escherichia coli.*[19] The enzymatic production of this amino-acid has now been demonstrated with extracts from seedlings of *Pisum sativum, Citrullus vulgaris,* and *Leucaena leucocephala.*[20]

The production of mimosine (9) from *O*-acetylserine and 3,4-dihydroxypyridine has previously been demonstrated in plants. In this connection it may therefore be of interest that 4-hydroxypyridine is converted into 3,4-dihydroxypyridine by an enzyme from an *Agrobacterium* sp.[21]

Willardiine [3-(2,4-dihydroxypyridine-1-yl)alanine] which can be produced from *O*-acetylserine and uracil has previously been found only in species of Leguminosae. Now this amino-acid, together with its γ-glutamyl derivative and γ-glutamylphenylalanylwillardiine, has been encountered in beech nuts (*Fagus silvatica* L.),[22,23] further extending the biological scope of the biosynthetic pathway to *N*-substituted alanines.

Ascorbalamic acid (10) is a compound widely occurring in higher plants, probably deriving from ascorbic acid and an alanine derivative.[24] *O*-Acetylserine seems a likely candidate for the alanine derivative indicating additional possibilities for the derivation of plant constituents from this compound. Ascorbalamic acid yields 3-(2-furoyl)-alanine (11) when treated with acid.[24] This latter amino-acid has recently been found in seeds of *Fagopyrum esculentum* Moench (buckwheat),[25] seeds of *Phaseolus radiatus,*[26] and in beech nuts,[22] and may therefore also be a further derivative from *O*-acetylserine.

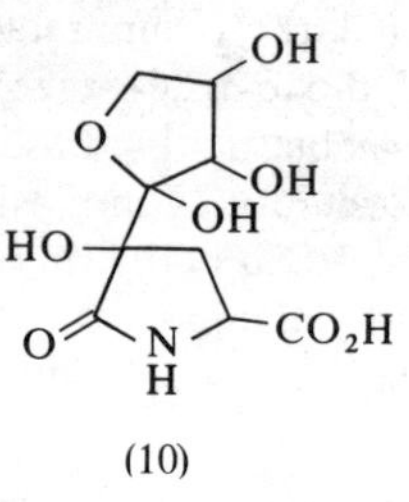

(10)

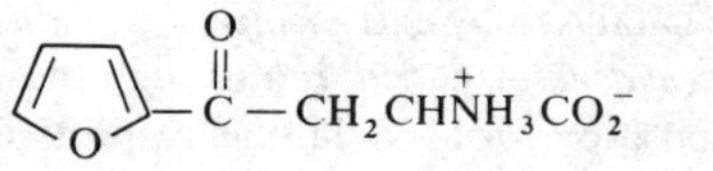

(11)

Selenium-containing Amino-acids.—Reviews on the chemistry and metabolism in plants and micro-organisms of selenium-containing amino-acids have appeared.[27,28] On incorporation of [^{75}Se]-L-selenomethionine, together with [*methyl*-^{3}H]-DL-selenomethionine, into selenomethylselenocysteine in *Astragalus bisulcatus,* more selenium radioactivity than ^{3}H-radioactivity is introduced, supporting previous findings that selenomethyl transfer does not take place. However, since one of the

[19] A. K. Williams, S. T. Cox, and R. G. Eagon, *Biochem. Biophys. Res. Comm.*, 1965, **18**, 250.
[20] I. Murakoshi, F. Kato, and J. Haginiwa, *Chem. and Pharm. Bull. (Japan)*, 1974, **22**, 480.
[21] G. K. Watson, C. Houghton, and R. B. Cain, *Biochem. J.*, 1974, **140**, 265.
[22] I. Kristensen, P. O. Larsen, and H. Sørensen, *Phytochemistry*, 1974, **13**, 2803.
[23] I. Kristensen and P. O. Larsen, *Phytochemistry*, 1974, **13**, 2799.
[24] R. Couchman, J. Eagles, M. P. Hegarty, W. M. Laird, R. Self, and R. L. M. Synge, *Phytochemistry*, 1973, **12**, 707.
[25] A. Ichihara, H. Hasegawa, H. Sato, M. Koyama, and S. Sakamura, *Tetrahedron Letters*, 1973, 37.
[26] T. Kasai, Y. Kishi, M. Sano, and S. Sakamura, *Agric. and Biol. Chem. (Japan)*, 1973, **37**, 2923.
[27] G. Zdansky, in 'Organic Selenium Compounds: Their Chemistry and Biology', ed. D. L. Klayman, Wiley, New York, 1973, p. 579.
[28] A. Shrift, in ref. 27, p. 760.

precursors used was racemic, whereas the other was the pure L-enantiomer, definite interpretation of the results is impossible.[29] The enzymatic production of dimethyl selenide by cleavage of *Se*-methyl selenomethionine has been demonstrated in leaves of *Brassica oleraceae*. *Se*-Methyl selenomethionine is formed in *B. oleraceae* when provided with selenite or selenate.[30, 31]

Homoserine, Homocysteine, Methionine, and Derivatives.—Homoserine occupies a key position as a precursor of the protein amino-acids threonine, methionine, and isoleucine, and of a number of non-protein amino-acids. It is derived from aspartic acid *via* β-aspartylphosphate and aspartic semi-aldehyde, the latter being also a precursor for lysine in the diaminopimelic acid pathway. The enzyme transforming aspartic acid to the phosphate, aspartic kinase, has been isolated from a number of plants.[32—37] The presence of the enzyme transforming β-aspartylphosphate into aspartic semialdehyde, aspartic semi-aldehyde dehydrogenase, has previously been established in *P. sativum* and the properties have been studied.[38, 39] The last enzyme in the sequence from aspartic acid to homoserine has also been isolated from *P. sativum* and its properties described.[36, 40] In agreement with the supposed precursor role of homoserine for methionine and other amino-acids, radioactivity from ^{14}C-labelled homoserine in *P. sativum* was incorporated mainly into threonine but also into isoleucine, aspartic acid, lysine, and pipecolic acid.[41]

The activation of homoserine has been studied in several plants.[42] *P. sativum*, in which *O*-acetylhomoserine is present, contains an enzyme which catalyses the formation of this compound from homoserine and acetyl-CoA. Correspondingly, *Lathyrus sativus*, in which *O*-oxalyl-homoserine is present,[43, 44] contains an enzyme catalysing the formation of this compound from homoserine and oxalyl-CoA. None of the other plants surveyed, including representatives of the green algae, horsetails, gymnosperms, and angiosperms, catalysed the synthesis of any *O*-acylhomoserine ester. On the other hand, synthesis of *O*-phosphohomoserine by the reaction catalysed by homoserine kinase was demonstrated in extracts of all plants examined, including the two exceptional legumes. This indicates that *O*-phosphohomoserine is the major activated homoserine derivative in plants.[42] In agreement, it has been found that *O*-phosphohomoserine is the natural substrate for cystathionine synthase in plants, whereas *O*-succinylhomoserine presumably is the substrate in algae, *E. coli* and *Salmonella typhimurium*, and *O*-acetylhomoserine is the substrate in *Bacillus subtilis* and in

29 C. M. Chow, S. N. Nigam, and W. B. McConnell, *Canad. J. Biochem.*, 1973, **51**, 489.
30 B. G. Lewis, C. M. Johnson, and T. C. Broyer, *Biochim. Biophys. Acta*, 1971, **237**, 603.
31 B. G. Lewis, C. M. Johnson, and T. C. Broyer, *Plant Soil*, 1974, **40**, 107.
32 P. A. Bryan, R. D. Cawley, C. E. Brunner, and J. K. Bryan, *Biochem. Biophys. Res. Comm.*, 1970, **40**, 1211.
33 R. M. Cheshire and B. J. Miflin, *Plant Physiol.*, 1973, **51**, s54.
34 K. F. Wong and D. T. Dennis, *Plant Physiol.*, 1973, **51**, 322.
35 K. F. Wong and D. T. Dennis, *Plant Physiol.*, 1973, **51**, 327.
36 H. Aarnes and S. E. Rognes, *Phytochemistry*, 1974, **13**, 2717.
37 H. Aarnes, *Physiol. Plant.*, 1974, **32**, 400.
38 K. Sasaoka, M. Kito, and H. Inagaki, *Agric. and Biol. Chem.* (*Japan*), 1963, **27**, 12.
39 K. Sasaoka, M. Kito, and H. Inagaki, *Agric. and Biol. Chem.* (*Japan*), 1963, **27**, 18.
40 K. Sasaoka, *Plant Cell Physiol.*, 1961, **2**, 231.
41 G. Goas, *Compt. rend.*, 1972, **275**, *D*, 2889.
42 J. Giovanelli, H. S. Mudd, and A. H. Datko, *Plant Physiol.*, 1974, **54**, 725.
43 J. Przybylska and J. Pawelkiewicz, *Bull. Acad. polon. Sci. Sér. Sci. Biol.*, 1965, **13**, 327.
44 J. Przybylska, *Acta Soc. Bot. Pol.*, 1972, **41**, 71.

Saccharomyces cerevisiae.[45] *O*-Acetylhomoserine is presumably also the natural substrate in *Corynebacterium glutamicum.*[46] γ-Cystathionase has been isolated from *Spinaceae oleraceae* and from *Zea mays* and the properties studied. Inhibition experiments with rhizobitoxine [2-amino-4-(2-amino-3-hydroxypropoxy)-*trans*-3-butenoic acid] (see below) suggest that methionine biosynthesis takes place *via* the cystathionine pathway but the simultaneous operation of the direct sulphuration pathway cannot be excluded.[47] No further information is available on the relative importance of the cystathionine pathway and the direct sulphuration pathway for the biosynthesis of homocysteine and methionine in plants, but cystathionine has been isolated from *Astragalus pectinatus*[48] and *Neptunia amplexicaulis.*[49] A review has appeared on the methylation of homocysteine to give methionine; in plants, a B_{12}-independent process is operating.[50]

The formation of dimethyl sulphide from *S*-methylmethionine in homogenates of seedlings of *Allium cepa* has been demonstrated,[51] as has the transformation of DL-methionine to methionine sulphoxide in *Penicillium citreoviride*, the latter oxidation taking place with hydrogen peroxide produced in the glucose oxidase reaction.[52]

An enzyme containing only a single molecular species and catalysing the transformation of L-methionine into methanethiol, ammonia, and 2-oxobutyric acid has been isolated from *Clostridium sporogenes.* DL-Homocysteine, L-ethionine, L-cysteine, L-cystine, DL-methionine sulphoxide, and DL-methionine sulphone are also substrates for the enzyme whereas 4-methylthio-2-oxobutyric acid is not.[53]

From *Chromobacterium violaceum* an enzyme system catalysing the synthesis of 2-amino-4-cyanobutyric acid, homocysteine, and thiocyanate from cyanide and homocystine has been described. The first step in this reaction may be the non-catalysed reaction between cyanide and homocystine to give 2-amino-4-thiocyanatobutyric acid and homocysteine, whereas the second, enzyme-catalyzed step is proposed to be the reaction between 2-amino-4-thiocyanatobutyric acid and cyanide to give thiocyanate and 2-amino-4-cyanobutyric acid.[54] 2-Amino-4-cyano-butyric acid has previously been found in *C. violaceum.*[55]

Canavanine, known previously only from species of Leguminosae, has been identified in *Allium cepa*, suggesting that the capacity to synthesize this compound is widely distributed.[56]

2-Amino-3-butenoic acid (vinylglycine) (12), the simplest βγ-unsaturated α-amino-acid, has previously been postulated as an intermediate, possibly enzyme-bound, in the conversion of homoserine into threonine[57] and 2-oxobutyric acid.[58] Hence, it is

45 A. H. Datko, J. Giovanelli, and S. H. Mudd, *J. Biol. Chem.*, 1974, **249**, 1139.
46 H. Kase and K. Nakayama, *Agric. and Biol. Chem.* (*Japan*), 1974, **38**, 2021.
47 J. Giovanelli, L. D. Owens, and S. H. Mudd, *Plant Physiol.*, 1973, **51**, 492.
48 S. N. Nigam and W. B. McConnell, *Phytochemistry*, 1972, **11**, 377.
49 P. J. Peterson and P. J. Robinson, *Phytochemistry*, 1972, **11**, 1837.
50 R. Ruediger and L. Jaenicke, *Mol. Cell. Biochem.*, 1973, **1**, 157.
51 T. Hattula and B. Granroth, *J. Sci. Food Agric.*, 1974, **25**, 1517.
52 M. Uyeda, S. Takenoby, and M. Hongo, *Agric. and Biol. Chem.* (*Japan*), 1974, **38**, 1797.
53 W. Kreis and C. Hession, *Cancer Res.*, 1973, **33**, 1862.
54 C. Ressler, O. Abe, Y. Kondo, B. Cottrell, and K. Abe, *Biochemistry*, 1973, **12**, 5369.
55 M. M. Brysk and C. Ressler, *J. Biol. Chem.*, 1970, **245**, 1156.
56 J. Mannhalter and H. Michl, *Monatsh.*, **105**, 889.
57 M. Flavin and C. Slaughter, *J. Biol. Chem.*, 1960, **235**, 1112.
58 B. I. Posner and M. Flavin, *J. Biol. Chem.*, 1972, **247**, 6402.

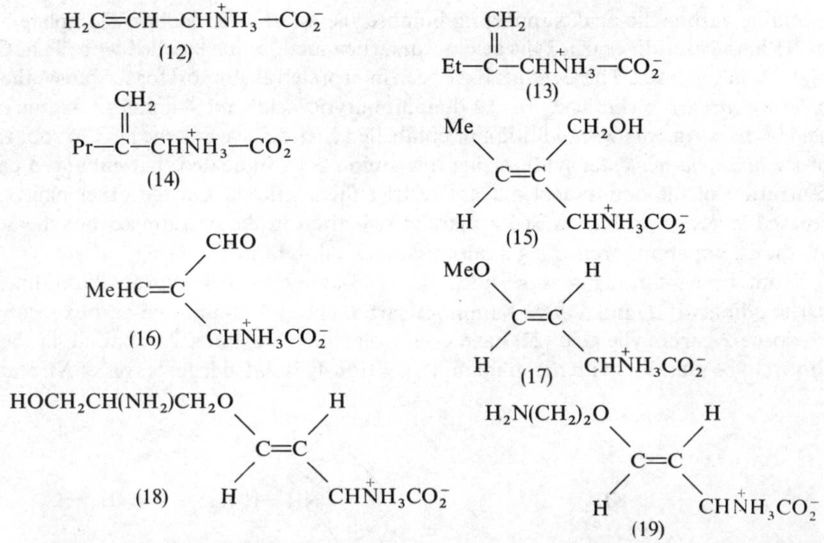

(12) (13) (14) (15) (16) (17) (18) (19)

of interest that 2-amino-3-butenoic acid has been isolated (as the partly racemic D-enantiomer) from the mushroom *Rhodophyllus nidorosus*.[59] Recently, a number of βγ-unsaturated amino-acids have been discovered in Nature, including β-methylene-L-norvaline (13) from the mushroom *Lactarius helvus*,[60] β-methylene-L-norleucine (14) from the mushroom *Amanita vaginata* var. *fulva*,[61] (*Z*)-L-2-amino-3-hydroxymethyl-3-pentenoic acid (15) and L-2-amino-3-formyl-3-pentenoic acid (16) from the mushroom *Bankera fuligineoalba*,[62] (*E*)-L-2-amino-4-methoxy-3-butenoic acid (17) from *Pseudomonas aeroginosa*,[63,64] (*E*)-L-2-amino-4-(2-amino-3-hydroxypropoxy)-3-butenoic acid (rhizobitoxine) (18), from *Rhizobium japonicum*,[65] and (*E*)-L-2-amino-4-(2-aminoethoxy)-3-butenoic acid (19) from a *Streptomyces sp.*[66] It may also be of interest that (*E*)-2-amino-4-methoxy-3-butenoic acid is an efficient, irreversible inhibitor of transaminases[67]; so too is 2-amino-3-butenoic acid which forms covalent links with the enzyme protein.[68]

Azetidine-2-carboxylic Acid and Derivatives.—Methionine labelled with ^{3}H at C-4 and ^{14}C at C-1 fed to *Convallaria majalis*, leads to labelled azetidine-2-carboxylic acid (20) with no decrease in the ^{3}H/^{14}C-ratio, thus excluding aspartic acid or aspartic semialdehyde as intermediates. Simultaneous feeding of ^{14}C-labelled methionine and unlabelled homoserine results in decreased incorporation of radioactivity into

59 G. Dardenne, J. Casimir, M. Marlier, and P. O. Larsen, *Phytochemistry*, 1974, **13**, 1897.
60 B. Levenberg, *J. Biol. Chem.*, 1968, **243**, 6009.
61 R. Vervier and J. Casimir, *Phytochemistry*, 1970, **2**, 2059.
62 R. R. Doyle and B. Levenberg, *Biochemistry*, 1968, **8**, 2457.
63 J. P. Scannell, D. L Pruess, T. C. Demny, L. H. Sello, T. Williams, and A. Stempel, *J. Antibiotics*, 1972, **25**, 122.
64 U. Sahm, G. Knobloch, and F. Wagner, *J. Antibiotics*, 1973, **26**, 389.
65 L. D. Owens, J. F. Thompson, R. G. Pitcher, and T. Williams, *J.C.S. Chem. Comm.* 1972, 714.
66 D. L. Pruess, P. J. Scannell, M. Kellett, H. A. Ax, J. Janecek, T. H. Williams, A. Stempel, and J. Berger, *J. Antibiotics*, 1974, **27**, 229.
67 R. R. Rando, *Nature*. 1974, **250**, 586.
68 R. R. Rando, *Biochemistry*, 1974, **13**, 3859.

azetidine carboxylic acid, supporting homoserine as an intermediate. Complete loss of ^{3}H in the azetidinecarboxylic acid occurs when methionine labelled with ^{3}H at C-2 and ^{14}C at C-1 is fed. These results have been interpreted as support for the biosynthesis of azetidinecarboxylic acid *via* 2,4-diaminobutyric acid and 4-amino-2-oxobutyric acid.[69] Experiments with addition of unlabelled azetidinecarboxylic acid to root tips of *Parkinsomia aculeata*, synthesizing this amino-acid, indicated that enhanced concentration of the amino-acid did not restrict the synthesis. On the other hand, increased levels of proline caused a marked reduction in the azetidinecarboxylic acid synthesis, apparently reflecting a false feedback inhibition.[3]

From beech nuts (*Fagus silvatica* L.), *N*-(3-amino-3-carboxypropyl)azetidine-2-carboxylic acid (21) and *N*-[*N*-(3-amino-3-carboxypropyl)-3-amino-3-carboxypropyl] azetidine-2-carboxylic acid (22) have been isolated.[70] The latter compound has been shown to be identical with nicotianamine, previously isolated from leaves of *Nicotiana*

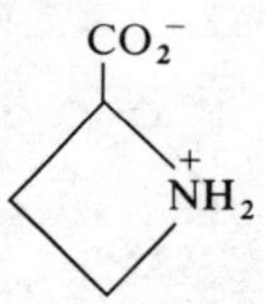

(20)

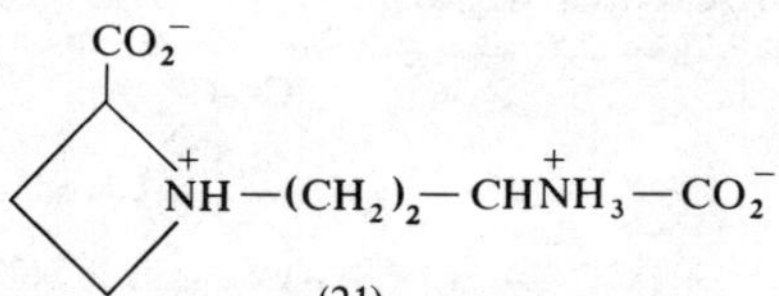

(21)

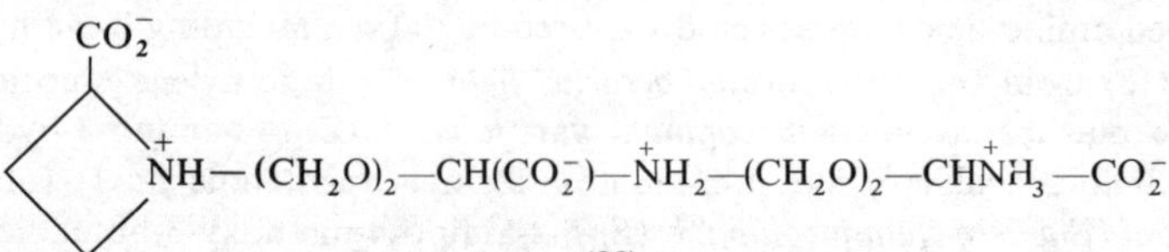

(22)

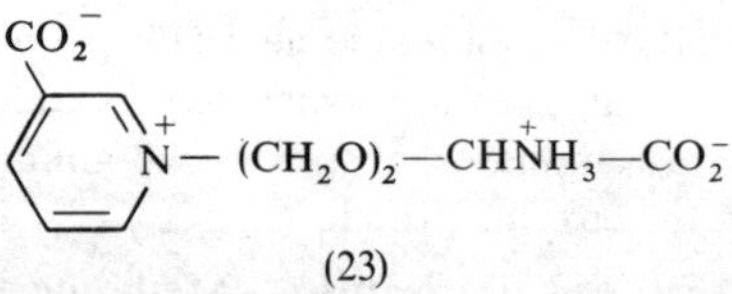

(23)

tabacum, but assigned a different structure.[71] The two amino-acids can be produced chemically from azetidinecarboxylic acid by heating in aqueous solution, the reaction being an electrophilically assisted nucleophilic ring-opening. However the amino-acids are not artefacts but genuine constituents of the beech nuts, and no free azetidinecarboxylic acid is present. The ring-opening also takes place with other nucleophiles. Thus, azetidinecarboxylic acid with NH_3 gives 2,4-diaminobutyric acid, with water it gives homoserine, with MeSH methionine, and with nicotinic acid the amino-acid nicotianine (23).[70] Compound (23) has also been identified in leaves from *Nicotiana tabacum*.[72] Hence, azetidinecarboxylic acid, previously found in Liliaceae and

[69] E. Leete, G. E. Davis, C. R. Hutchinson, K. K. Woo and M. R. Chedekel, *Phytochemistry*, 1974, **13**, 427.
[70] I. Kristensen and P. O. Larsen, *Phytochemistry*, 1974, **13**, 2791.
[71] M. Noma, M. Noguchi, and E. Tamaki, *Tetrahedron Letters*, 1971, 2017.
[72] M. Noguchi, H. Sakuma, and E. Tamaki, *Phytochemistry*, 1968, **7**, 1861.

Agavaceae,[73] Leguminosae,[74], and Chenopodiaceae,[75] is presumably present also in Fagaceae and Solanaceae. Azetidinecarboxylic acid is a proline antagonist and toxic.[76] Its widely observed occurrence may reflect an almost ubiquitous distribution within the class of higher plants, possibly attributable to 'metabolic noise,'[3] and the production of nicotianine and nicotianamine may be methods for detoxification. In any case, it seems a justified conclusion that these latter amino-acids are biosynthesized from azetidinecarboxylic acid. Likewise it is conceivable that 2-amino-4-(isoxazolin-5-one-2-yl)butyric acid, present in *Lathyrus odoratus*,[15] derives from azetidinecarboxylic acid by reaction with the heterocyclic ring. In connexion with the discussion of γ-substituted α-aminobutyric acids it should perhaps be mentioned that 3-(3-amino-3-carboxypropyl)uridine, which is a constituent of tRNA in *Escherichia coli*,[77,78] is produced through the intermediacy of *S*-adenosylmethionine.[79] In any event, it is obvious that azetidinecarboxylic acid is itself an activated C_4-intermediate, even though it apparently possesses a notable stability, in *Convallaria majalis*.[80] It is possible for various higher plants, including *Urtica dioica*, *Scirpus lacustris*, *Eleocharis palustris*, and *Juncus effusus*, to degrade L- (but not D-) azetidinecarboxylic acid.[81] *Agrobacterium* sp. can transform L-azetidine-2-carboxylic acid *via*-2-hydroxybutric acid into malic semialdehyde.[82] In the algae *Cyanidium caldarium* ring-labelled L-azetidinecarboxylic acid gives rise to labelled methionine and serine.[83]

Straight-chain C_5-Amino-acids. (2*S*,4*S*)-, and (2*S*,4*R*)-2-Amino-4-hydroxypentanoic acid have been isolated from the mushroom *Boletus satanas* Lenz.[84] The (2*S*, 4*S*)-isomer has previously been isolated from *Lathyrus odoratus*.[85] L-2-Amino-4-chloro-4-pentenoic acid has been isolated from the mushroom *Amanita pseudoporphyria*.[86]

The biosynthesis of 5-aminolevulinic acid (24) in plants has been studied. It accumulates in *Cucumis sativus* L. seedlings and in leaves of *Hordeum vulgare* and *Phaseolus*

$$H_3\overset{+}{N}CH_2CO(CH_2)_2CO_2^-$$

(24)

vulgaris, when levulinic acid is supplied. Levulinic acid is known to be a specific, competitive inhibitor of 5-aminolevulinic acid dehydrase, rendering the amino-acid an acceptable chlorophyll precursor in higher plants, so that chlorophyll biosynthesis is regulated at the level of the formation of 5-aminolevulinic acid.[87] Incorporation of labelled compounds into the plants demonstrated that both glycine and

[73] L. Fowden, *Biochem. J.*, 1956, **64**, 323.
[74] M. Sung and L. Fowden, *Phytochemistry*, 1969, **8**, 2095.
[75] L. Fowden, *Phytochemistry*, 1972, **11**, 2271.
[76] L. Fowden, D. Lewis, and H. Tristram, *Adv. Enzymol.*, 1967, **29**, 89.
[77] Z. Ohashi, M. Maeda, J. A. McCloskey, and S. Nishimura, *Biochemistry*, 1974, **13**, 2620.
[78] S. Friedman, H. J. Li, K. Nakanishi, and G. Van Lear, *Biochemistry*, 1974, **13**, 2932.
[79] S. Nishimura, Y. Taya, Y. Kushino, and Z. Ohashi, *Biochem. Biophys. Res. Comm.*, 1974, **57**, 702.
[80] L. Fowden and M. Bryant, *Biochem. J.*, 1959, **71**, 210.
[81] R. Schwarz, 'Zur Ökochemie der Azetidin-2-carbonsäure', Dissertation, Göttingen, 1973.
[82] P. M. Dunnill and L. Fowden, *Phytochemistry*, 1965, **4**, 445.
[83] R. F. Troxler and A. S. Brown, *Biochim. Biophys. Acta*, 1974, **366**, 341.
[84] P. Matzinger, P. Catalfomo, and C. H. Eugster, *Helv. Chim. Acta*, 1972, **55**, 1478.
[85] L. Fowden, *Nature*, 1966, **209**, 807.
[86] S. Hatanaka, S. Kaneko, Y, Niimura, F. Kinoshita, and G. Soma, *Tetrahedron Letters*, 1974, 3931.
[87] S. I. Beale and P. A. Castelfranco, *Plant Physiol.*, 1974, **53**, 291.

succinic acid were poor precursors, whereas glutamic acid, glutamine, and 2-oxoglutaric acid served well. The carboxy and methylene groups of glycine were incorporated into 5-aminolevulinic acid to about equal extents; the carboxy-carbon of glutamate was incorporated almost as efficiently as the internal carbons of the same compound. These results indicate that 5-aminolevulinic acid is not produced by the action of succinyl-CoA-glycine succinyl transferase (δ-aminolevulinic acid synthetase), but rather that a direct route, involving the complete carbon skeleton of glutamic acid, is operating.[88]

The biosynthesis of glutamine has been studied in *Vicia faba*,[89] in *Pisum sativum*,[90—92] and in *Spinaceae oleraceae*.[91] It has been shown recently that the main route for nitrogen into amino-acids in higher plants is through glutamine synthetase and a ferredoxin-dependent glutamine: 2-oxoglutarate aminotransferase producing two molecules of glutamic acid. Asparagine was not a substrate for the enzyme, found in chloroplasts of *Pisum sativum*.[93] Similar conclusions have been reached by studies with tissue cultures from *Acer pseudoplatanus* and *Pisum sativum*, although here asparagine also was able to act as an amino-group donor;[94] similar results were obtained with carrot-tissue cultures.[95]

The enzymes of the 4-aminobutyric acid pathway, *i.e.* glutamate decarboxylase, 4-aminobutyrate: 2-oxoglutarate aminotransferase, and succinic semialdehyde dehydrogenase were found in *Raphanus sativus*[96] and in *Lupinus albus*.[97] [1-^{14}C]-4-Aminobutyric acid gave succinic acid as the first labelled product in the mushroom *Agaricus bisporus*.[98]

Straight-chain C_6-Amino-acids. (2*S*,4*S*)- and (2*S*,4*R*)-2-Amino-3-hydroxy-4-hexynoic acid have been isolated from the mushroom *Tricholomopsis rutilans*,[99] and 2-amino-5-hydroxyhexanoic acid from *Crotalaria* seeds.[100]

Experiments with labelled precursors in *Hordeum vulgare* have demonstrated that lysine is produced by the diaminopimelic acid pathway. Labelled 2-aminoadipic acid can also give rise to labelled lysine, but this is not reflecting a biosynthetic pathway of any significance, though 2-aminoadipic acid is widely distributed in plants.[101] Demonstration of lysine-sensitive aspartate kinase in various plants,[32—35,37] and of dihydrodipicolinic acid synthase in *Zea mays*[33] also provides evidence for the diaminopimelic acid pathway.

The previously established transformations of D-lysine into pipecolic acid and of L-lysine into alkaloids have been substantiated and discussed. L-Lysine is a precursor for the alkaloids sedamine, *N*-methylpelletierine, and *N*-methylallosedendrine in

88 S. I. Beale and P. A. Castelfranco, *Plant Physiol.*, 1974, **53**, 297.
89 A. Haystead, *Planta*, 1973, **111**, 271.
90 D. O'Neal and K. W. Joy, *Arch. Biochem., Biophys.*, 1973, **159**, 113; *Plant Physiol.*, 1974, **54**, 773.
91 D. O'Neal and K. W. Joy, *Nature New Biol.*, 1973, **246**, 61.
92 A. V. Pushkin, Z. G. Evstigneeva, and V. L. Krestovich, *Biokhimiya*, 1974, **39**, 533.
93 P. J. Lea and B. J. Miflin, *Nature*, 1974, **251**, 614.
94 M. W. Fowler, W. Jessup, and G. Sarkissian, *F.E.B.S. Letters*, 1974, **46**, 340.
95 D. K. Dougall, *Biochem. Biophys. Res. Comm.*, 1974, **58**, 639.
96 J. G. Streeter and J. F. Thompson, *Plant Physiol.*, 1972, **49**, 579.
97 I. Andres, P. Gonzales, and A. Santos-Ruiz, *Physiol. Chem. Phys.*, 1973, **5**, 357.
98 P. Baldy and J. C. Latche, *Compt rend.*, 1972, **275**, *D*, 2123.
99 Y. Niimura and S. Hatanaka, *Phytochemistry*, 1974, **13**, 175.
100 R. Pant and H. M. Fales, *Phytochemistry*, 1974, **13**, 1626.
101 B. L. Møller, *Plant Physiol.*, 1974, **54**, 638.

two *Sedum* species, and for anabasine in *Nicotiana glauca*, whereas pipecolic acid is derived from D-lysine in all of these plants.[102] In this connection it is of interest that L-lysine has been established as a precursor of pipecolic acid in rumen ciliate protozoans[103] and that diaminopimelic acid is a precursor of both lysine and pipecolic acid in these protozoans, which are unable to metabolize 2-aminoadipic acid.[104] Furthermore, in the fungus *Rhizoctonia leguminicola* it has been demonstrated that L-lysine is a precursor of L-pipecolic acid, presumably *via* 2-oxo-6-aminohexanoic acid and Δ^1-piperideine-2-carboxylic acid, whereas D-lysine is degraded *via* N^6-acetyllysine in the same species. L-Pipecolic acid is a precursor of piperidine alkaloids in this fungus.[105]

The presence of N^6-methyllysine has been established in *Sedum acre* and *S. sarmentosum.* Radioactivity from [*methyl*-^{14}C]methionine and from [6-^{3}H]lysine is incorporated into this amino-acid, which however is not an intermediate in the biosynthesis of the alkaloid *N*-methylpelletierine in these species.[106] It has also been shown that N^6-methyllysine is not a precursor for sedamine in *Sedum acre*.[107]

Lathyrine (25), first isolated from *Lathyrus tingitanus* and widely distributed within *Lathyrus* species,[108] derives from homoarginine and 4-hydroxyhomoarginine.[109] It has now been shown that 4,5-dehydrohomoarginine is not incorporated into lathyrine in *L. tingitanus.* On basis of this evidence, together with chemical studies, it is proposed that lathyrine is derived from homoarginine starting with a 1,5-hydrogen shift followed by reaction of the alkyl radical with oxygen to give 4-hydroxyhomoarginine, oxidation to 4-oxohomoarginine, and, finally, cyclization.[110] Both homoarginine and 4-hydroxyhomoarginine are widely distributed within the *Lathyrus* species.[108]

Carnitine (26) in *Neurospora crassa* is biosynthesized from lysine *via* N^6-trimethyllysine betaine and γ-butyrobetaine,[111] as in mammalian systems.[112]

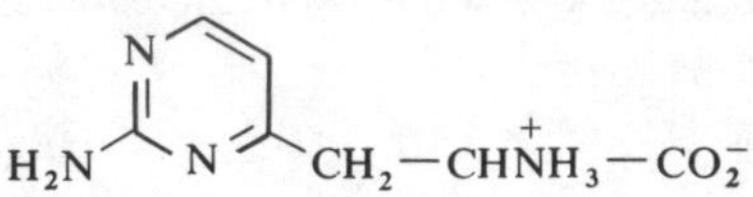

(25)

$(Me)_3\overset{+}{N}CH_2CHOHCH_2CO_2^-$

(26)

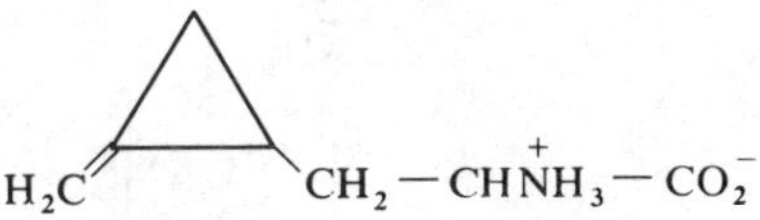

(27)

$Bu^iCH\overset{+}{N}H_3CHOHCH_2CO_2^-$

(28)

[102] E. Leistner, R. N. Gupta, and I. A. Spenser, *J. Amer. Chem. Soc.*, 1973, **95**, 4040.
[103] R. Onodera and M. Kandatsu, *Agric. and Biol. Chem. (Japan)*, 1972, **36**, 1989.
[104] R. Onodera and M. Kandatsu, *Agric. and Biol. Chem. (Japan)*, 1974, **38**, 913.
[105] F. P. Guengerich and H. P. Broquist, *Biochemistry*, 1973, **12**, 4270.
[106] E. Leistner and I. D. Spenser, *J. Amer. Chem. Soc.*, 1973, **95**, 4715.
[107] P. Korzan and T. J. Gilbertson, *Phytochemistry*, 1974, **13**, 435.
[108] E. A. Bell, in 'Chemotaxonomy of the Leguminosae', ed. J. B. Harborne, D. Boulter, and B. L. Turner, Academic Press, London and New York, 1971, p. 179.
[109] E. A. Bell and J. Przybylska, *Biochem. J.*, 1965, **94**, 35P.
[110] R. C. Hider and D. I. John, *Phytochemistry*, 1973, **12**, 119.
[111] D. W. Horne and H. P. Broquist, *J. Biol. Chem.*, 1973, **248**, 2170.
[112] H. T. Haighler and H. P. Broquist, *Biochem. Biophys. Res. Comm.*, 1974, **56**, 676; R. A. Cox and C. L. Hoppel, *Biochem. J.*, 1974, **143**, 699.

Branched-chain Amino-acids.—In *Acer pseudoplatanus* only small incorporation of labelling from [^{14}C]isoleucine into hyptoglycin A (27) took place. Higher, though still low, incorporation percentages were found with [*methyl*-^{14}C]methionine and [^{14}C]acetate. No incorporation of radioactivity from ^{14}C-labelled leucine was observed. These preliminary results may indicate that hypoglycin A is produced by condensation of three acetate units to give a C_6-skeleton, followed by formation of the cyclopropane ring by formal addition of the methyl group of methionine at a position of 4,5-unsaturation. If this is so, there is no common origin for hypoglycin A and 2-amino-4-methylhex-4-enoic acid found in related plants, the latter amino-acid being derived from isoleucine.[113] A discussion of the biosyntheses of the branched chain amino-acids found in Sapindaceae and Hippocastancaceae has been presented.[3]

(3*S*, 4*S*)-4-Amino-3-hydroxy-6-methylheptanoic acid (28) is a constituent of the antibiotic pepstatin.[114] Studies of its biosynthesis in *Streptomyces testaceus* indicate that C-1 and C-2 are derived from malonate, whereas the remaining six C-atoms originate from leucine.[115]

(2*S*, 3*S*)-3-Hydroxy-4-methyleneglutamic acid has been isolated from *Gleditsia caspica* where it co-occurs with (2*S*, 4*R*)-4-methylglutamic acid and (2*S*, 3*S*, 4*R*)-3-hydroxy-4-methylglutamic acid.[116] This supports a metabolic relation between 4-methyl- and 4-methylene-substituted glutamic acids even though such a relation has not been demonstrated by biosynthetic studies. The *O*-β-D-galactoside of (2*S*, 4*R*)-4-hydroxy-4-isobutylglutamic acid has been isolated from *Reseda odorata* L.[117] L-3-(3-Carboxyfuran-4-yl)alanine is a constituent of the mushroom *Phyllotopsis nidulans*.[118]

Imino-acids.—*Agrobacterium tumefaciens* degrades lysopine (29) to lysine and pyruvate by a membrane-bound enzyme system.[119,120] The presence of octopine (30) has been demonstrated in normal tissue cultures of *Nicotiana tabacum*, *Helianthus annuus*, and pinto bean, and the amounts were increased by induction with *Agro-*

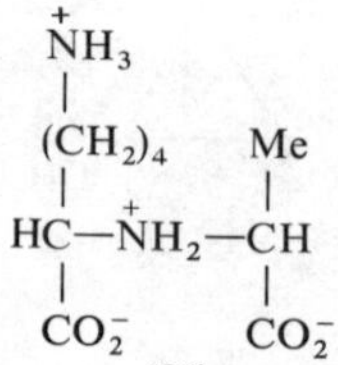

(29)

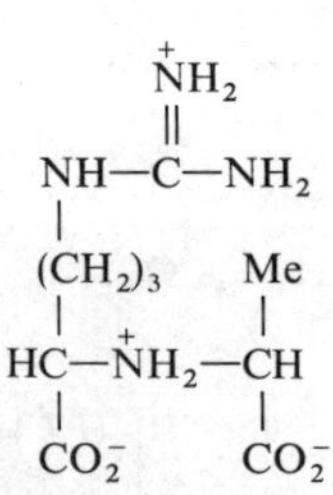

(30)

[113] L. Fowden and H. M. Pratt, *Phytochemistry*, 1973, **12**, 1677.
[114] H. Morishima, T. Takita, and H. Umezawa, *J. Antibiotics*, 1973, **26**, 115.
[115] H. Morishima, T. Sawa, T. Takita, T. Aoyagi, T. Takeuchi, and H. Umezawa, *J. Antibiotics*, 1974, **27**, 267.
[116] G. Dardenne, J. Casimir, and H. Sørensen, *Phytochemistry*, 1974, **13**, 2195.
[117] P. O. Larsen, H. Sørensen, D. W. Cochran, E. W. Hagaman, and W. Wenkert, *Phytochemistry*, 1973, **12**, 1713.
[118] R. R. Doyle and B. Levenberg, *Phytochemistry*, 1974, **13**, 2813.
[119] M. F. Jubier and B. Lejeune, *Physiol. Veg.*, 1972, **10**, 653.
[120] M. F. Jubier, *F.E.B.S. Letters*, 1972, **28**, 129.

bacterium tumefaciens.[121] In other studies with tissue cultures of *Nicotiana tabacum* and *Crepis capillaris*, octopine could not be found in healthy plants or in non-infected tissue cultures, but only in the cultures transformed by *A. tumefaciens.* However, small amounts of octopine were found in root tips of pea and bean seedlings.[122]

Details have been presented of the previously reported biosynthetic studies on lycomarasmine (31). Aspartic acid and glycine are good precursors for the C_4- and the C_2-units, respectively. Glucose is a better precursor for the C_3-unit than glycine, alanine, serine, or aspartic acid.[123] However, studies with different strains of *Fusarium oxysporum melonis* indicated that the normal biosynthetic route to lycomarasmine is *via* aspargillomarasmine A (32) and B(33), thus implying that the C_2-moiety in lycomarasmine originates in a C_3-unit.[124]

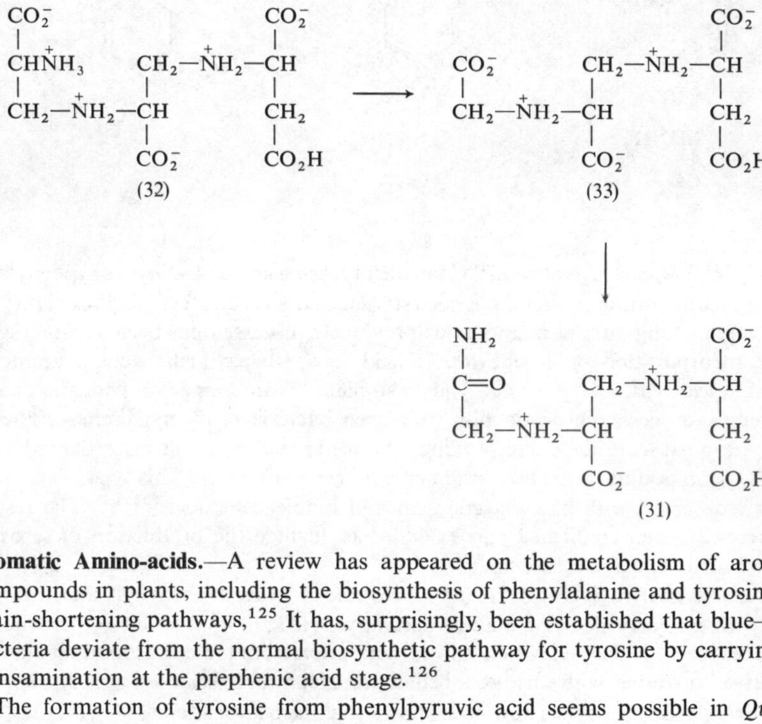

Aromatic Amino-acids.—A review has appeared on the metabolism of aromatic compounds in plants, including the biosynthesis of phenylalanine and tyrosine and chain-shortening pathways,[125] It has, surprisingly, been established that blue–green bacteria deviate from the normal biosynthetic pathway for tyrosine by carrying out transamination at the prephenic acid stage.[126]

The formation of tyrosine from phenylpyruvic acid seems possible in *Quercus pedunculata.* However, direct hydroxylation of phenylpyruvic acid or phenylalanine does not seem to be involved; more complex mechanisms, leading to partial or complete degradation of phenylpyruvic acid, seem at play.[127] This agrees with the

[121] R. Johnson, R. H. Guderian, F. Eden, M. D. Chilton, M. P. Gordon, and E. W. Nester, *Proc. Nat. Acad. Sci. U.S.A.*, 1974, **71**, 536.
[122] M. F. Wendt-Gallitelli and I. Dobrigkeit, *Z. Naturforsch.*, 1973, **28c**, 768.
[123] C. R. Popplestone and A. M. Unrau, *Canad. J. Chem.*, 1973, **51**, 3943.
[124] P. Camporota, A. Trouvelot, and M. Barbier, *Compt. rend.*, 1973, **276**, *D*, 1903.
[125] H. A. Stafford, *Ann. Rev. Plant Physiol.*, 1974, **25**, 459.
[126] S. L. Stenmark, D. L. Pierson, R. A. Jensen, and G. I. Glover, *Nature*, 1974, **247**, 290.
[127] P. Gadal, A. Boudet, and G. Robreau, *Plant Sci. Letters*, 1974, **3**, 55.

general supposition that tyrosine is not derived from phenylalanine in plants.[128] Radioactivity from ^{14}C-labelled shikimic acid and tyrosine was effectively incorporated into 3',4'-dihydroxyphenylalanine in *Vicia faba*, whereas phenylalanine did not serve as a precursor.[128]

Stizolobic acid (34) and stizolobinic acid (35) have been identified in the mushroom *Amanita pantherina*.[129] These amino-acids, presumably derived from 3',4'-dihydroxyphenylalanine, have previously been encountered only in species of Leguminosae.[108] Degradation of aromatic amino-acids by ring-opening reactions has attracted increased attention.[125,130] 3',4'-Dihydroxyphenylalanine is a precursor for betalamic

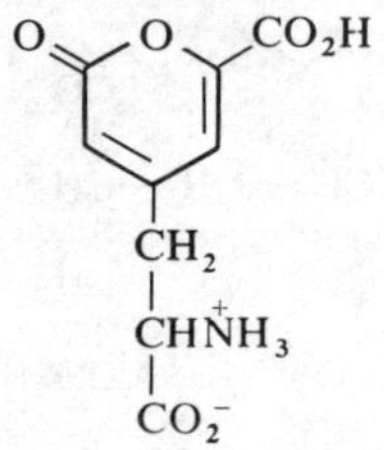

(34)

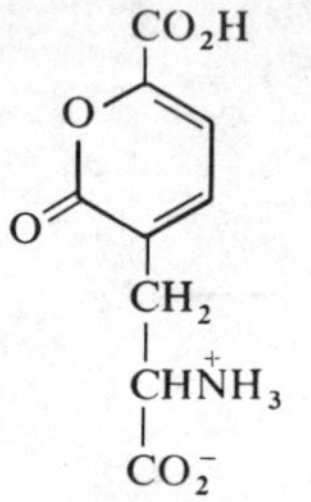

(35)

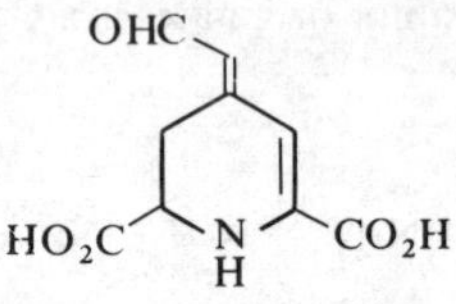

(36)

acid (36).[131] The cleavage of ring-^{14}C labelled tyrosine and 3',4'-dihydroxyphenylalanine with the liberation of $^{14}CO_2$ has been studied in tissue cultures of various plants.[132]

The chain elongation of amino-acids previously discussed has been substantiated by the incorporation of 2-benzylmalic acid into 3-benzylmalic acid, 2-amino-4-phenylbutyric acid, and phenylethylglucosinolate.[133] In connexion with the chain-lengthening process, the production by rumen bacteria of phenylalanine, tyrosine, and tryptophan from the corresponding arylacetic acids without prior degradation to C_6–C_1-compounds or smaller fragments deserves attention. This type of reaction was first observed with phenylacetic acid and indoleacetic acid.[134,135] The results have recently been confirmed and extended to include the production of tyrosine from *p*-hydroxyphenylacetic acid.[136]

(3-Carboxyphenyl)glycine (37) and (3-carboxy-4-hydroxyphenyl)glycine (38) co-occur with the corresponding alanine derivatives [(39), (40)] in species of Cruciferae, Resedaceae, and Iridaceae, with a known derivation from the corresponding alanine derivative. In studies with various, labelled precursors in *Reseda lutea* L., *R. odorata* L. and *Iris* x *hollandica* it has now been shown that the mandelic acids (41), (42) and the glyoxylic acids (43), (44) are intermediates in these transformations whereas the corresponding cinnamic acids and phenylacetic acids are not. The transformation

[128] T. Griffith and E. E. Conn, *Phytochemistry*, 1973, **12**, 1651.
[129] W. S. Chilton, C. P. Hsu, and W. T. Zdybak, *Phytochemistry*, 1974, **13**, 1179.
[130] B. E. Ellis, *Lloydia*, 1974, **37**, 168.
[131] N. Fischer and A. S. Dreiding, *Helv. Chim. Acta*, 1972, **55**, 649.
[132] B. E. Ellis, *Planta*, 1973, **111**, 113.
[133] D. Dörnemann, W. Löffelhardt, and H. Kindl, *Canad. J. Biochem.*, 1974, **52**, 916.
[134] M. J. Allison, *Biochem. Biophys. Res. Comm.*, 1965, **18**, 30.
[135] M. J. Allison and I. M. Robinson, *Biochem. J.*, 1967, **102**, 36P.
[136] S. Kristensen, *British J. Nutrition*, 1974, **31**, 357.

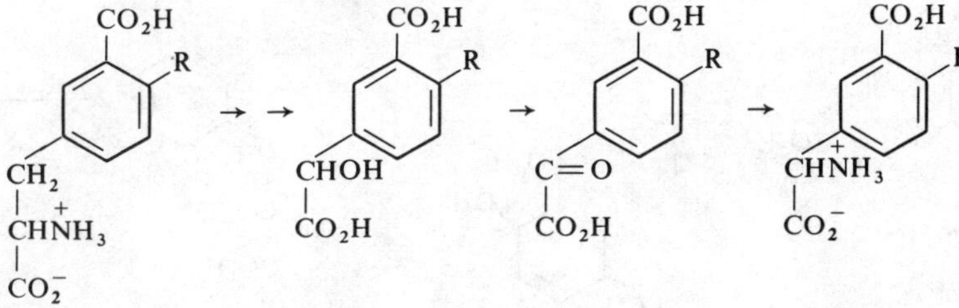

(39) R = H (41) R = H (43) R = H (37) R = H
(40) R = OH (42) R = OH (44) R = OH (38) R = OH

of the glyoxylic acids into the glycine derivatives seems rather unspecific, since *Iris*, when supplied with ^{14}C-labelled 3-carboxy-4-hydroxyphenylglyoxylic acid (44), produces labelled (3-carboxy-4-hydroxyphenyl)glycine although this amino-acid is not normally present in *Iris* species. Normal transamination may not be involved since the glycine derivatives occur in the plants as partly racemized D-amino acids.[137] Although other phenylglycines occur in plants and in peptide antibiotics, only limited information is available about their biosynthesis. It has been shown that in *Streptomyces griseoviridus* *N*-methylphenylglycine, present in etamycin, is derived from phenylalanine without rearrangement of the carbon-skeleton. A low but definite incorporation of radioactivity from [*carboxy*-^{14}C]phenylalanine into *N*-methylphenylglycine, with the radioactivity occurring only in C-1, was also observed.[138] It may be of interest that *p*-hydroxymandelic acid is a metabolite of DL-phenylalanine, presumably with phenylacetic acid and *p*-hydroxyphenylacetic acid as intermediates, in *Aspergillus niger*.[139] The same reaction pathway is probably operating in the fungus *Polyporus hispidus*. Here, tyrosine also gives *p*-hydroxyphenylacetic acid.[140] β-Tyrosine (45) is produced directly from tyrosine (46) by a tyrosine αβ-mutase in *Bacillus brevis*.[141]

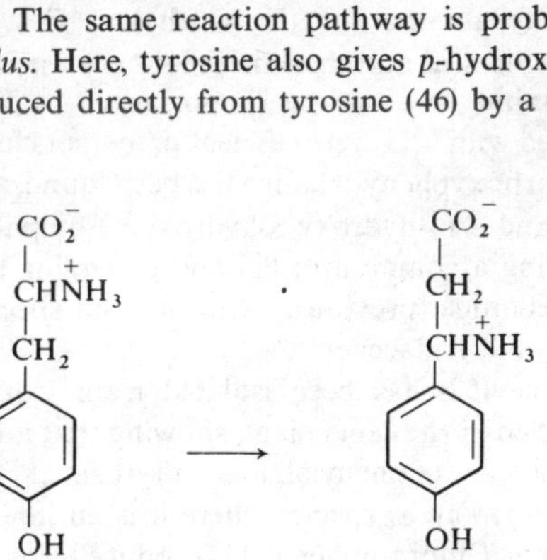

(46) (45)

Detailed evidence has appeared regarding the incorporation of stereospecifically labelled shikimic acid into 3-(3-carboxyphenyl)alanine (39), supporting a derivation of this amino-acid from chorismic acid (47), *via* isochorismic acid (48), isoprephenic

[137] P. O. Larsen and E. Wieczorkowska, *Biochim. Biophys. Acta*, 1975, **381**, 409.
[138] D. J. Hook and L. C. Vining, *Canad. J. Biochem.*, 1973, **51**, 1630.
[139] G. Kishore, M. Sugumaran, and C. S. Vaidyanathan, *Biochem. Biophys. Res. Comm.*, 1974, **56**, 851.
[140] P. W. Perrin and G. H. N. Towers, *Phytochemistry*, 1973, **12**, 583.
[141] Z. Kurylo-Borowska and T. Abramsky, *Biochim. Biophys. Acta*, 1972, **264**, 1.

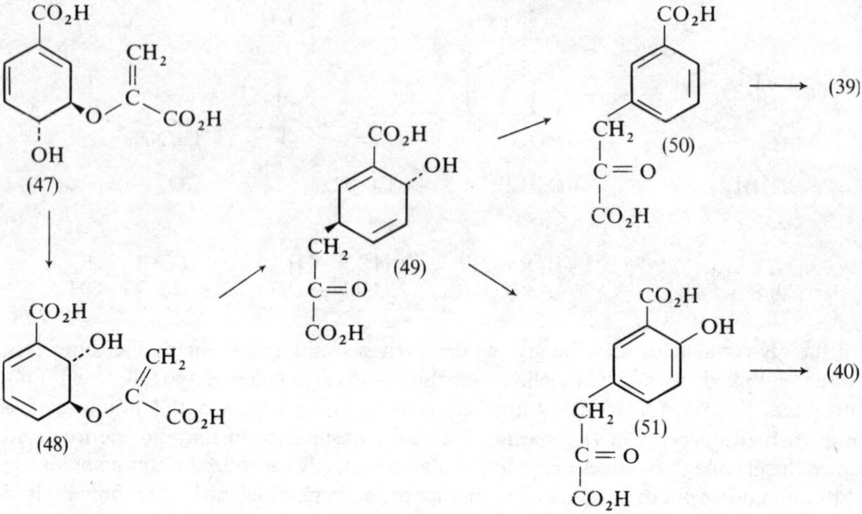

acid (49), and 3′-carboxyphenylpyruvic acid (50), in *Reseda odorata* L. and *R. lutea* L. The $^3H/^{14}C$-ratio in 3-(3-carboxyphenyl)alanine, when labelled by incorporation of (−)-(6*S*)-[6-^{3}H]shikimic acid, mixed with (−)-[G-^{14}C]shikimic acid, was substantially higher than that of the precursor. The isotope effect must operate by competition between pathways from isoprephenic acid to 3′-carboxyphenylpyruvic acid and to 3′-carboxy-4′-hydroxyphenylpyruvic acid (51). This indicates that 3-(3-carboxy-4-hydroxyphenyl)alanine (40) (in which no ^{3}H was retained) is derived directly from isoprephenic acid and that the previously established hydroxylation of 3-(3-carboxyphenyl)alanine is of minor importance.[142] The two phenylpyruvic acids [(50), (51)], labelled with ^{14}C, were efficient precursors for the alanine derivatives (39) and (40).[137] 3-(3-Carboxyphenyl)alanine has been found, along with 3-(3-hydroxymethylphenyl)alanine and 3-(4-hydroxy-3-hydroxymethylphenyl)alanine, in *Caesalpinia tinctoria*, suggesting a common origin and extending the range of occurrence of 3-(3-carboxyphenyl)alanine (previously known from species of Cruciferae, Resedaceae, Cucurbitaceae, and Iridaceae).[143]

p-Aminophenylalanine (52) has been isolated from *Vigna vexillata*.[144] Its biosynthesis has been studied in the same plant, showing that it is derived from shikimic acid without the intermediacy of phenylalanine or tyrosine.[145] *p*-Aminophenylalanine is also known from *Streptomyces* species where it is an intermediate in the biosynthesis of chloramphenicol. Chloramphenicol is produced from chorismic acid without the intermediacy of phenylalanine, tyrosine, or prephenic acid. The nitrogen of the aromatic amino-group is derived from the common pool of nitrogen. Experiments with shikimic acid, stereospecifically labelled with ^{3}H at C-6, have demonstrated that the pro-6*R*-hydrogen atom is lost whereas the pro-6*S*-hydrogen atom is re-

142 P. O. Larsen, D. K. Onderka, and H. G. Floss, *Biochim. Biophys. Acta*, 1975, **381**, 397.
143 R. Watson and L. Fowden, *Phytochemistry*, 1973, **12**, 617.
144 G. A. Dardenne, M. Marlier, and J. Casimir, *Phytochemistry* 1972, **11**, 2567.
145 G. A. Dardenne, P. O. Larsen, and E. Wieczorkowska, *Biochim. Biophys. Acta*, 1975, **381**, 416.

tained in the position *ortho* to the side-chain in chloramphenicol.[146–149] The biosynthesis of chloramphenicol seems to be closely related to those of anthranilic acid, tryptophan, *p*-aminobenzoic acid (53), and *p*-hydroxybenzoic acid.[150] On this basis, the biosynthesis of *p*-aminophenylalanine has been proposed to proceed from chorismic acid (47) through 4-amino-3-enolpyruvylcyclohexa-1,5-dienecarboxylic acid (54), 3-(4-amino-1-carboxycyclohexa-2,5-dienyl)pyruvic acid (55), and 4′-aminophenylpyruvic acid (56). Compound (54) has also been proposed as an inter-

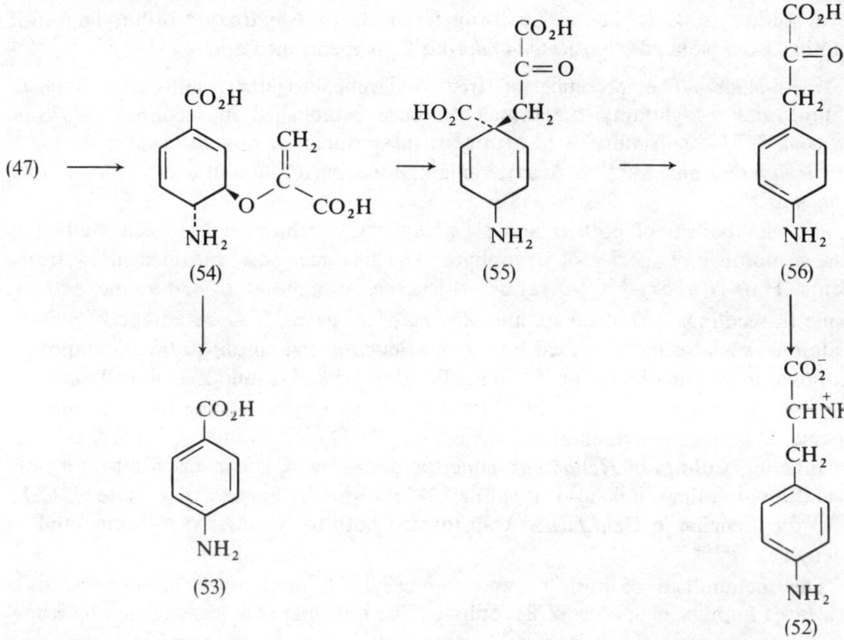

mediate—possibly enzyme-bound—in the biosynthesis of *p*-aminobenzoic acid (53). This pathway, together with pathways (i) for the biosynthesis of *p*-hydroxybenzoic acid, 3,4-dihydro-3,4-dihydroxybenzoic acid, phenylalanine, and tyrosine from chorismic acid, (ii) for the biosynthesis of 3-(3-carboxyphenyl)-alanine, 3-(3-carboxy-4-hydroxyphenyl)alanine, salicylic acid, and 2,3-dihydro-2,3-dihydroxybenzoic acid from chorismic acid *via* isochorismic acid, and (iii) for anthranilic acid and 2-amino-3-hydroxy-2,3-dihydroxybenzoic acid from chorismic acid *via* 2-amino-3-enolpyruvylcyclohexa-4,6-dienecarboxylic acid, has been proposed as part of a coherent framework for chorismic acid derivatives.[135] Similar proposals for the intermediates in the biosynthesis of anthranilic acid, *p*-aminobenzoic acid, and *p*-aminophenylalanine have

[146] L. C. Vining, V. S. Malik, and D. W. S. Westlake, *Lloydia*, 1968, **31**, 355.
[147] W. P. O'Neill, R. F. Nystrom, K. L. Rinehart, and D. Gottlieb, *Biochemistry*, 1973, **12**, 4775.
[148] A. Emes, H. G. Floss, D. A. Lowe, D. W. S. Westlake, and L. C. Vining, *Canad. J. Microbiol.*, 1974, **20**, 347.
[149] A. Jones and D. W. S. Westlake, *Canad. J. Microbiol.*, 1974, **20**, 1599.
[150] B. Skarka, D. W. S. Westlake, and L. C. Vining, *Chem. Zvesti*, 1970, **24**, 66.

been put forward in a recent monograph on shikimic acid and its derivatives.[151] *p*-Aminophenylalanine and *p*-aminophenylserine have been shown to be intermediates in the formation of the nitroaromatic unit of the antibiotic aureotine in *Streptomyces thioluteus.*[152]

***O*-Glycosides of Amino-acids.**—A detailed report has appeared on the isolation of *O*-(β-D-glucopyranosyl)-L-tyrosine and 2-[3-(β-D-glucopyranosyloxy)-4-hydroxyphenyl]ethylamine from seeds of *Entada pursaetha.*[153] The metabolism of 3′,4′-dihydroxyphenylalanine, and of the glucosides of this amino-acid and tyrosine, has been studied in *Vicia faba.*[154] A β-D-galactoside of 4-hydroxy-4-isobutylglutamic acid has been isolated from *Reseda odorata* L. as mentioned above.[117]

D-Amino-acids.—The presence of free D-alanine, together with *N*-malonyl-D-alanine and γ-L-glutamyl-D-alanine, has been established in seedlings of *Pisum sativum.*[155–157] D-Alanine is also present in seedlings of *Lolium* sp., *Zea mays,*[158] and *Helianthus annuus.*[159] D-Alanyl-D-alanine has been found in leaves of *Nicotiana tabacum.*[160]

The metabolism of both D and L-[*methyl*-^{14}C]methionine has been studied in a large number of species of Bryophyta. The L-isomer was transformed by trans-plants. Thus [*carboxy*-^{14}C]-D-valine, -D-leucine, -D-alanine, -D-methionine, and -D-lysine in seedlings of *Lolium* sp. and *Zea mays* all gave $^{14}CO_2$. Radioactivity from D-alanine was also incorporated into valine, leucine, and an unidentified compound proposed to be 2-methylserine.[161] In seedlings of *Lolium* sp. and *Zea mays,* [*carboxy*-^{14}C]-D-leucine is transformed into not only *N*-malonyl-D-leucine but also into L-leucine, 2-oxo-4-methylpentanoic acid, and $^{14}CO_2$.[158] Similarly, [*carboxy*-^{14}C]-D-valine in seedlings of *Helianthus annuus* and *Zea mays,* is converted into not only *N*-malonyl-D-valine but also L-valine, 2-oxo-3-methylbutyric acid, and $^{14}CO_2$.[159, 162] D-Leucine in *Vicia faba* is transformed both to *N*-malonyl-D-leucine and to D-leucine.[163, 164]

The metabolism of both D- and L-[*Methyl*-^{14}C]methionine has been studied in a large number of species of Bryophyta. The L-isomer was transformed by transamination, whereas the D-isomer in some species was transformed into the same compounds as the L-isomer, in others into the *N*-acetyl derivative. These results have been considered in a taxonomic context, based on the known malonylation of D-amino-acids in vascular plants, the acetylation in fungi and some lichens, and the identical metabolism of D- and L- amino-acids in bacteria and algae.[165]

[151] E. Haslam, 'The Shikimate Pathway', Butterworths, London, 1974.
[152] R. Cardillo, C. Fuganti, D. Ghiringhelli, D. Giangrasso, and P. Grasselli, *Tetrahedron Letters*, 1972, 4875.
[153] P. O. Larsen, E. Pedersen, H. Sørensen, and P. Sørup, *Phytochemistry*, 1973, **12**, 2243.
[154] R. Longo, A. Castellani, P. Sberze, and M. Tibolla, *Phytochemistry*, 1974, **13**, 167.
[155] T. Ogawa, M. Fukuda, and K. Sasaoka, *Biochim. Biophys. Acta*, 1973, **297**, 60.
[156] M. Fukuda, A. Tokumura, and T. Ogawa, *Phytochemistry*, 1973, **12**, 2593.
[157] M. Fukuda, T. Ogawa, and K. Sasaoka, *Biochim. Biophys. Acta*, 1973, **304**, 363.
[158] R. W. Aldag and J. L. Young, *Planta*, 1970, **95**, 187.
[159] R. W. Aldag, J. L. Young, R. Kickuth, and F. Vardakis, *Biochem. Physiol. Pflanzen*, 1972, **163**, 363.
[160] M. Noma, M. Noguchi, and E. Tamaki, *Agric. and Biol. Chem.* (*Japan*), 1973, **37**, 2439.
[161] R. W. Aldag and J. L. Young, *Agron. J.*, 1970, **62**, 184.
[162] R. W. Aldag and J. L. Young, *Z. Pflanzenphysiol.*, 1974, **71**, 83.
[163] F. Führ and W. Steffens, *Landwirt. Forsch.*, 1972, **25**, 226.
[164] W. Steffens, W. Thiemann, W. Darge, and F. Führ, *Landwirtsch. Forsch., Sonderh.*, 1973, **28**, 61.
[165] M. Porkorny. *Phytochemistry*, 1974, **13**, 965.

O-Malonylaminobenzoic acid has been isolated from *Arachis hypogeae* L. thus demonstrating that aromatic amino-groups also may be transformed into malonyl derivatives.[166] A transaminase isolated from germinating *Arachis hypogeae* was specific for D-methionine.[167] A transaminase from seedlings of *Pisum sativus*, (devoid of racemase activity), was specific for D-amino-acids and may be responsible for the presence of rather large amounts of D-alanine in pea seedlings.[168]

3 Cyanogenic Principles

General.—Hydrogen-cyanide-releasing constituents of living matter continue to draw attention. Knowledge acquired within the two-year period since the last Report,[1] has added significantly to our understanding, yet not profoundly changed the general picture. Cyanolipids, and cyanogenic and pseudo-cyanogenic glycosides still constitute the recognized classes of natural hydrogen cyanide progenitors.

Individual determination of the four known types of cyanolipids[1] in established and novel sapindaceous seed oils has been facilitated by ^{1}H n.m.r. spectroscopy.[169] The occurrence of (57) to the extent of 50%, as the sole cyanolipid in the seed oil of *Cardiospermum hirsutum* Willd. is remarkable.[169]

A useful review has appeared, covering traditional and novel methods for detection, isolation, and identification of the 26 cyanogenic and pseudocyanogenic glycosides known at the moment,[170] Acacipetalin, a cyanogenic glucoside of long standing, has been assigned the structure (58)[170a] possessing a known cyanolipid counter-

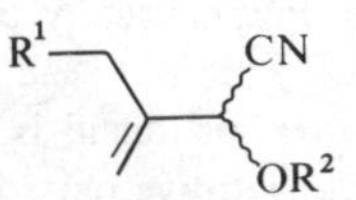

(57) R^1 = $OCO(CH_2)_{18}Me$; R^2 = $CO(CH_2)_{14-18}Me$

(59) R^1 = H; R^2 = $CO(CH_2)_{18}Me$

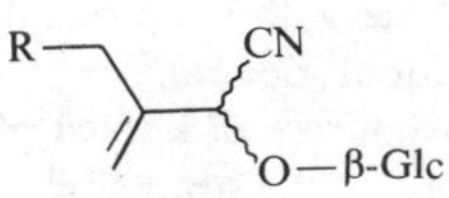

(58) R = H

(60) R = OH

part (59),[1] whereas cardiospermin (60) constitutes a novel cyanogenic glucoside, encountered in the vegetative parts of *Cardiospermum hirsutum*;[171] the high contents of the cyanolipid (57) in the seed oil of this species were commented upon above. The glucoside holocalin (61) has been shown by two groups[172,173] to possess the (*R*)-configuration at the benzylic centre; consequently, zierin, a diastereoisomer, belongs to the (*S*)-series. The interesting structure (62) has been assigned to simmondsin,[174] isolated from seeds of the jojoba plant, *Simmondsia californica* Nutt. (Buxaceae),[175] along with its 4- and 5-demethoxy-analogues, and the *cis*- and *trans*-2′-ferulates.[176]

166 Y. Kimura, M. Saito, T. Inoue, and S. Tamura, *Agric. and Biol. Chem. (Japan)*, 1974, **38**, 1507.
167 J. I. Durham, P. W. Morgan, J. M. Prescott, and C. M. Lyman, *Phytochemistry*, 1973, **12**, 2123.
168 T. Ogawa, M. Fukuda, and K. Sasaoka, *Biochem. Biophys. Res. Comm.*, 1973, **52**, 998.
169 D. Seigler, *Phytochemistry*, 1974, **13**, 841.
170 D. S. Seigler, *Phytochemistry*, 1975, **14**, 9.
170a C. S. Butterfield, E. E. Conn, and D. S. Seigler, *Phytochemistry*, 1975, **14**, 993.
171 D. S. Seigler, C. Eggerding, and C. Butterfield, *Phytochemistry*, 1974, **13**, 2330.
172 S. R. Jensen and B. J. Nielsen, *Acta Chem. Scand.*, 1973, **27**, 2661.
173 A. Nahrstedt, *Phytochemistry*, 1973, **12**, 2799.
174 C. A. Elliger, A. C. Waiss, jun., and R. E. Lundin, *J. Org. Chem.*, 1974, **39**, 2930.
175 C. A. Elliger, A. C. Waiss, jun., and R. E. Lundin, *J. Chem. Soc. Perkin I*, 1973, 2209.
176 C. A. Elliger, A. C. Waiss, jun., and R. E. Lundin, *Phytochemistry*, 1974, **13**, 2319.

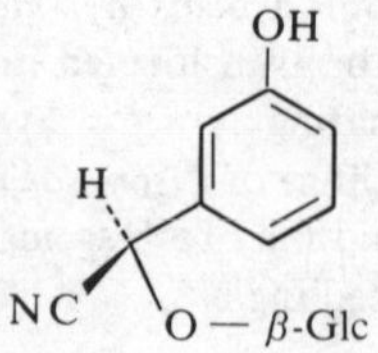

(61)

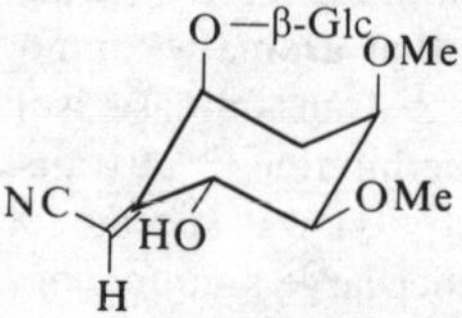

(62)

The non-cyanogenic *Simmondsia* glucosides bear structural resemblance to the αβ-unsaturated nitriles encountered in certain oils of Sapindaceae, others of which contain true cyanogenic cyanolipids.[1] Dehydration of the corresponding cyanohydrins, as yet unreported, or, alternatively, condensation of a nitrile-furnishing entity onto a cyclohexane unit, *e.g.* dehydroquinic acid, are conceivable, though unproven, biosynthetic pathways to cyclic glucosides of type (62).[175] Another intriguing novelty is the detection of *N*-cyano-compounds in natural sources, *viz.* as derivatives of alkaloids of the strychnine series in dried leaves of certain *Strychnos* species.[176] A cyanogenic, as yet unidentified, psychrophilic basidiomycete, commonly known as snow mould, type B, strain W2, has been found to produce cyanohydrins of glyoxylic and pyruvic acid, R·C(OH)(CN)·CO_2H (R = H, and Me), when cultivated on suitable media. The synthesis, claimed in earlier studies, of the glucosides linamarin and lotaustralin by the mould could not be confirmed.[178] The cyanide-accumulating fungus *Rhizoctonia solani* contains 2-aminobutyronitrile, supposedly formed upon reaction of cyanide with propionaldehyde and ammonia and enzymatically converted into 2-aminobutyric acid.[179]

New sources of known glycosides have been revealed: prunasin is a minor constituent of bitter almonds[180] and is also present in the vegetative parts of wild cherry, with amygdalin predominating in the generative parts.[181] Again, prunasin, the sole leaf glycoside of several *Cotoneaster* species, is accompanied in the fruits by amygdalin.[182] Sambunigrin, a repeatedly reported constituent of common elder (*Sambucus nigra*), co-occurs in certain collections with prunasin and the *m*-hydroxylated analogues zierin and holocain.[172] *Scheuchzeria palustris* (Juncaginaceae) has been reported as a new source of triglochinin.[183]

The ability to synthesize cyanogenic glycosides may be a general property of vascular plants, even if accumulation may often appear restricted and erratic. As recently pointed out, however, three major groups of glycosides, typified by: (i) prunasin, (ii) taxiphyllin, and (iii) linamarin, with an *in vivo* derivation from phenylalanine, tyrosine, and valine/isoleucine, respectively, are not randomly distributed.[184] Thus, ferns elaborate glycosides of the prunasin type, the gymnosperms glycosides which are exclusively taxiphyllin;[185] the class of angiosperms appears less uniform.

177 N. G. Bisset, A. K. Choudhury, and M. D. Walker, *Phytochemistry*, 1974, **13**, 255.
178 B. A. Tapper and M. A. Macdonald, *Canad. J. Microbiol.*, 1974, **20**, 563.
179 B. P. Mundy, F. H. S. Liu, and G. A. Strobel, *Canad. J. Biochem.*, 1973, **51**, 1440.
180 U. Schwarzmaier, *Phytochemistry*, 1972, **11**, 2358.
181 A. Nahrstedt, *Phytochemistry*, 1972, **11**, 3121.
182 A. Nahrstedt, *Phytochemistry*, 1973, **12**, 1539.
183 H. W. L. Ruijgrok, *Phytochemistry*, 1974, **13**, 161.
184 R. Hegnauer, *Biochem. System.*, 1973, **1**, 191.
185 B. Bijl-van Dijk, A. C. van der Plas-de Wijs, and H. W. L. Ruijgrok, *Phytochemistry*, 1974, **13**, 159,

Here, the potential of synthesizing any one, or even mixtures, of the three major glycoside types seems rather widely and evenly distributed, with the notable exception of the Magnoliidae subsection (comprising Polycarpicae and Papaveraceae) which, quite strikingly, shares with the monocotyledons a synthetic capacity restricted to glycosides of the taxiphyllin type. This circumstance has been adduced in support of a Magnoliidae-derived ancestry of the monocotyledons.[184]

Biosynthesis.—Studies of the *in vivo* synthesis of cyanogenic principles, still essentially limited to the glycosides, have continued. Two useful reviews have appeared which summarize the state of affairs up to 1972—73.[186,187] These, or the previous Report,[1] should be consulted for an account of the established anabolic pathways leading from certain protein amino-acids (valine, isoleucine, phenylalanine, and tyrosine) to the glycosidic end-products, *via* intermediates such as oximes, or 2-hydroxyoximes, nitriles, and cyanohydrins. Efforts have lately been directed towards a more refined and detailed understanding of certain of the individual steps in the general sequence, including their generality. We shall briefly discuss some major recent achievements.

The sequences aldoxime → nitrile → cyanohydrin → glucoside, or aldoxime → hydroxyaldoxime → cyanohydrin → glucoside, established by feeding and 'trapping' experiments for linamarin and prunasin biosynthesis in flax and cherry laurel shoots. respectively, have also been shown to obtain for dhurrin in sorghum shoots. Here, tyrosine, 4-hydroxyphenylacetaldoxime, and 4-hydroxyphenylacetonitrile all serve well. Support for the oxime as a true intermediate has been provided by 'trapping' experiments, which, however, failed in the case of the nitrile, most likely because of toxicity problems.[188] Studies with $^{18}O_2$ have ascertained that in linen flax seedlings the cyanohydrins of acetone and methyl ethyl ketone, the immediate natural precursors of the glucosides linamarin and lotaustralin, are produced by direct introduction of molecular oxygen into the corresponding nitriles or aldoximes, in the latter case followed by dehydration.[189] The stereochemical course of oxygenation has been studied in detail in the case of lotaustralin (63), previously known from *Trifolium repens.* Exposure of (±)-2-hydroxy-2-methylbutyronitrile to glucosylation, *in vivo* or *in vitro*, afforded 55—60% of lotaustralin and 45—50% of *epi*-lotaustralin (64). The exclusive occurrence in flax of (63) hence reflects the natural availability

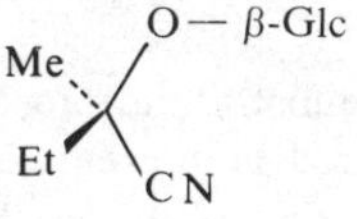

(63)

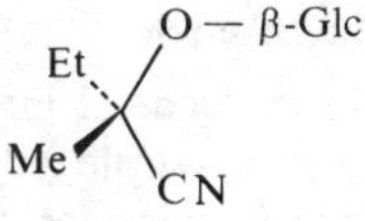

(64)

of only the (*R*)-cyanohydrin. From this conclusion, supplemented with the demonstration that L-isoleucine in excised flax shoots is converted into (63) and L-*allo*-isoleucine predominantly into (64), it follows that the introduction of oxygen takes place with retention of configuration, in keeping with the presumed

[186] E. E. Conn, *Biochem. Soc. Symp.*, 1973, **38**, 277.
[187] H. R. Schütte, in 'Fortschritte der Botanik', Springer Verlag, Berlin, 1973, Vol. 35, p. 103.
[188] K. J. F. Farnden, M. A. Rosen, and D. R. Liljegren, *Phytochemistry*, 1973, **12**, 2673.
[189] H. Zilg, B. A. Tapper, and E. E. Conn, *J. Biol. Chem.*, 1972, **247**, 2384.

involvement of a mixed function oxidase. The ability of flax shoots to convert 2-hydroxy-2-ethylbutyronitrile and 1-hydroxycyclopentanecarbonitrile into the corresponding β-D-glucosides contrasts with their failure to produce these non-natural glucosides, or deidaclin (65), a natural product, on feeding with the appropriate α-amino-acids.[190] A UDP-glucose:aldehyde cyanohydrin β-glucosyltransferase has been isolated from *Sorghum* seedlings and partially purified. Though exhibiting the highest activity for aromatic cyanohydrins, the enzyme will also accept a series of monophenolic substrates, whereas no significant transfer was observed to the cyanohydrins of acetone and acetaldehyde, both of which were rapidly glucosylated by the flax β-glucosyltransferase. Contrasting sharply with the latter in its stereospecificity, the *Sorghum* enzyme selectively converts (±)-4-hydroxymandelonitrile into dhurrin (66), virtually uncontaminated by taxiphyllin (67).[191]

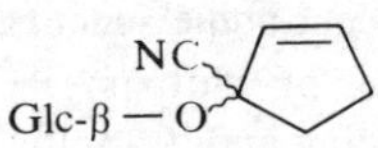

(65)

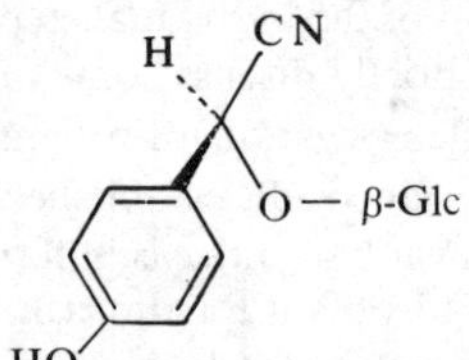

(66)

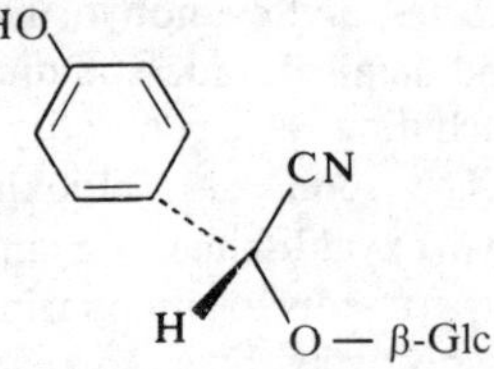

(67)

Cyanogenesis is not limited to the plant kingdom. The long-known production of hydrogen cyanide and benzaldehyde by polydesmid millipedes has been subjected to a detailed study in the species *Harpaphe haydeniana.* Convincing evidence has been adduced for mandelonitrile being the stored source of HCN and benzaldehyde, utilized as a defensive secretion. Extensive tracer studies have established the operation of an anabolic pathway from phenylalanine to mandelonitrile identical with, or very similar to, that established for higher plants. Again, a striking parallelism exists between the ability of the millipedes to metabolize HCN to form β-cyanoalanine, and the same potential of several species of higher plants.[192] Further studies of animal cyanogenesis are awaited with interest.

4 Glucosinolates

General.—Even if the general interest in the chemistry and biochemistry of glucosinolates seems unabated, no addition is required at present to the collection of 71 individual natural glucosides[193] known at the time of the last Report.[1] However, new sources of known glucosinolates have been reported. Thus, within the family Capparaceae: isopropyl-, and most likely also 2-butyl-glucosinolate, are present in *Boscia senegalensis* in addition to methylglucosinolate which is widely found within the family,[194] and (*R*)-2-hydroxy-3-butenylglucosinolate is a constituent of

[190] H. Zilg and E. E. Conn, *J. Biol. Chem.*, 1974, **249**, 3112.
[191] P. F. Reay and E. E. Conn, *J. Biol. Chem.*, 1974, **249**, 5826.
[192] S. S. Duffey, E. W. Underhill, and G. H. N. Towers, *Comp. Biochem. and Physiol.*, 1974, **47B**, 753.
[193] A. Kjær, in 'Chemistry in Botanical Classification', ed. G. Bendz and J. Santesson, Academic Press, New York and London, 1974, p. 229.
[194] A. Kjær, A. Schuster, P. Delaveau, and B. Koudogbo, *Phytochemistry*, 1973, **12**, 725.

Zilla spinosa.[195] In members of the Cruciferae: about 15 known, and several unidentified, glucosinolates occur in species of the genus *Cakile*; [196] a similar number of known compounds have been observed within the genus *Thelypodium*.[197]

Isothiocyanates found in Nature have invariably been regarded as secondary products of glucosinolate derivation. Hence, it may occasion surprise that isothiocyanates are now be to counted as natural products in their own right, after both sesqui-[198,199] and di-terpenoid[199] isothiocyanates have been recorded as constituents of marine sponges, most likely originating from the accompanying isonitriles.

Much progress can be recorded in our understanding of myrosinases, a collection of isoenzymes catalyzing the hydrolytic detachment of glucose from glucosinolates. Thus, the chromatographic separation of three isoenzymes from white mustard seed (*Sinapis alba*),[200] and of four major isoenzymes from rape seed (*Brassica napus*)[201] has been reported. Though widely varying in physicochemical and enzymic properties, including their susceptibility to ascorbic acid activation, the isoenzymes exhibited only minor differences in their substrate specificity.[202] Japanese workers similarly separated and purified four isoenzymes with myrosinase activity from mustard seed, differently activated by ascorbic acid.[203] Amino-, mercapto-, and histidyl groupings seem essential for enzymic activity.[204] Ascorbic acid activation of the plant myrosinases seems mediated through conformational changes rather than dissociation–association of the genuine enzyme.[205]

Much interest is still associated with the repeatedly observed curious difference between the types of products resulting from hydrolysis of glucosinolates catalyzed by endogenous enzymes (autolysis) and those produced when extraneous myrosinases are implied.[206] In the latter case, isothiocyanates (69), resulting from rearrange-

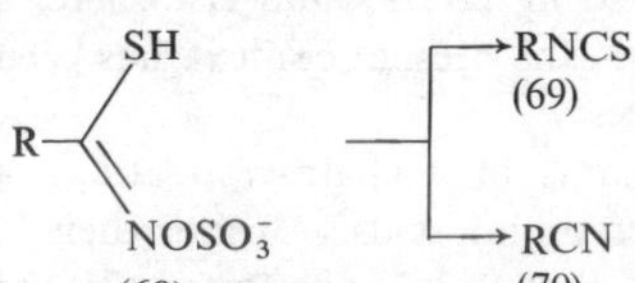

(68) (69) (70)

ment, and nitriles (70), arising by fragmentation of the aglucone (68), are typical products. A few years ago, however, it was discovered that unheated seed meals of rape, or *Crambe abyssinica*, when subjected to autolysis, converted much of the parent (*R*)- and (*S*)-2-hydroxy-3-butenylglucosinolate, respectively, into the racemic, diastereomeric 1-cyano-2-hydroxy-3,4-epithiobutanes (71). A comparative study of the product composition as a function of extraneous and endogenous enzyme sources in rape-seed meal has now appeared.[206] Episulphide formation from un-

[195] M. S. Karawya, G. M. Wassel, and B. S. Elmenshawi, *Acta Pharm. Suec.*, 1973, **10**, 477.
[196] J. E. Rodman, *Contrib. Gray Herbar.*, 1974, No. 205.
[197] J. A. Al-Shehbaz, *Contrib. Gray Herbar.*, 1973, No. 204.
[198] F. Cafieri, E. Fattorusso, S. Magno, C. Santacroce, and D. Sica, *Tetrahedron*, 1973, 4259.
[199] B. J. Burreson and P. J. Scheuer, *J.C.S. Chem. Comm.*, 1974, 1035.
[200] R. Björkman and J.-C. Janson, *Biochim, Biophys. Acta*, 1972, **276**, 508.
[210] B. Lönnerdal and J.-C. Janson, *Biochim, Biophys. Acta*, 1973, **315**, 421.
[202] R. Björkman and B. Lönnerdal, *Biochim. Biophys. Acta*, 1973, **327**, 121.
[203] M. Ohtsuru and T. Hata, *Agric. and Biol. Chem. (Japan)*, 1972, **36**, 2495.
[204] M. Ohtsuru and T. Hata, *Agric. and Biol. Chem. (Japan)*, 1973, **37**, 269.
[205] M. Ohtsuru and T. Hata, *Agric. and Biol. Chem. (Japan)*, 1973, **37**, 1971.
[206] V. K. Srivastava and D. C. Hill, *Phytochemistry*, 1974, **13**, 1043.

saturated glucosinolates during autolysis may be general. Thus, it was recently noted that within a collection of turnip rape strains (*Brassica campestris*), certain cultivars of the Indian Yellow Sarson type possessed the ability to convert 3-butenylglucosinolate into (±)-1-cyano-3,4-epithiobutane (72), rather than 3-butenyl iso-

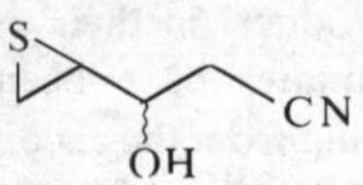

(71)

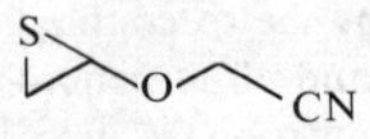

(72)

thiocyanate, on autolysis. After heating for 30 min at 115 °C, followed by addition of enzyme, seed meals of all cultivars afforded isothiocyanate as the major product.[207] For an understanding of the deviating reaction course during autolysis, the separation from *Crambe* seed-meal extracts of a protein factor, devoid of thioglucosidase activity but promoting the formation of the 1-cyano-2-hydroxy-3,4-epithiobutanes (71) in the presence of endogenous thioglucosidase, deserves attention. Alone, the thioglucosidase proper converts the parent glucosinolate essentially into 1-cyano-2-hydroxy-3-butene.[208] Following this observation, renewed consideration might be given to another problem of long standing, namely the formation of allyl and benzyl thiocyanate, along with, or instead of, the corresponding isothiocyanates, in selected crucifer species.[208,209]

Biosynthesis.—The common *in vivo* pathway, leading from α-amino-acids *via* oximes to glucosinolates in higher plants, can now be regarded as accepted. The evidence on which it rests has been reviewed in Vol. 2,[1] and elsewhere.[187,210] The general role of oximes, obviously pertinent in the present context, has been the subject of another recent review.[211]

The operation in higher plants of a chain-elongating mechanism, successively transforming certain protein α-amino-acids (73) into their higher homologues (76) through the intermediacy of 2- (74) and 3-substituted (75) malic acids, has formerly been suggested and, in part, experimentally supported. Previous studies on *Nasturtium officinale*, a source of 2-phenylglucosinolate, which established phenylalanine (73, R = $PhCH_2$), 3-benzylmalic acid (75, R = $PhCH_2$), and 2-amino-4-phenylbutyric acid (76, R = $PhCH_2$) as efficient biochemical precursors, have now been supplemented with a demonstration of 2-benzylmalic acid (74, R = $PhCH_2$) in the same role, both in *Nasturtium* and *Barbarea vulgaris*.[133]

Efforts to elucidate the biochemical background for the low glucosinolate contents in certain rape-seed strains, such as the Bronowski cultivar, have continued. A postulated biosynthetic block in the *in vivo* synthesis of 3-butenylglucosinolate, situated somewhere between the parent 2-amino-6-(methylthio)caproic acid (77) and desulpho-3-butenylglucosinolate (79), has now been traced to a step prior to 5-methylthiopentanal oxime (78).[212]

[207] J. T. O. Kirk and C. G. Macdonald, *Phytochemistry*, 1974, **13**, 2611.
[208] H. L. Tookey, *Canad. J. Biochem.*, 1973, **51**, 1654.
[209] M. Saarivirta, *Planta Medica*, 1973, **24**, 112.
[210] E. W. Underhill, L. R. Wetter, and M. D. Chisholm, *Biochem. Soc. Symp. Cambridge, Eng.*, 1973, **38**, 303.
[211] S. Mahadevan, *Ann. Rev. Plant. Physiol.*, 1973, **24**, 69.
[212] E. Josefsson, *Physiol. Plant.*, 1973, **29**, 28.

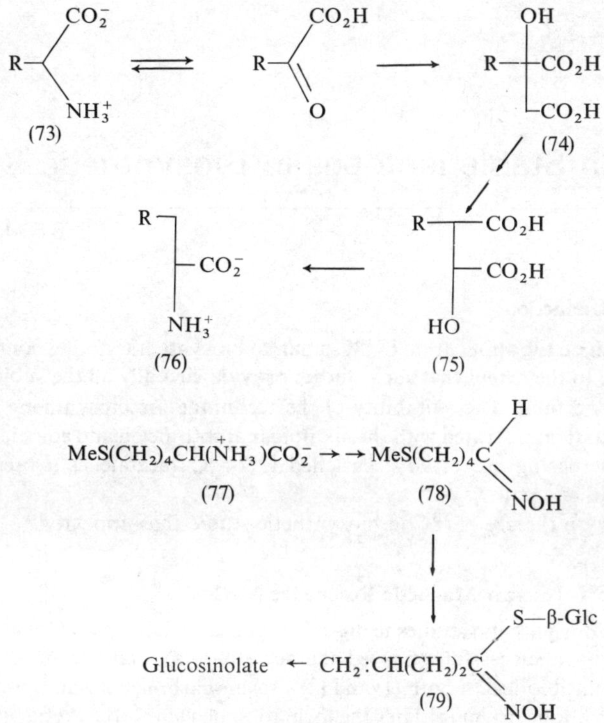

Hydroxy-groups in aliphatic side-chains, as present in *e.g.* the (*R*)-2-hydroxy-3-butenylglucosinolate in rape seed, appear to be introduced at a late stage in the biosynthetic sequence. Low incorporation of desulpho-3-butenylglucosinolate (79) into the hydroxy-glucosinolate within the Bronowski cultivar, as compared with that observed in an ordinary rape strain, suggests the existence of at least one biosynthetic block in the low-glucosinolate strain situated after the formation of the oxime (78).[212]

The demonstration, based on chromatographic evidence, that horse-radish plants are capable of converting $Na_2{}^{75}SeO_4$ into a selenium analogue of allylglucosinolate, yet with an unknown oxidation level of the Se-label, represents an interesting novelty warranting more detailed studies.[213]

Improved insight into the detailed biosynthetic pathways, leading to glucosinolates, provides a good occasion for reconsidering the potential value of this collection of plant constituents as biochemical markers in a taxonomic and evolutionary context. Attempts in this direction have been reported.[193,196,197]

[213] J. M. Stewart, S. N. Nigam, and W. B. McConnell, *Canad. J. Biochem.*, 1974, **52**, 144.

7

N.M.R. with Stable Isotopes in Biosynthetic Studies

BY M. TANABE

1 Introduction

In 1974, research on the application of ^{13}C n.m.r. to biosynthetic studies continued at a very brisk pace, to the extent that such studies provide virtually all the subject-matter for the present account. The suitability of the technique for biosynthetic studies in higher plants was demonstrated with the alkaloids camptothecin and colchicine. There has been an increasing use of 90% enriched [1,2-$^{13}C_2$]acetate as a precursor for defining biosynthetic pathways.

A new review on the use of ^{13}C in biosynthetic studies has appeared.[1]

2 ^{13}C Nuclear Magnetic Resonance Method

Rifamycin S.—Biosynthetic studies using ^{13}C n.m.r. on rifamycin S (1)[2] and streptovaricin D (2)[3] have successfully proved the acetate–propionate origin of the 'ansa' chains of these antibiotics. In both (1) and (2) a seven-carbon meta-amino-compound of unknown metabolic origin initiates the formation of a single polyketide chain composed of acetate and propionate units as shown in Scheme 1. Investigations on the origin of the C_7 unit are of interest because of the unusual *meta*-aminobenzoic acid substitution pattern and the reported non-incorporation of [^{14}C]shikimic acid.[4]

^{13}C N.m.r. studies indicated that [1-^{13}C]glucose (15% enriched) labelled C-1 and C-10 of this aromatic chromophore.[5] The relative enrichment factors for C-1 and C-10, as calculated from the ratio of intensities in the enriched and natural abundance spectra, were 3.4 and 1.9 respectively. These studies also showed that [1-^{13}C]glycerate (90% enriched) labelled C-8 and C-3 of rifamycin S, where the relative enrichment factors were 4.4 and 2.0 respectively.

The results were interpreted as evidence for a shikimate-type origin of the aromatic chromophore. The apparent contradiction in the very poor and non-specific incorporation of [^{14}C]shikimic acid was rationalized by proposing that the C_7 unit is derived from an earlier intermediate of aromatic biosynthesis such as 5-dehydroquinic acid (3) or 5-dehydroshikimic acid (4). The metabolic route shown in Scheme 2 would explain this labelling pattern. The higher enrichment at C-1 with [1-^{13}C]glucose indicated that it was derived from the methylene carbon of phosphoenolpyruvate

[1] U. Séquin and A. I. Scott, *Science*, 1974, **186**, 101.
[2] (*a*) R. J. White, E. Martinelli, G. G. Gallo, G. C. Lancini, and P. J. Benyon, *Nature*, 1973, **243**, 273; (*b*) E. Martinelli, R. J. White, and G. G. Gallo, *Tetrahedron Letters*, 1974, 1367.
[3] B. Milavetz, K. Kakinuma, K. L. Rinehart, J. P. Rolls, and W. J. Haak, *J. Amer. Chem. Soc.*, 1973, **95**, 5793.
[4] A. Karlsson, G. Sartori, and R. J. White, *European J. Biochem.*, 1974, **47**, 251.
[5] R. J. White and E. Martinelli, *F.E.B.S. Letters*, 1974, **49**, 233.

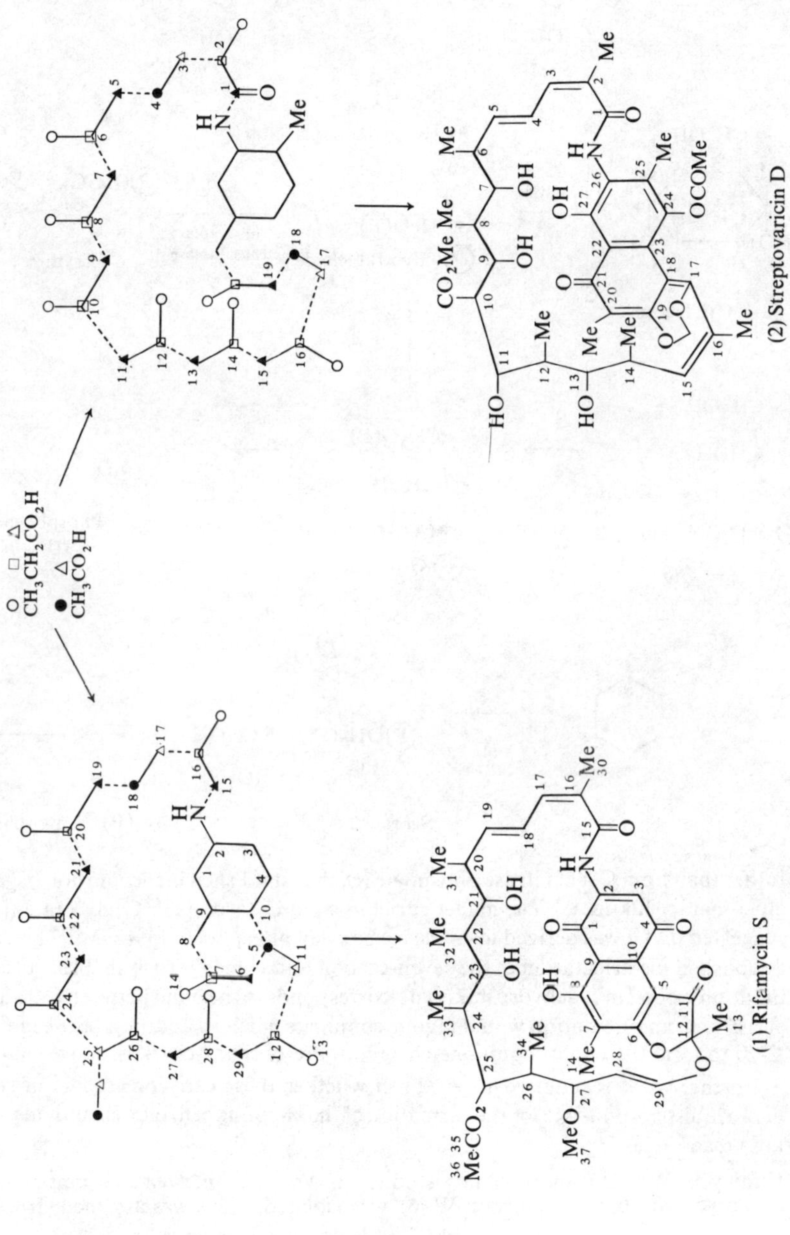

Scheme 1

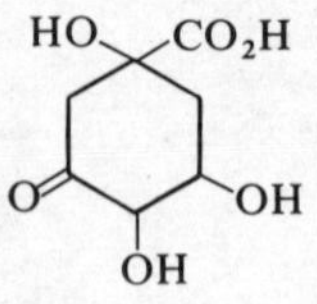

(3)

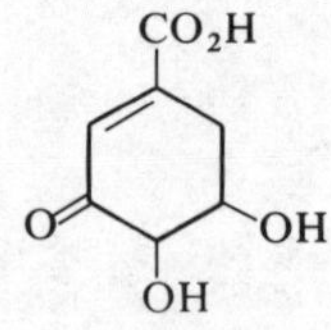

(4)

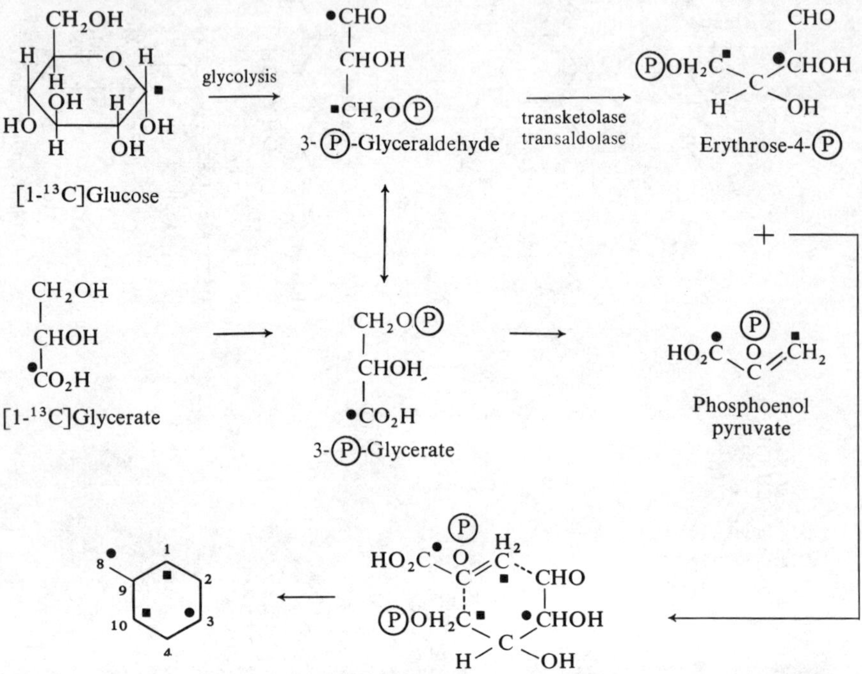

Scheme 2 Ⓟ = pyrophosphate

rather than from C-4 of tetrose phosphate (*cf.* the established incorporation of [1-^{14}C]-glucose into shikimate). The higher enrichment of C-8 with [1-^{13}C]glycerate similarly suggested that it was derived from the carboxyl of phosphoenolpyruvate. These results established the orientation of the seven-carbon unit, which is such that the location of the double bond in 5-dehydroshikimate corresponds to the bond between C-9 and C-10 of (1). Such an orientation would favour conjugate Michael addition between C-8 and C-10 to form the naphthoquinone chromophore of rifamycin S, as shown in (5).

Further work is required to establish whether these early precursors of shikimic acid are also responsible for the formation of the *m*-aminobenzoic acid unit in the other ansa macrolides.

Rifamycin W.—During screening studies on *Nocardia mediterranei* mutants, a new ansamycin antibiotic, rifamycin W (6), was isolated.[6] This was assigned structure (6)

[6] E. Martinelli, G. G. Gallo, P. Antonini, and R. J. White, *Tetrahedron*, 1974, **30**, 3087.

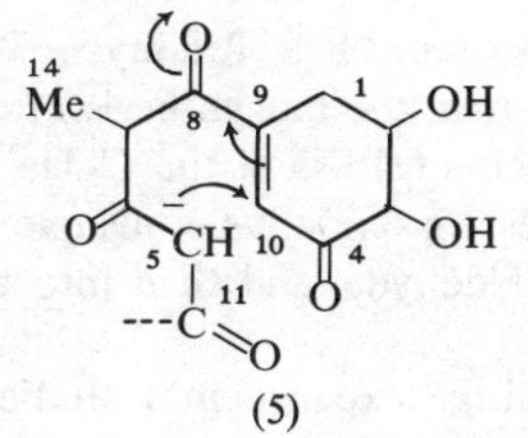

(5)

by u.v., i.r., m.s., and ^{1}H and ^{13}C n.m.r. spectroscopic studies and by comparison of its properties with the model compound rifamycin S (1). Rifamycin W is a new structural

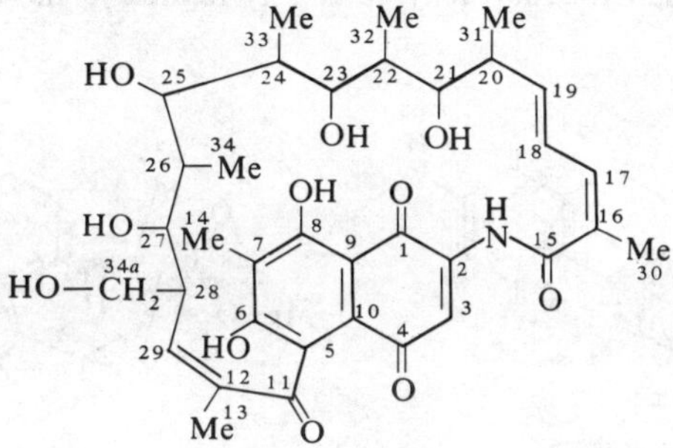

(6) Rifamycin W

variation in the rifamycin se: ies and has a continuous ansa carbon chain not interrupted by an oxygen atom, corresponding to that of streptovaricin D (2), the other naphthalenic ansa-macrolide.

Studies on the incorporation of ^{13}C-enriched precursors into rifamycin W revealed the expected biosynthetic pattern for ansa-chain formation.[7] The chain was initiated by the same seven-carbon unit, with the formation of a single linear polyketide chain composed of eight propionate and two acetate units.

C-6, -15, -19, -21, -23, -25, and -27 of rifamycin W were labelled by [1-^{13}C]propionate and C-7, -12, -16, -20, -22, -24, -26, and -28 by [2-^{13}C]propionate. [3-^{13}C]Propionate led to enrichment at C-13, -14, -30, -31, -32, -33, -34, and -34a, and [1-^{13}C]acetate enriched C-11 and -17. The alignment of these units in rifamycin W accounted for its biosynthesis and was identical to that previously proposed for the biosynthesis of rifamycin S and streptovaricin D (Scheme 3).

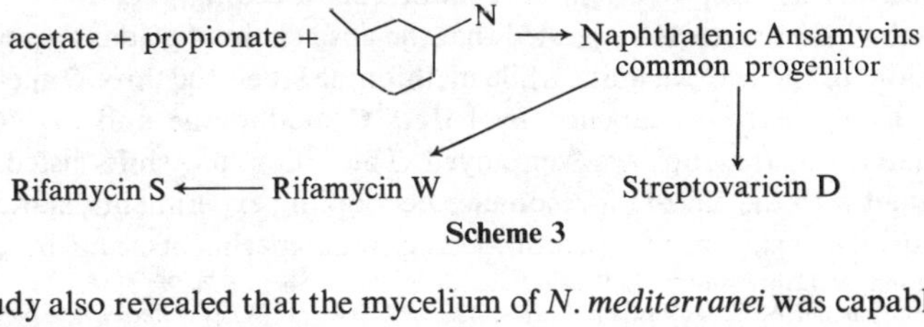

Scheme 3

This study also revealed that the mycelium of *N. mediterranei* was capable of oxidative removal of the C-34a methyl group and inserting oxygen into the ansa chain to

[7] R. J. White, E. Martinelli, and G. Lancini, *Proc. Nat. Acad. Sci. U.S.A.*, 1974, **71**, 3260.

transform rifamycin W to rifamycin S. Rifamycin W was thus considered to be the missing common progenitor in the biosynthesis of other known rifamycins with an oxygen atom in the ansa chain. Loss of the C-34a methyl group in the rifamycins presumably involves successive oxidative conversions into hydroxymethyl (as found in rifamycin W), into the aldehyde, and then into the carboxylic acid, followed by decarboxylation.

In the absence of definitive experimental studies, several possible mechanisms can be considered for the introduction of oxygen between C-12 and C-29, such as by epoxidation of the $\Delta^{12(29)}$ double bond of rifamycin W and subsequent cleavage of the C-12—C-29 bond of the epoxide, or by a modified Bayer–Villiger oxidation. Oxidative loss of the C-34a methyl group may be closely related to the oxygen-insertion, as shown in Scheme 4.

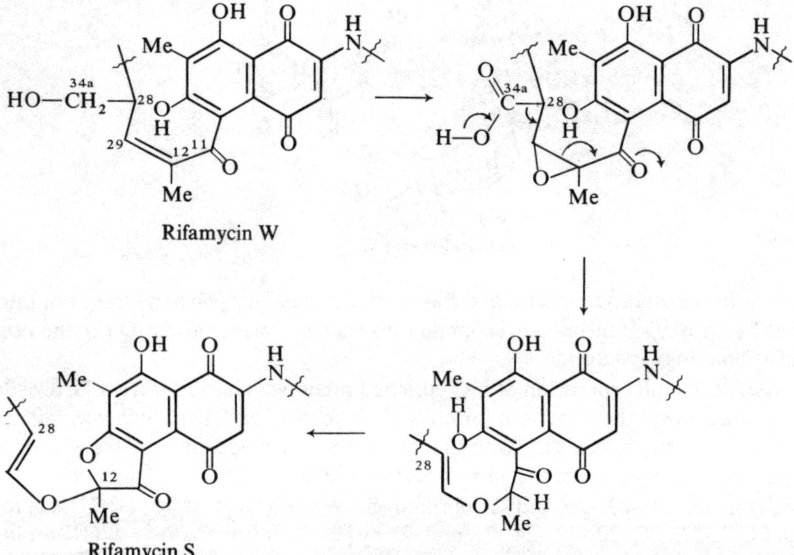

Scheme 4

Geldanamycin.—Biosynthetic results on the only benzoquinone in the ansamycin series, geldanamycin (7), revealed that this is formed through the same type of biogenetic pathway as streptovaricin and rifamycin.[8] Feeding experiments conducted with ^{14}C-labelled precursors suggested that the ansa chain of geldanamycin is formed from propionate and acetate units, while methionine labels the three *O*-methyl groups.

To identify the carbons labelled by [*Me*-^{13}C]methionine and [1-^{13}C]propionate, ^{13}C n.m.r. spectroscopy was employed. The ^{13}C n.m.r. shifts listed in Table 1 were assigned with the aid of off-resonance decoupling experiments, standard chemical-shift correlations,[9] specific proton-decoupling experiments, and by comparison

[8] R. D. Johnson, A. Haber, and K. L. Rinehart, *J. Amer. Chem. Soc.*, 1974, **96**, 3316.

[9] (*a*) L. F. Johnson and W. C. Jankowski, 'Carbon-13 NMR Spectra', Wiley–Interscience, New York, 1972; (*b*) J. B. Stothers, 'Carbon-13 NMR Spectroscopy', Academic Press, New York, 1972; (*c*) G. C. Levy and G. L. Nelson, 'Carbon-13 Nuclear Magnetic Resonance for Organic Chemists', Wiley–Interscience, New York, 1972.

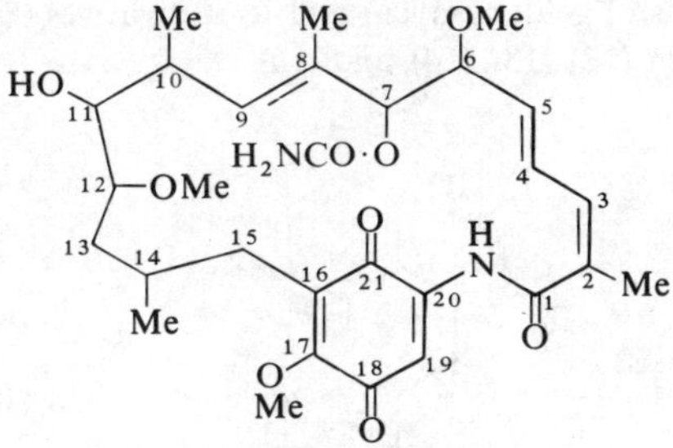

(7) Geldanamycin

Table 1 ^{13}C *N.m.r. data for geldanamycin and derivatives*

Group	*Carbon*	*Chemical Shift*[a] (7)[b]	(8)[c]	(9)[d]	(10)[e]
—C═O	C-21	183.7[f]	183.3[g]	184.2[h]	183.2[i]
	C-18	183.1[f]	182.6[g]	180.5[h]	183.0[i]
	C-1	169.1	169.2	167.7	170.0
	C-7a	156.0	156.0	155.8	156.1
═C—O—	C-17	156.4	154.7	158.1	156.4
═C—	C-20	139.6	140.9	148.8	140.4
	C-2	133.2[j]	133.5[k]	132.3[l]	133.9[m]
	C-8	132.6[j]	132.5[k]	131.4[l]	133.7[m]
	C-16	128.1	117.1	127.4	127.6
═CH—	C-5	137.8[s]	137.3	134.2	134.5
	C-9	131.9[s]	132.1	130.2	129.1
	C-4	128.4	128.2	129.1	128.4
	C-3	125.7[s]	125.9	125.5	126.4
	C-19	110.9[s]	108.5	97.3	109.8
—CH—O—	C-12	81.6[n]	81.1[o]	79.7[p]	78.6[q]
	C-7	80.6[s]	80.6	78.6[p]	78.2[q]
	C-6	80.3[n]	80.4[o]	77.2[p]	77.9[q]
	C-11	71.9	71.8	72.9[p]	74.4[q]
—OCH_3	C-17a	61.0	—	61.0	61.2
	C-12a	56.5	56.5	55.9	56.0
	C-6a	56.0	56.0	55.9	56.0
—CH—	C-10	32.1	32.2	35.1	32.3
	C-14	26.6[s]	26.8	28.4	29.6
—CH_2—	C-15	31.7	31.2	34.7	31.8
	C-13	31.0[s]	30.9	30.9	30.4
—CH_3	C-14a	23.2[s]	23.1	19.3[r]	21.8
	C-10a	13.0[s]	13.2	17.1[r]	13.2
	C-8a	12.5	12.5	13.9[r]	14.3
	C-2a	12.2[s]	12.3	12.2	12.2
				51.8[t]	170.0[u]
					15.1[v]

[a] p.p.m. downfield from TMS; [$^2H_6$]DMSO = 39.5 p.p.m.; [b] (7) = geldanamycin; [c] (8) = des-*O*-methylgeldanamycin; [d] (9) = geldanamycin methanolysis product; [e] (10) = geldanamycin acetate; [f–r] These values may be interchanged; [s] Assignment made by specific proton decoupling; [t] Methoxy at C-1; [u] Acetate carbonyl carbon; [v] Acetate methyl carbon.

of the ^{13}C n.m.r. shifts of geldanamycin and its derivatives (8), (9), and (10) with the model compounds (11), (12), (13), (14), and (15).

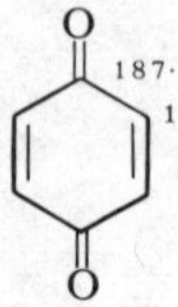

(11)

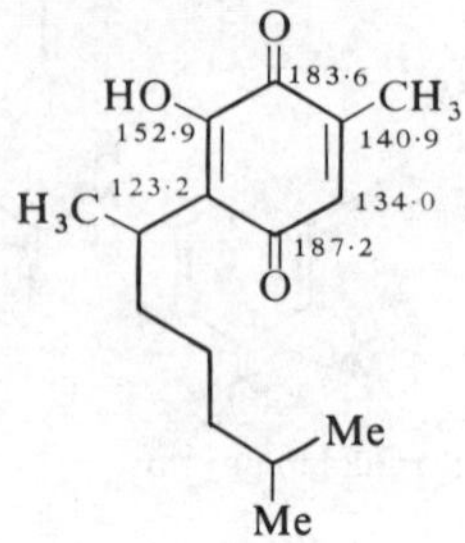

(12)

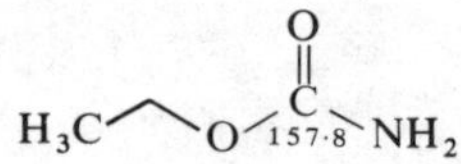

(13)

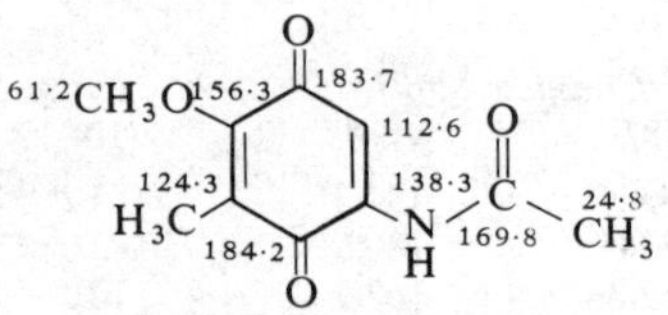

(14)

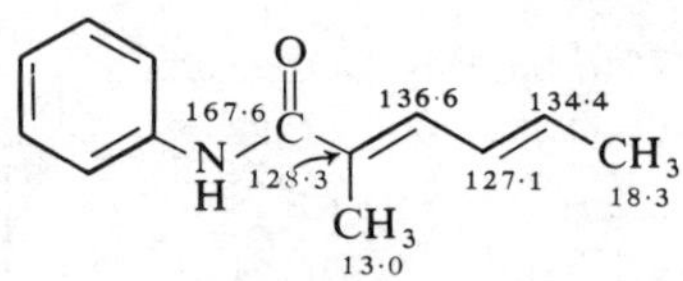

(15)

Mass spectrometry of geldanamycin labelled by [*Me*-^{13}C]methionine showed 81% unlabelled, 10% mono-, 5% di-, and 4% tri-labelled (an average of 11% at each of three labelled carbons). The ^{13}C n.m.r. spectrum clearly showed that only the three methoxy-peaks at 56.0, 56.5, and 61.0 p.p.m. were enriched by ^{13}C, with relative intensities of 1.08:1.08:0.86. The peak at 61.0 p.p.m. was assigned to C-17a since it was missing in the ^{13}C n.m.r. spectrum of (8). The 56.5 p.p.m. signal was assigned to C-12a since it showed an upfield shift in the spectrum of geldanamycin 11-acetate, presumably caused by steric interaction between the C-11 acetoxyl- and the C-12a methoxy- groups. The remaining peak at 56.0 p.p.m. was then assigned to C-6a.

Geldanamycin enriched by [1-^{13}C]propionate was 65% unlabelled, 18% mono-, 10% di-, 5% tri-, and 2% tetra-labelled, an average of 15% labelled at each of the four labelled carbons. In the ^{13}C n.m.r. spectrum only four highly enriched peaks were observed at 31.0, 80.6, 131.9, and 169.1 p.p.m. (relative intensities of 1.17:0.90:1.14:0.79). The 31.0, 80.6, and 131.9 p.p.m. resonances were identified by specific proton decoupling of the attached protons at $\delta = 1.45$, 4.90, and 5.52. The remaining absorption at 169.1 p.p.m. was identified as the amide carbonyl since it resonated at nearly the same position as the amide carbonyl of streptovaricin D. The labelling at C-1, C-7, C-9, and C-13 corresponds perfectly to a clockwise direction of growth to form the ansa chain of propionate and acetate units as shown in Scheme 5. The reverse direction of growth commencing with the C-1 amide group was excluded. The clockwise formation of the ansa chain of geldanamycin is thus the same as that found for rifamycin and streptovaricin.

The benzoquinone unit with attached carbon (C-15 to C-21) was not labelled by propionate or methionine. The formation of the ansa chain of geldanamycin is therefore presumably also initiated by the same C_7 *m*-aminobenzoic acid unit as in rifamycin and streptovaricin. Further experimental verification is needed in geldanamycin

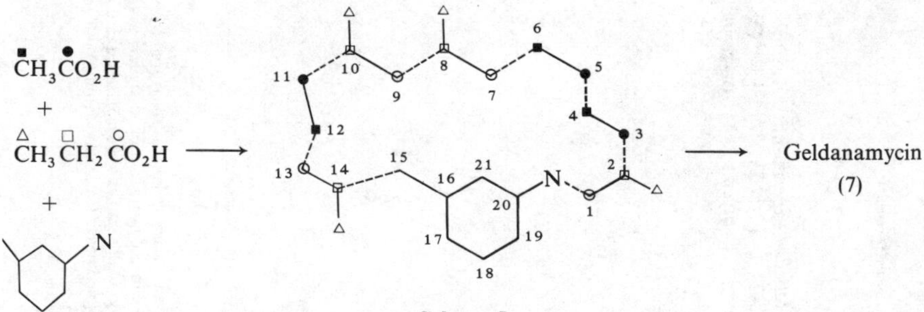

Scheme 5

biosynthesis to ascertain if precursors such as [^{13}C]glucose and [^{13}C]glyceric acid are in fact incorporated as they are in rifamycin S.

Metacycloprodigiosin and Undecylprodigiosin.—Earlier ^{13}C n.m.r. studies established that prodigiosin (16), the red pigment produced by *Serratia marcescens*, is biosynthesized from acetate, alanine, serine, proline, and methionine by a unique polypyrrole-forming process.[10] Similar ^{13}C n.m.r. biosynthetic studies have now been reported[11] for two related tripyrrole metabolites, metacycloprodigiosin (17) and undecylprodigiosin (18) which are produced as co-metabolites in *Streptomyces longisporus ruber* (M-3).

The ^{13}C n.m.r. chemical shifts of (17) and (18) are shown in Table 2, along with incorporation values from each of the ^{13}C-labelled precursors fed. Definitive chemical shift assignments for the methylene carbons of the undecyl side-chain of (17) and (18)

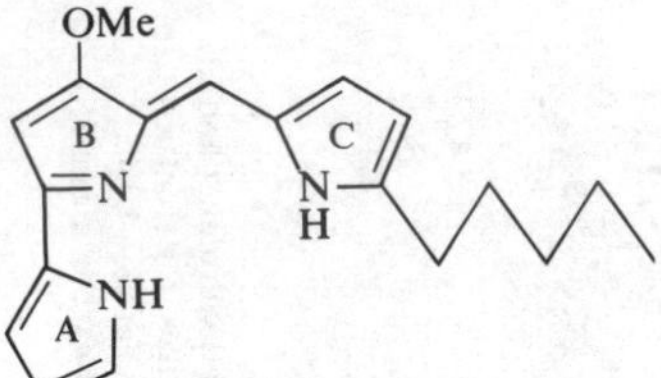

(16)

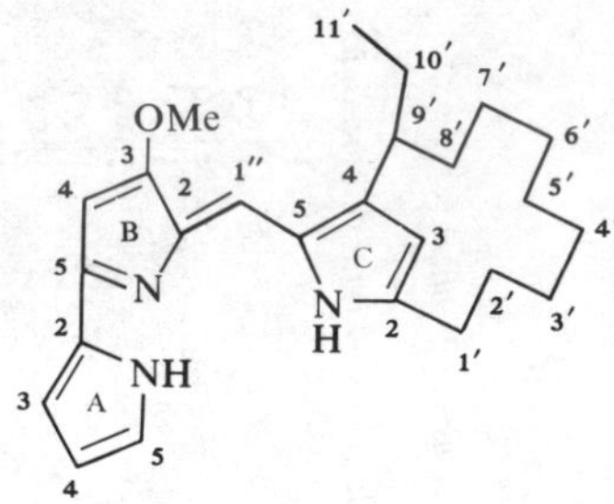

(17)

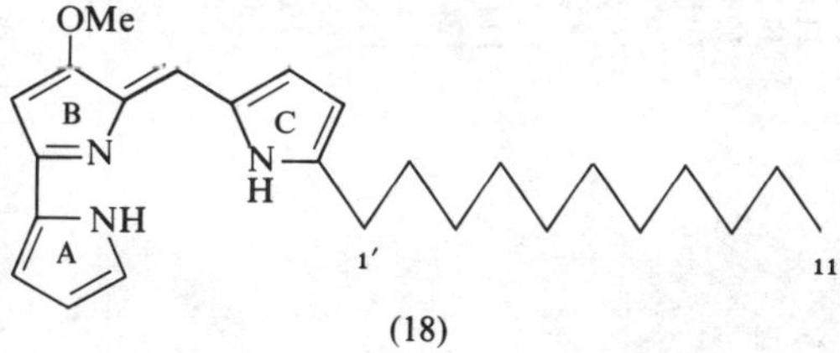

(18)

[10] (*a*) R. J. Cushley, D. R. Anderson, S. R. Lipsky, R. J. Sykes, and H. H. Wasserman, *J. Amer. Chem. Soc.*, 1971, **93**, 6284; (*b*) H. H. Wasserman, R. J. Sykes, P. Peverada, C. K. Shaw, R. J. Cushley, and S. R. Lipsky, *J. Amer. Chem. Soc.*, 1973, **95**, 6874.

[11] H. H. Wasserman, C. K. Shaw, R. J. Sykes, and R. J. Cushley, *Tetrahedron Letters*, 1974, 2787.

Table 2 ^{13}C *N.m.r. data for (17) and (18)*

	Chemical shifts[a,e]		*Percent excess of* ^{13}C *in the labelled experiment*[b,c]				
	Undecylprodigiosin (18)	*Metacycloprodigiosin* (17)	[1-^{13}C]-*acetate*	[2-^{13}C]-*acetate*	[1-^{13}C]-DL-*proline*	[2-^{13}C]-*glycine*	[3-^{13}C]-DL-*serine*
A2	131.2	131.6	—	1(1)[c]			
A3	115.7	115.0	—	1(1)	—	—	1
A4	112.8	112.6	—	4(4)	—	—	2
A5	125.6	125.0	4(3)[c]				
B2	141.4	140.4	—	—	—	2	—
B3	171.9	171.7	2(1)				
B4	98.0	98.1	—	5(4)	—	—	2
B5	163.2	162.1	1(1)	—	22(23)[c]	—	—
C2	147.4	147.2	3(2)				
C3	111.4	111.9	—	6(6)	—	—[d]	2[d]
C4	124.0	144.0	3(3)				
C5	131.2	130.2	—	—	—	3	—
1″	118.8	116.1	—	—	—	2	1
OMe	61.0	61.0	—	—	—	6	2
11′	16.8	15.3	—	4(3)	—	—	1
10′	25.4	—	3				
9′	34.7	42.1	—	4	—	—	1
8′	—	36.9	3(1)				
1′	30.1	—	—	(4)			

[a] Chemical shifts are given relative to HMDS (hexamethyldisilane) and are calculated from the solvent peak $CHCl_3$ (δc 80.0); [b] Percent excess of label is measured from peak heights and is rounded off to the nearest 1%; [c] Values in parentheses are derived from undecylprodigiosin; [d] In the case of undecylprodigiosin, enriched signals were observed at B2, C5, 1″, and OMe for the glycine feedings and at A3, A4, B4, C3, 1″, OMe, 11′, and 9′ for the serine feedings. However, since very small samples were isolated, quantitative data were difficult to obtain; [e] Unassigned peaks for carbon atoms 1′—7′ and 10′ in (17): 32.8, 30.7, 30.3, 29.4, 27.9, 27.5, 25.4. Unassigned peaks for carbon atoms 2′—8′ in (18): 32.4—32.1.

could not be made. Incorporation of [1-^{13}C]proline,[1-^{13}C]acetate, and [2-^{13}C]acetate establishes the close biosynthetic relationship between (17), (18), and (16), particularly in the formation of the methoxybipyrrole in rings A and B common to these metabolites.

The high level of incorporation of [1-^{13}C]proline at carbon B5 (22—25%) verifies that ring A and carbon B5 are derived from an intact proline unit. As in the biosynthesis of prodigiosin, carbons B3 and B4 of (17) and (18) are derived from an acetate unit. Similarities in the biosynthesis of all three compounds included the incorporation of [3-^{13}C]serine at the 1″-position. However, substantial simultaneous conversion of serine into acetate also occurs in this organism, as noted by the incorporations at carbons A3, B4, C3, C11′, and C9′. Conversion of serine into the 'one-carbon' pool also occurs as demonstrated by incorporation of label at the methoxy-carbon.

Incorporations of the label from [2-^{13}C]glycine were observed at B2, 1″, C5, and OMe in (17) and (18). The labelling at B2 and B1″ from glycine parallels results in the biosynthesis of (16) and may be accounted for by the well-established glycine–serine interconversion. A major difference between the biosynthesis of (17) and (18) compared with (16) is the appearance of glycine label at C5. This result agrees with the different mode of acetate incorporation into ring C of (17) and (18) shown in Scheme 6.

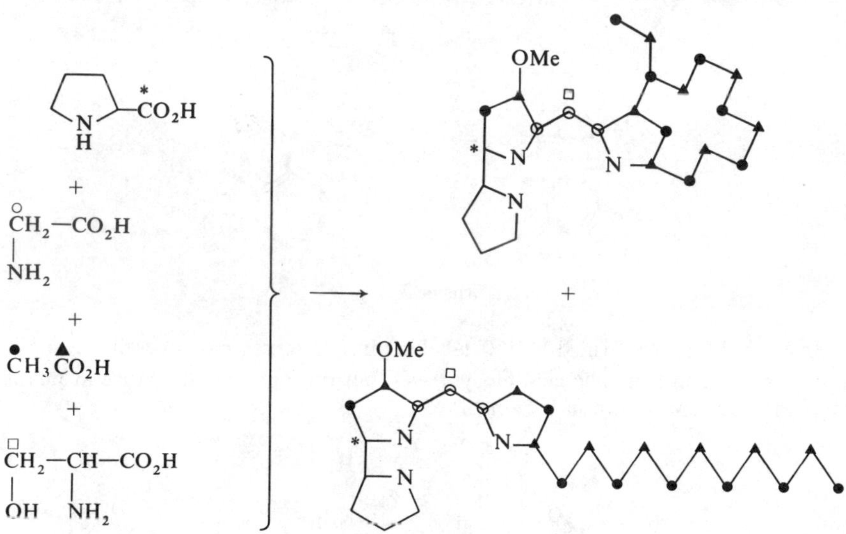

Scheme 6

The formation of ring C in undecylprodigiosin (18) would seem to involve the initial formation of a C_{14} polyketide chain with eventual incorporation of three of the carbons into ring C, with the fourth ring C-carbon and the nitrogen being derived from glycine. The remaining carbons of the polyketide chain would then constitute the C_{11} side-chain of (18). Formation of metacycloprodigiosin (17) appears to involve oxidative cyclization of C4 to form the *meta*-bridged fused-ring system.

Penicillic Acid.—The unique ability of the doubly labelled acetate method to yield details of biosynthetic processes has been used to good advantage in clarifying the mechanism of aromatic ring cleavage processes in polyketide biosynthesis. The method depends on the detection of carbon–carbon coupling in the ^{13}C n.m.r. spectrum of a metabolite labelled by 90% enriched [1,2-$^{13}C_2$]acetate. Such carbon–carbon couplings are only observed when a doubly labelled acetate unit remains intact during subsequent biosynthetic transformations. This method was used to elucidate the details of the oxidative ring-fission in the biosynthesis of the mould metabolite penicillic acid (19).[12] Birch[13] had previously conducted [1-^{14}C]acetate feeding studies on *Penicillium cyclopium* Westling to show that three such labels were incorporated into penicillic acid, and had suggested the biosynthetic sequence shown in Scheme 7*. Cleavage of the aromatic ring in (20) would afford the keto-dicarboxylic acid (21) which on decarboxylation would be transformed to penicillic acid (19).

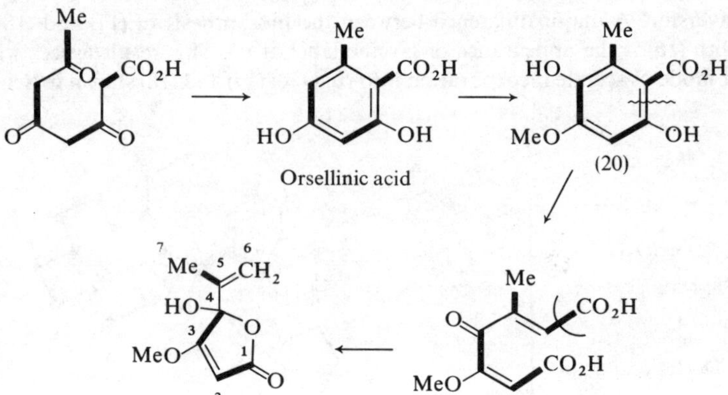

(19) (21)

Scheme 7

Mosbach[14] later confirmed by ^{14}C labelling in *P. baarnense* that orsellinic acid is an intermediate in penicillic acid biosynthesis, but through the alternative aromatic ring-fission process shown in Scheme 8.

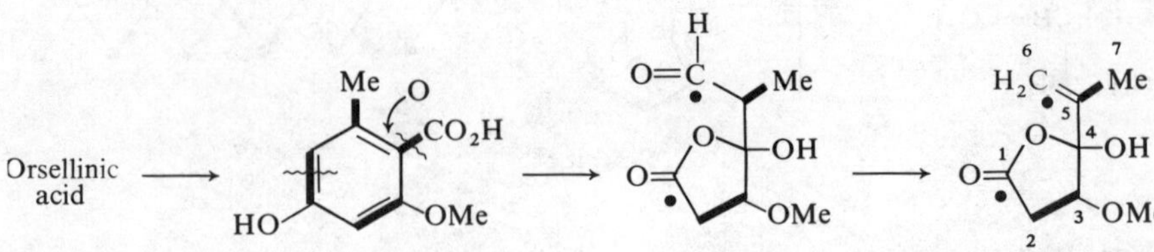

(19)

Scheme 8

[12] H. Seto, L. W. Cary, and M. Tanabe, *J. Antibiotics*, 1974, **27**, 558.

[13] A. J. Birch, G. E. Blance, and H. J. Smith, *J. Chem. Soc.*, 1958, 4582.

[14] K. Mosbach, *Acta Chem. Scand.*, 1960, **14**, 457.

* In this and subsequent schemes the bold lines represent bonds derived from $^{13}CH_3$–$^{13}CO_2H$ and indicate that these adjacent carbons are spin–spin coupled.

These two proposals were readily differentiated by the use of doubly labelled 90% enriched [1,2-$^{13}C_2$]acetate. If penicillic acid were labelled by the route of Scheme 7 it would have three pairs of carbon–carbon couplings, between C-1 and C-2, C-3 and C-4, and C-5 and C-7 whereas two carbon–carbon pairs would be cleaved in the Mosbach proposal of Scheme 8, and this would result in only two pairs of carbon–carbon couplings, between C-2 and C-3, and C-5 and C-7. N.m.r. results obtained,[12] shown in Table 3, were assigned by off-resonance decoupling and by comparison with

Table 3 ^{13}C *N.m.r. data for penicillic acid*

Carbon	*Chemical shift* (δc) *p.p.m. from* TMS	*Coupling constant* J_{C-C}/Hz	*Multiplicity on off-resonance decoupling*[a]
C-1	171.52	—	s
C-2	89.99	78	d
C-3	178.17	77	s
C-4	117.58	—	s
C-5	139.86	44	s
C-6	117.58	74	t
C-7	17.30	45	q
C-8	59.65	—	q

[a] s singlet, d doublet, t triplet, and q quartet

model compounds. Off-resonance decoupling confirmed the coincidental overlap of the C-4 and C-6 signals since the C-6 signal appeared as a triplet superimposed on the singlet C-4 signal. The proton-coupled spectrum of penicillic acid further confirmed the position of the quaternary C-4 resonance that remained as a singlet. The lowest field resonance at 178.17 p.p.m. was assigned to the vinylic C-3 carbon since a similarly substituted carbon in 4-methoxy-Δ^3-pyrrolin-2-one had a similar chemical shift. The spectrum of the labelled penicillic acid had carbon–carbon couplings between C-2 and C-3, with J = 77 Hz, which is of the magnitude expected for the coupling of two sp^2 carbons with an electronegative substituent, as well as between C-5 and C-7, J = 44 Hz, for the coupling between sp^2 and sp^3 carbons.[9] The presence of only two pairs of carbon–carbon couplings confirmed the conclusions of Mosbach, *i.e.* that the aromatic ring-fission process occurs as outlined in Scheme 8.

Satellite peaks less intense than those at C-2, -3, -5, and -7 were observed with the C-6 signal, J = 74 Hz, and were due to coupling to C-5. The complementary additional satellite signals of reduced intensity at C-5 due to spin-spin coupling to C-6 were also visible in the original spectrum. Carbon-6 was enriched with ^{13}C since it was originally derived from a doubly labelled acetic acid unit that underwent carbon–carbon bond cleavage.

Multicolic Acid and Multicolosic Acid.—The benzene-ring fission process that occurs in the formation of multicolic acid (22) and multicolosic acid (23) in *Penicillium multicolor* was also examined by the [1,2-$^{13}C_2$]acetate method.[15] These metabolite structures were established by ^{1}H and ^{13}C n.m.r., and u.v. spectroscopic studies as well as by chemical correlations, particularly the oxidative conversion of methyl

[15] J. A. Gudgeon, J. S. E. Holker, and T. J. Simpson, *J.C.S. Chem. Comm.*, 1974, 636.

O-methylmulticolate (24) into a product (25) identical with the product from the methylation of multicolosic acid (23).

The ^{13}C-assignments shown in Table 4 for methyl *O*-methylmulticolate (24) are

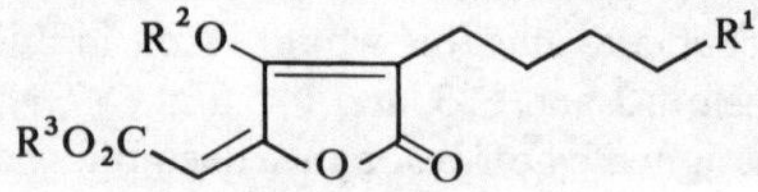

	R^1	R^2	R^3
(22)	CH_2OH	H	H
(23)	CO_2H	H	H
(24)	CH_2OH	Me	Me
(25)	CO_2H	Me	Me

Table 4 ^{13}C *N.m.r. data for methyl O-methyl multicolate* ($CDCl_3$)

Carbon	δc	$J_{^{13}C-^{13}C}$/Hz
1	168.3[a]	—
2	109.7	48
3	160.6[a]	—
4	150.2	90
5	23.4	48
6	29.7	35
7	25.5	35
8	32.2	35
9	62.2	35
10	100.9	90
11	163.8	—
Me—O_2CR	52.0	—
Me—OR	59.5	—

[a] These shifts may be interchanged.

based on measurements of the proton noise-decoupled and off-resonance decoupled ^{13}C n.m.r. spectra of (24) along with its transformation products as model compounds. Table 4 shows an ambiguity with a possible reversal of assignments between C-1 and C-3. It should be noted that in a similar chemical shift environment in penicillic acid (19), the β-carbon in the β-methoxy-α,β-unsaturated-γ-lactone was assigned the more deshielded resonance.

Feeding experiments with [1-^{13}C]- and [2-^{13}C]-acetate established the labelling pattern of (24) indicated in Scheme 9. This label distribution is only compatible with

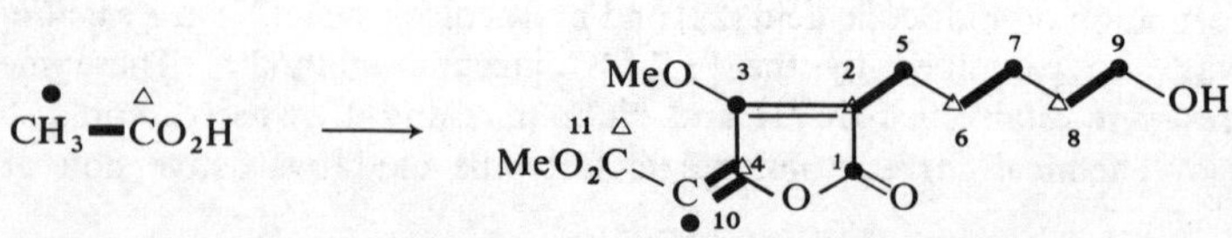

Scheme 9

oxidative cleavage of a polyketide-derived aromatic precursor such as 6-pentylresorcylic acid (26) between C-4 and C-5, as indicated in path a of Scheme 10. Direct confirmation of this specific cleavage process was derived from methyl *O*-methylmulticolate (24) labelled by [1,2-$^{13}C_2$]acetate. In the ^{13}C n.m.r. spectrum, this material showed four pairs of carbon–carbon couplings with *J* values of expected magnitude, between C-8 and C-10 (90 Hz, sp^2–sp^2), C-2a and C-5 (48 Hz, sp^2–sp^3), C-6 and C-7 (35 Hz, sp^3–sp^3), and C-8 and C-9 (35 Hz, sp^3–sp^3). The absence of couplings at C-1, C-3, and C-11 excludes the alternative possibility of C-1—C-2 cleavage (path b in Scheme 10); the intermediacy of any symmetrical aromatic intermediate such as 5-pentylresorcinol in the biosynthesis of (22) is also precluded.

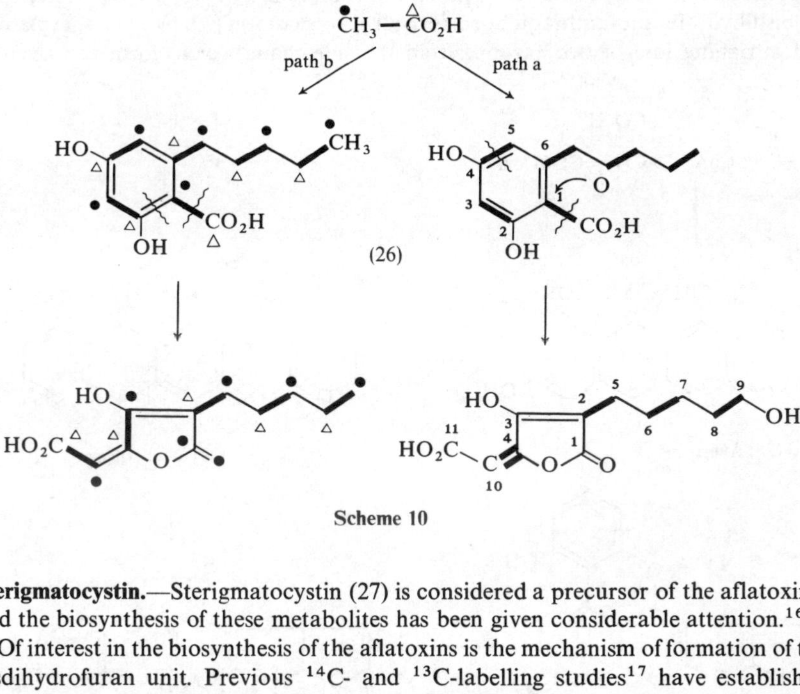

Scheme 10

Sterigmatocystin.—Sterigmatocystin (27) is considered a precursor of the aflatoxins, and the biosynthesis of these metabolites has been given considerable attention.[16]

Of interest in the biosynthesis of the aflatoxins is the mechanism of formation of the bisdihydrofuran unit. Previous ^{14}C- and ^{13}C-labelling studies[17] have established that the carbon skeleton of all these fungal metabolites is completely derived from acetate units and that methionine contributes to the *O*-methyl groups. This result led to the proposal that the bisdihydrofuran unit is formed by rearrangement of a naphthacene *endo*-peroxide that originates by cyclization of a C-18 linear polyketide chain (Scheme 11). Studies on the biosynthesis of aflatoxin B_1 (32) using C^{14}-labelled compounds have shown, however, that this route is unlikely and that averufin (28)[18] and sterigmatocystin (27)[19] are likely intermediates. The overall biogenetic relation-

[16] W. B. Turner, 'Fungal Metabolites,' Academic Press, New York, p. 176.
[17] (*a*) M. Biollaz, G. Büchi, and G. Milne, *J. Amer. Chem. Soc.*, 1970, **92**, 1035; (*b*) J. S. E. Holker and L. J. Mulheirn, *Chem. Comm.*, 1968, 1576; (*c*) M. Tanabe, T. Hamasaki, and H. Seto, *Chem. Comm.*, 1970, 1539.
[18] (*a*) M. T. Lin and D. P. H. Hsieh, *J. Amer. Chem. Soc.*, 1973, **95**, 1668; (*b*) M. T. Lin, D. P. H. Hsieh, R. C. Yao, and J. A. Donkersloot, *Biochemistry*, 1973, **12**, 5167.
[19] D. P. H. Hsieh, M. T. Lin, and R. C. Yao, *Biochem. Biophys. Res. Comm.*, 1973, **52**, 992.

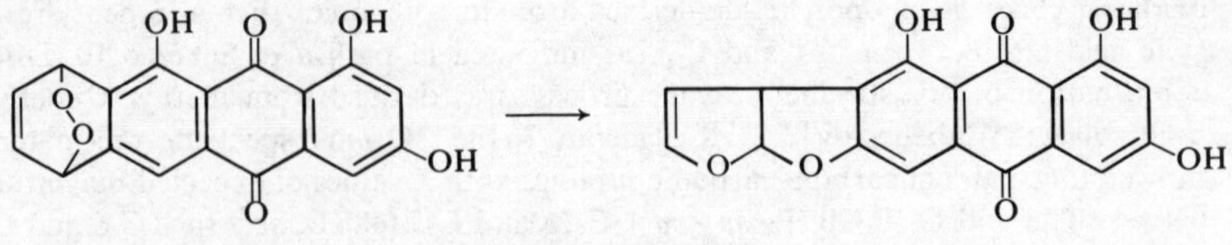

Scheme 11

ship within these bisdihydrofuran metabolites appears to be as shown in Scheme 12. The C-20 linear polyketide is formed by the initial condensation of one acetyl-Co-A with nine malonyl-Co-A units. Cyclization yields averufin (28), which is converted into the bisdihydrofurano-anthraquinone derivative versicolorin (29) by a yet unexplained and intriguing loss of two carbons from the side-chain. Versicolorin can then be

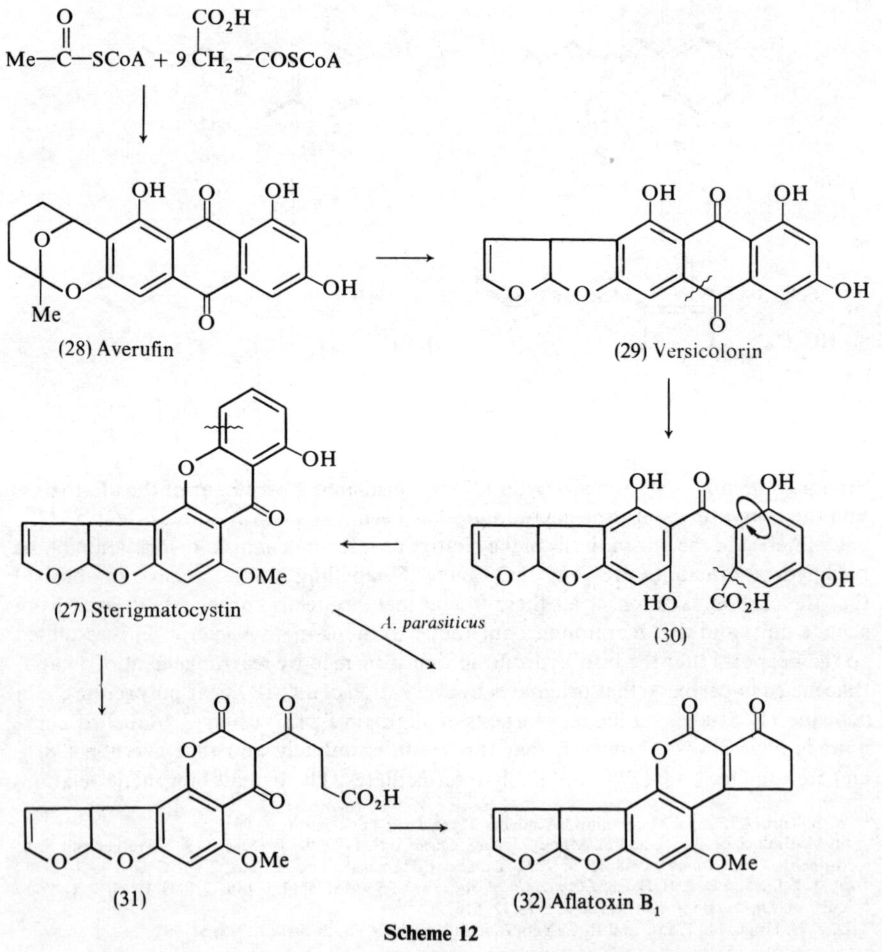

Scheme 12

cleaved at a quinone carbonyl in a Baeyer–Villiger type oxidative process to a hydroxy-acid (30). Decarboxylation of (30) followed by xanthone formation affords sterigmatocystin (27), the final level of oxidation in *A. versicolor*. In *A. parasiticus*, further oxidative cleavage of the aromatic ring in (27) would yield the keto-acid (31), which can aldol-condense, eliminate water, and decarboxylate to yield aflatoxin B (32).

Sterigmatocystin biosynthesis in *A. versicolor* was studied with [1,2-$^{13}C_2$]acetate (90% enriched) to obtain more detailed information about the process.[20] The C-20 linear polyketide chain can be folded in a counter-clockwise manner to afford averufin, and subsequently sterigmatocystin, with a labelling pattern from [1,2-$^{13}C_2$]acetate shown in (27a) of Scheme 13. Alternatively, the C-20 polyketide can be folded in a clockwise direction in a manner which will afford an [1,2-$^{13}C_2$]acetate-enriched averufin and sterigmatocystin with a different labelling pattern as indicated in (27b) of Scheme 13. These pathways can be differentiated since they clearly possess different spin–spin coupling relationships between the carbons.

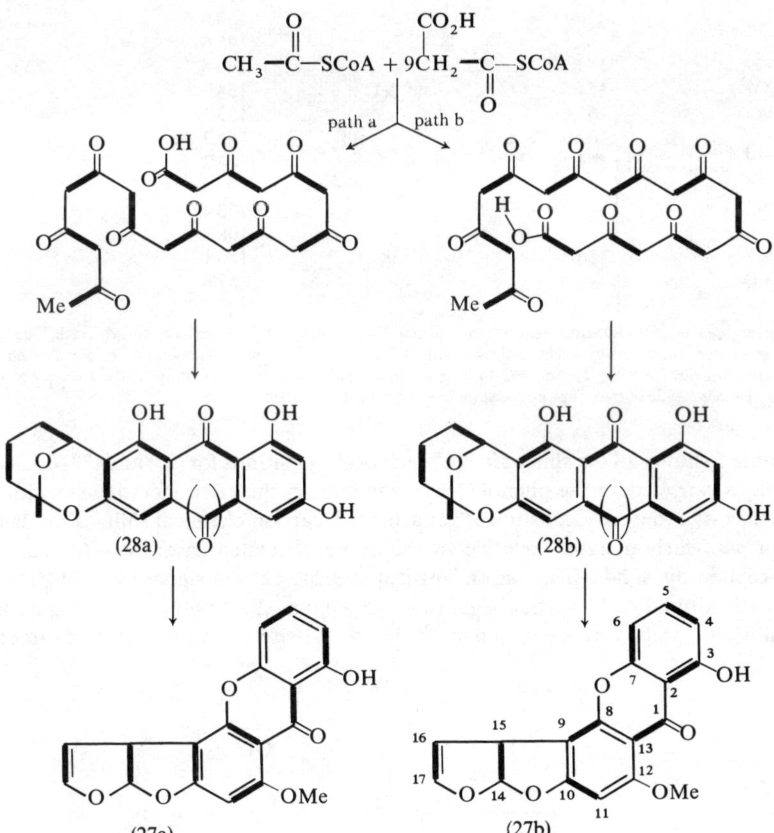

Scheme 13

[20] H. Seto, L. W. Cary, and M. Tanabe, *Tetrahedron Letters*, 1974, 4491.

The carbon chemical shifts of sterigmatocystin were assigned as shown in Table 5, by the use of off-resonance proton decoupling, specific single-frequency proton decoupling, and by comparison with the chemical shifts of model compounds. 6-Methoxysterigmatocystin (33) was a valuable model in the shift assignments of (27).

Table 5 ^{13}C *N.m.r. data*[20] *for sterigmatocystin and 6-methoxysterigmatocystin*

	Sterigmatocystin		*6-Methoxysterigmatocystin*	
Carbon	δc[a]	^{13}C–^{13}C *Coupling constants*/Hz	δc	*Calcd.*
15	47.2	34	47.2	—
OMe	56.7	—	56.7	—
11	90.9	72	90.9	—
16[b]	102.5	76	102.2	—
4	106.4	58	109.0	108.8
6	110.7	70	139.0	142.0
14	113.3	34	113.3	—
5	136.1	59	120.6	121.1
17	145.5	76	145.6	—
7	153.3	—[c]	143.8	144.5
12	154.5	singlet	154.0	—
3	161.4	—[c]	153.1	154.6
8, 10	162.9	—[c]	162.7	—
	164.2	—[c]	164.2	—
2, 9, 13	—	—	105.2	—
	—	—	106.4	—
	—	—	108.7	—
1	180.3	58	180.1	—
6-OMe	—	—	57.3	—

[a] Determined in [$^2H_6$]DMSO, relative to internal TMS, saturated solution. Hz, pulse width; 20; [b] This assignment was confirmed by selective decoupling at δ 5.5 in the ^{1}H n.m.r. spectrum; therefore, the previous assignment[3] given for these carbons (C-14 and C-16) should be reversed. [c] Owing to the overlap of these signals, the magnitude of the coupling constant could not be determined.

As Table 5 shows, all the shift values at each of the positions for (27) and (33) coincide, except, as expected, in the phenolic ring which bears the extra methoxy-substituent. The effect of a methoxy-substituent on aromatic-carbon chemical shifts is known to shift a *para*-carbon signal (relative to the unmethoxylated case) by −6.8 p.p.m., a *meta*-carbon by +2.4 p.p.m., an unsubstituted *ortho*-carbon signal by −15.0 p.p.m., and a substituted *ortho*-carbon signal by −8.8 p.p.m. The methoxy-bearing carbon signal itself is shifted by +31.3 p.p.m.[20a] By use of these substituent-shift parameters,

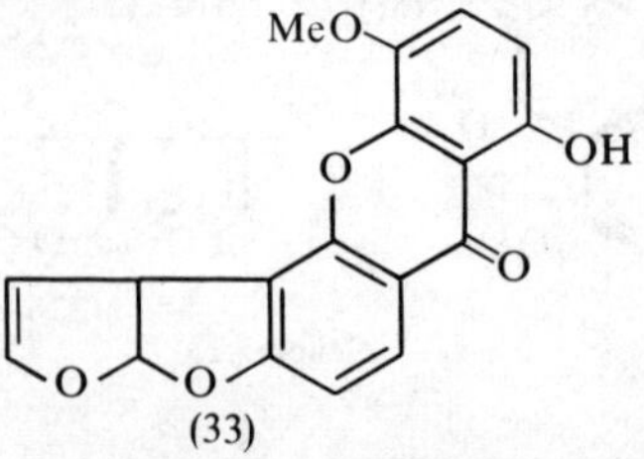

(33)

[20a] R. H. Levin, J. Y. Lallemand, and J. D. Roberts, *J. Org. Chem.*, 1973, **38**, 1983.

the calculated and observed shift values for (27) are compared in Table 5. The values are in very good agreement and these results further confirm the validity of the shift assignments.

Intense satellite signals caused by ^{13}C–^{13}C spin–spin coupling were observed with most of the carbon signals from the sterigmatocystin enriched by [1,2-$^{13}C_2$]acetate as indicated in Table 5. All the observed coupling constants were in good agreement with the degree of *s*-hybridization on each carbon as well as its substitution pattern, the two major factors influencing the magnitude of ^{13}C–^{13}C couplings.[9b]

Detection of couplings between C-4—C-5 and C-6—C-7 in the enriched sterigmatocystin clearly indicated that the C_{20} linear polyketide chain is folded in the clockwise fashion outlined in path b of Scheme 13.* The use of [1,2-$^{13}C_2$]acetate in polyketide biosynthesis thus promises to be a useful method for distinguishing between different possible folding patterns of polyketide chains.

The observed ^{13}C–^{13}C couplings in the bisdihydrofurano-unit of enriched (27) at C-14—C-15 and C-16—C-17 established that these carbons are derived from intact acetic acid units. Any proposed mechanism for the conversion of the C_6 side-chain of averufin into the C_4 bisdihydrofuran unit of sterigmatocystin must account for this labelling pattern. An interesting mechanism[21] which does just this is shown in Scheme 14. In this, averufin is first opened up to a hydroxy methyl ketone which is then

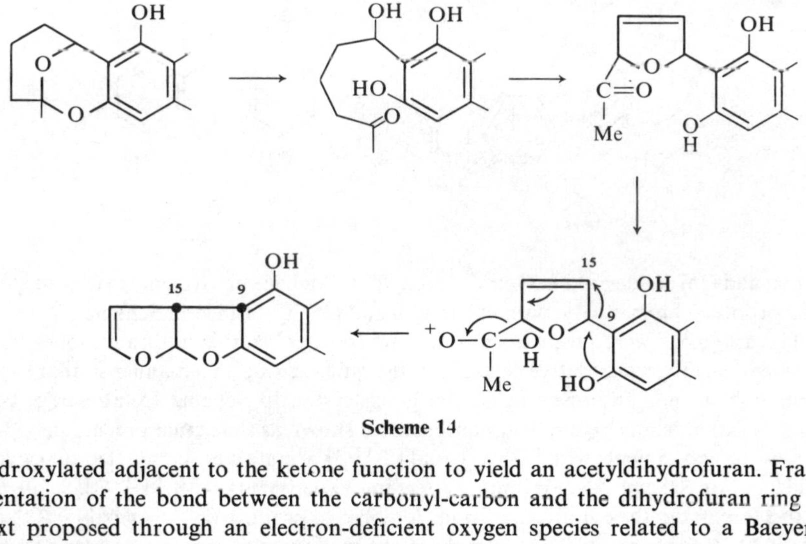

Scheme 14

hydroxylated adjacent to the ketone function to yield an acetyldihydrofuran. Fragmentation of the bond between the carbonyl-carbon and the dihydrofuran ring is next proposed through an electron-deficient oxygen species related to a Baeyer–Villiger type intermediate. Concerted shift of the bonds leads directly to the bisdihydrofuran system.

* Note added in proof: A recent assignment of the ^{13}C spectrum of sterigmatocystin (K. G. R. Pachler, P. S. Steyn, R. Vleggaar, and P. L. Wessels, *J.C.S. Chem. Comm.*, 1975, 355) favours route *a* of Scheme 13. Work by the same group on aflatoxin B_1 (P. S. Steyn, R. Vleggaar, P. L. Wessels, and D. B. Scott, *J.C.S. Chem. Comm.*, 1975, 193) supports this pathway and also the Thomas mechanism for the later steps in aflatoxin biosyntheses.

[21] R. Thomas, cited by M. O. Moss in 'Phytochemical Ecology,' ed. J. B. Harborne, Academic Press, London, 1972, 140.

Tajixanthone.—The biosynthesis of another xanthone, tajixanthone (34), was investigated with the aid of ^{13}C n.m.r.[22] In addition to (34), *Aspergillus variecolor* produces a number of minor metabolites including shamixanthone (35) and arugosins A (36), B (37), and C (38). These metabolites constitute a closely related biogenetic group suggesting that formation of (35) occurs by cyclodehydration of the dihydroxy-benzophenone system present as a hemiketal in (36) and (37). The dihydropyran rings

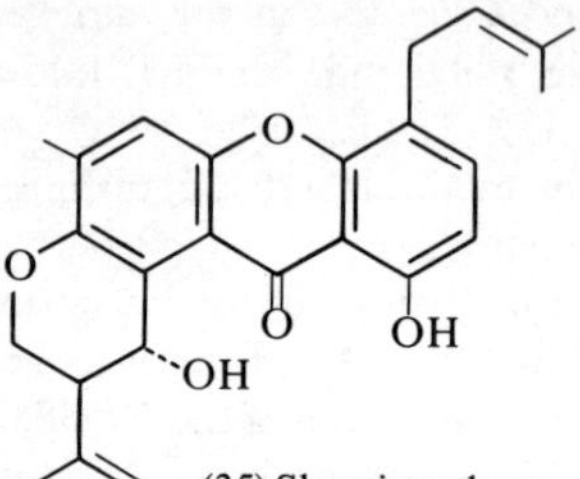

(35) Shamixanthone

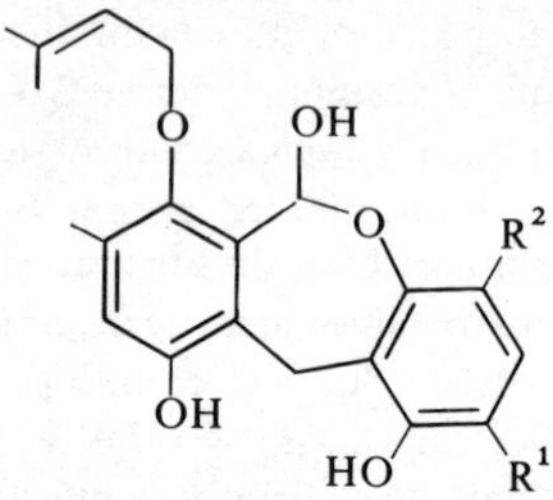

(36) $R^1 = CH_2CH{=}CHMe_2$, $R^2 = H$
(37) $R^1 = H$, $R^2 = CH_2CH{=}CHMe_2$

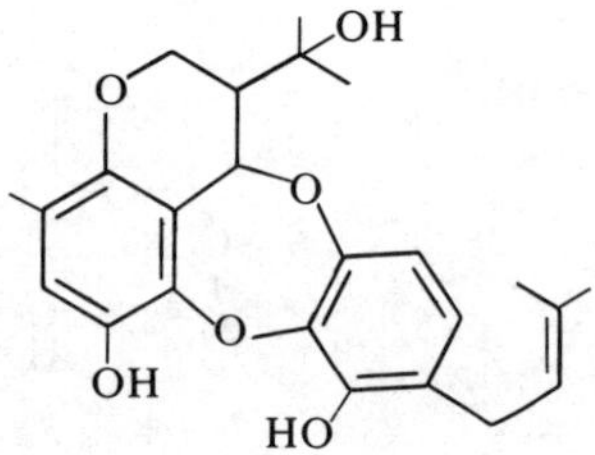

(38)

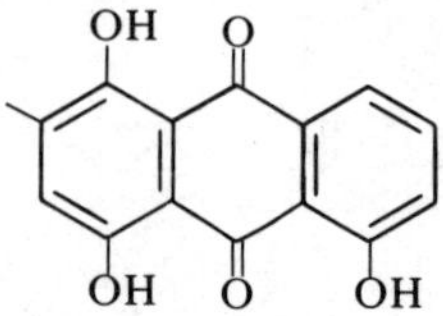

(39)

of (34) and (35) are derived from the *o*-prenyloxy aldehyde equivalents in (36) and (37). The proposed biosynthetic pathway to (34) and (35) is outlined in Scheme 15.

The arugosins were proposed to originate from polyketide anthraquinones, such as islandicin (39), by oxidative cleavage of the quinone ring, in a manner similar to the formation of the xanthone ring in sterigmatocystin. In Scheme 15, an earlier precursor of islandicin, chrysophenol anthrone, is shown as the actual precursor.

Feeding experiments with [1-^{13}C]- and [2-^{13}C]-acetate were conducted to provide evidence for Scheme 15. Preliminary feeding experiments with [1-^{14}C]acetate revealed that under the culture conditions required to produce (34), 1.0 g of [1-^{13}C]acetate (92.1%) for every 25.0 g of sucrose in the medium was required to reach the required level of ^{13}C-enrichment. This concentration of acetate was found to have no metabolic effects on the organism or on the yields of (34) produced.

Extensive studies on the ^{13}C n.m.r. shifts of tajixanthone (34) (see Table 6) were conducted on at least ten related derivatives, by the use of off-resonance decoupling and by carrying out lanthanide-induced shifts (LIS) studies.

Assignment of the sp^3-hybridized carbon resonances at less than 80 p.p.m. was

[22] J. S. E. Holker, R. D. Lapper, and T. J. Simpson, *J.C.S. Perkin I*, 1974, 2135.

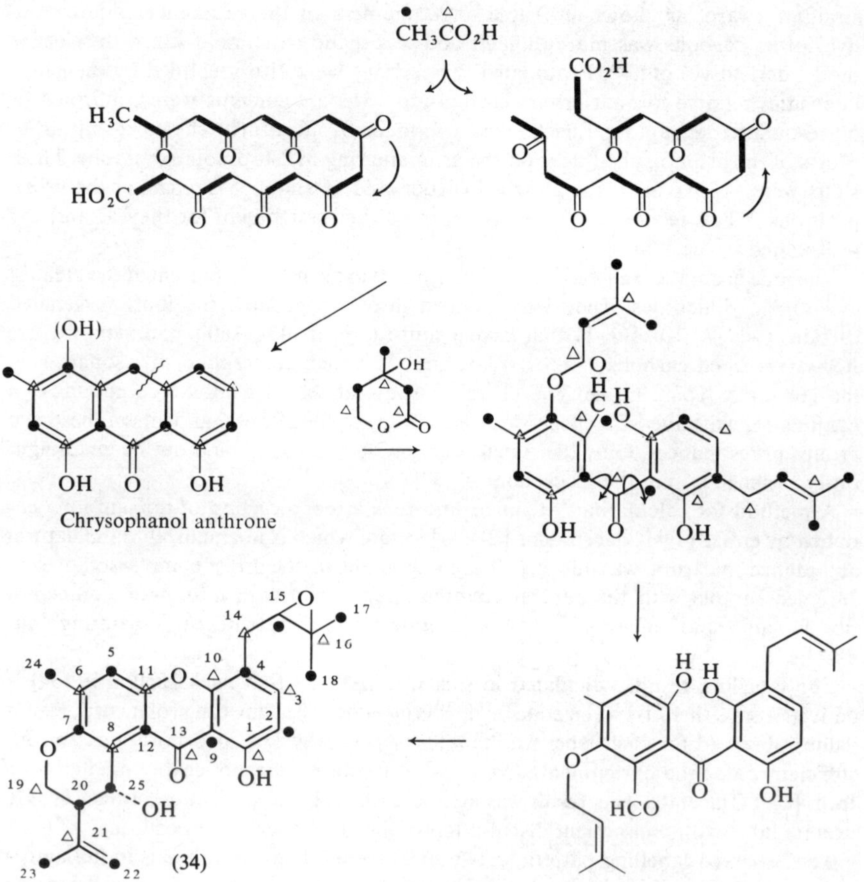

Scheme 15

Table 6 ^{13}C *N.m.r. data for tajixanthone* (34) *and derivatives*

Carbon	*Incorporation values*[a]	*Carbon*	*Incorporation values*[a]
1	152.3	14	28.5(t)
2	109.7(d)	15	63.0(d)
3	136.4(d)	16	58.5
4	114.8	17	19.0(q)
5	118.7(d)	18	24.7(q)
6	137.8	19	64.3(t)
7	149.0	20	44.8(d)
8	120.6	21	412.0
9	108.7	22	111.8(t)
10	159.7	23	22.4(q)
11	151.4	24	17.3(q)
12	116.4	25	63.0(d)
13	183.6		

[a] d, doublet; t, triplet; q, quartet

straightforward, as shown in Table 6. Assignment of the resonances due to sp^2-hybridized carbons was more difficult and was made by comparison with selective model derivatives of (34). Protonated sp^2-carbons were also identified by their high intensities relative to quarternary carbons, and the assignments were confirmed by off-resonance decoupling. Effective use was made of substituent chemical shift (SCS) effects of the bromo-substituent on the aromatic ring in 2-bromotajixanthone. These shifts were -6 p.p.m. at the attached carbon and at about $+3$ p.p.m. at the *ortho*-positions.[9c] This technique yielded the specific shift assignment for the C-2 and C-3 protonated sp^2-carbons.

The nine aromatic ring-carbons not bearing hydrogen atoms presented the greatest assignment difficulties. They were divided into two groups: the four oxygenated carbons C-1, -7, -10, and -11 that have a shift range of 148—160 p.p.m.; and the five non-oxygenated carbons C-4, -6, -8, -9, and -12 which correspond to resonances in the 108—138 p.p.m. range. The chemical shifts of selective model compounds, in conjunction with the SCS effect, were used to make specific assignments in these two groups of resonances. Only the signals due to C-8 and C-12 could not be unambiguously assigned within the 25 carbons of (34).

A method for calculating ^{13}C-incorporations at each carbon of tajixanthone enriched by either [1-^{13}C]acetate or [2-^{13}C]acetate which is normalized to the natural abundance spectrum, was adopted. The method allows the direct comparison of each enriched sample with the natural abundance spectrum with automatic compensation for any randomization of [2-^{13}C]acetate label or instrumental operating conditions.

The labelling results calculated in this way indicate that each carbon in (34) is derived from either [1-^{13}C]acetate or [2-^{13}C]acetate. The wide range of incorporation values observed for each label was caused by generally low incorporations and insufficient data; the enrichments from [1-^{13}C]acetate were noticeably greater than from [2-^{13}C]acetate. This result was attributed to the randomization of the [2-^{13}C]-acetate label with consequent distribution of the label over more positions.

The observed labelling pattern shown in Table 6 for tajixanthone is in full agreement with Scheme 15, with derivation of the xanthone system from an anthrone or anthraquinone precursor and introduction of the prenyl units from mevalonate. The origin of C-25 from the acetate methyl group supports its formation from the methylene group of an anthrone precursor. Formation of the C-20—C-25 bond by linkage of two carbons derived from the methyl group of acetate is also consistent with the proposed cyclization of the *o*-prenyloxy aldehyde system in the precursor xanthone. As shown by Scheme 15 the precise cyclization pattern of the polyketide chain remains uncertain, though it would be considerably clarified by the double-labelling technique.

Ochrephilone.—A new azaphilone metabolite, ochrephilone (40), was isolated from cultures of *Penicillium multicolor* NRRL 2060 along with the co-metabolites sclerotiorin (41) and citrinin (42).[23] From structural similarity between ochrephilone (40) and rotiorin (43), it appeared that (40), like (43),[24] was derived from a polyketide. To

[23] H. Seto and M. Tanabe, *Tetrahedron Letters*, 1974, 651.

[24] A. J. Birch, A. Cassera, P. Fitton, J. S. E. Holker, H. Smith, G. A. Thompson, and W. B. Whalley, *J. Chem. Soc.*, 1962, 3583.

obtain definitive structural as well as biosynthetic information, chrephilone was labelled with 90% enriched [1,2-$^{13}C_2$]acetate. The ^{13}C n.m.r. data are shown in Table 7. The specific position of the ^{13}C–^{13}C spin–spin couplings observed in the ochrephilone carbon skeleton as well as the magnitude of these coupling constants are both in accord with structure (40).

The enrichment in (40) of approximately 2.5-fold at carbons C-2, -4, -6, -8, -10, -12, -14, -16, -18, and -20 by [2-^{13}C]acetate, along with the [1,2-$^{13}C_2$]acetate labelling result, agrees with the biosynthetic pathway outlined in Scheme 16. As with sclerotiorin, and rotiorin, the three C_1-units are presumably derived from formate or methionine.

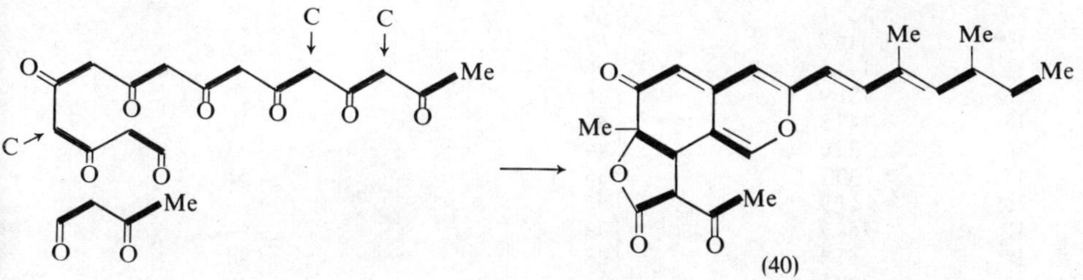

(40)

Scheme 16

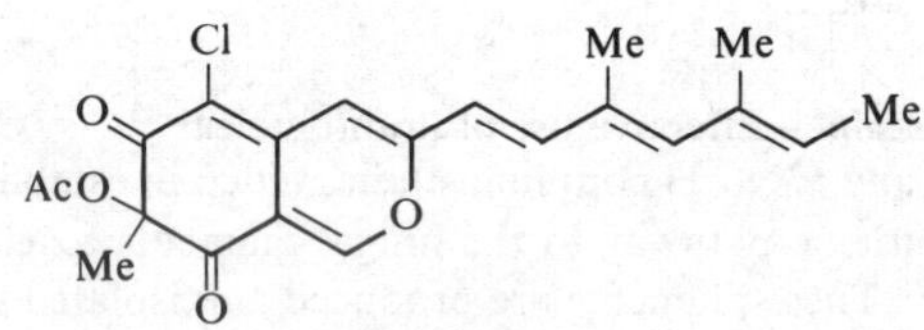

(41) Sclerotiorin

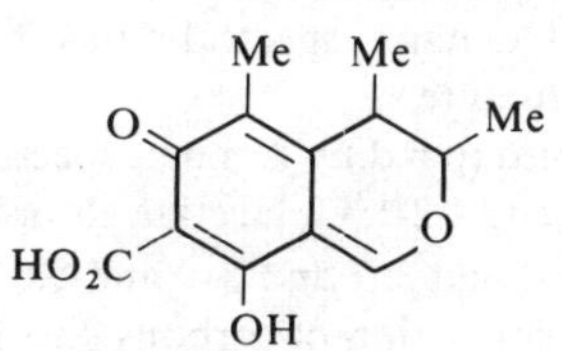

(42) Citrinin

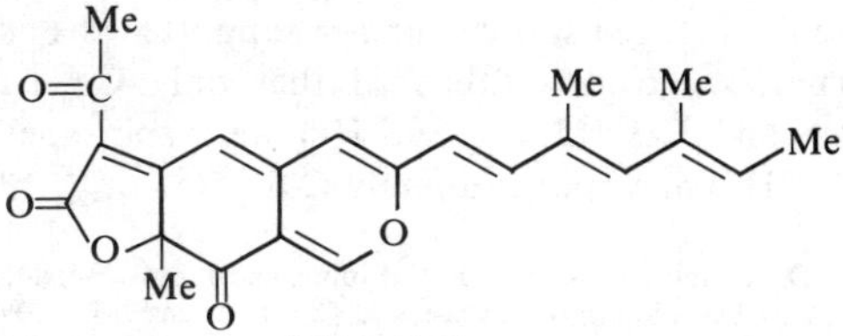

(43) Rotiorin

Table 7 ^{13}C *N.m.r. data for ochrephilone*

Carbon	δc	$J_{^{13}C-^{13}C}$/Hz *From* 1,2-^{13}C-*acetate* (90%)	$J_{^{13}C-^{13}C}$/Hz *From* 1:1, 1-^{13}C-*acetate* (90%): 2-^{13}C-*acetate* (90%)
1	147.0	78	—
2^a	114.0	78	42
3	42.9	34	42
4^a	82.9	34	42
5	191.0	61	42
6^a	106.1	61	72
7	144.7	—	72
8^a	107.6	—	72
9	157.1	69	72
10^a	116.2	69	70
11	141.5	56	70
12^a	131.9	56	70
13	147.3	44	70
14^a	35.0	44	35
15	30.1	36	35
16^a	11.9	36	—
17	168.7	48	—
18^a	57.4	48	40
19	200.1	44	40
20^a	30.2	44	
21	23.2		
22	20.2		
23	12.3		

^{a} Labelled by [2-^{13}C]acetate.

Tenellin and Bassianin.—Effective use of homonuclear ^{13}C–^{13}C coupling and heteronuclear ^{15}N–^{13}C and ^{15}N–^{1}H coupling studies aided in establishing the structure as well as the biosynthetic pathway to the fungal pigments tenellin (44) and bassianin (45) (Scheme 17).[25] These pigments are produced and isolated from the insect pathogenic fungi *Beauveria tenella* (Delacroix) Siem. and *Beauveria bassiana* (Bals.) Vuill.

Structure (44) was established for tenellin by degradative chemical reactions and spectroscopic studies; the ^{13}C n.m.r. spectral data (shown in Table 8) are fully consistent with the assigned structure.

The proton noise decoupled (p.n.d.) ^{13}C n.m.r. spectrum of the tenellin produced in the presence of 90% enriched [1,2-^{13}C$_2$]acetate showed five pairs of carbon–carbon couplings between carbons 2 and 3, 7 and 8, 9 and 10, 11 and 12, and 13 and 14. This result is in agreement with formation of carbons 2 to 14 through a linear hexaketide unit. The magnitude of the $J_{^{13}C-^{13}C}$ values (see Table 8) also agrees with the degree of *s*-hybridization on these carbons expected for structure (44).

The novel use of ^{15}N-labelled (44), obtained from cultures with K^{15}NO$_3$ (95% enriched) as the only nitrogen source, further supported the assigned structure (44). Its p.n.d.-^{13}C n.m.r. spectrum established that only C-6 ($^1J_{^{15}N-C}$ 15.0 Hz), C-2 ($^1J_{^{15}N-H}$ 11.0 Hz), and C-3 ($^2J_{^{15}N-C}$ 9.2 Hz) were spin–spin coupled to ^{15}N. In the corresponding ^{1}H n.m.r. spectrum, only C-6-*H* ($^2J_{^{15}N-H}$ 1.0 Hz) was coupled to

[25] (*a*) A. G. McInnes, D. G. Smith, C. K. Wat, L. C. Vining, and J. L. C. Wright, *J.C.S. Chem. Comm.*, 1974, 281; (*b*) A. G. McInnes, D. G. Smith, J. A. Walter, L. C. Vining, and J. L. C. Wright, *J.C.S. Chem. Comm.*, 1974, 282.

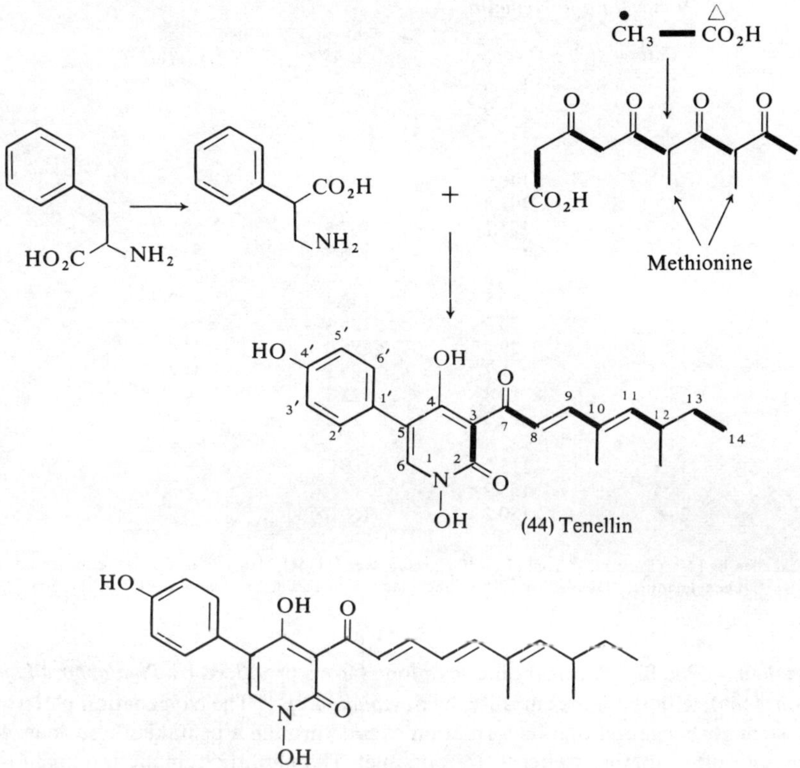

Scheme 17

^{15}N. These data confirm that the nitrogen in (44) is geminal to C-6-H and C-3 and directly bonded to C-2 and C-6, thus providing further evidence for a hydroxamic acid group.

Parallel studies on bassianin yielded structure (45) for this metabolite. Feeding experiments showed that the C-15 and C-16 methyl groups on the polyketide chain are derived, as expected, from L-[*methyl*-^{13}C]methionine. The alternate carbons of the polyketide chain were also labelled with either [1-^{13}C]acetate or [2-^{13}C]acetate.

Further studies also showed that DL-[1-^{13}C]phenylalanine labelled C-4 in the heterocyclic ring, and that C-6 was labelled by DL-[2-^{13}C]phenylalanine. Thus (44) is formed, as outlined in Scheme 17, by condensation of the methylated polyketide chain with a biogenetic unit that contains all the carbons of phenylalanine. At some stage in the biosynthesis, the carboxyl carbon of phenylalanine migrates to the benzylic carbon atom. Formal precedent for this type of carboxyl rearrangement process exists in the biosyntheses of tropic and pulvic acids from phenylalanine.[26]

[26] (*a*) M. L. Louden and E. Leete, *J. Amer. Chem. Soc.*, 1962, **82**, 4507; (*b*) W. S. G. Maass and A. C. Neish, *Canad. J. Bot.*, 1967, **45**, 59.

Table 8 ^{13}C *N.m.r. data for tenellin*

Carbon	δc	$^1J_{^{13}C-H}$/Hz	$^1J_{^{13}C-^{13}C}$/Hz
2[a]	157.5	—	75.3
3[b,c]	105.9	—	75.3
4[d]	173.0	—	—
5	110.9	—	—
6[e]	140.0	183.6	—
7[a]	193.8	—	55.9
8[b,c]	123.1	167.5	55.9
9[a]	149.7	155.9	53.6
10[b,c]	132.6	—	53.6
11[a]	150.8	154.4	43.2
12[b,c]	34.6	125.4	43.2
13[a]	29.4	126.0	34.4
14[b,c]	11.7	125.2	34.4
15[f]	19.8	125.8	—
16[f]	12.3	126.2	—
1′	122.8	—	—
3′–5′	115.0	160.0	—
4′	156.9	—	—
2′–6′	130.2	160.0	—

[a] Labelled by [1-^{13}C]acetate; [b] from [1,2-$^{13}C_2$]acetate labelled (44); [c] labelled by [2-^{13}C]acetate; [d] labelled by [1-^{13}C]phenylalanine; [e] labelled by [2-^{13}C]phenylalanine; [f] labelled by [*methyl*-^{13}C]-methionine.

Scytalone.—The fungal metabolite scytalone (46) is produced by *Phialophora lagerbergii*[27] and, with the same chirality, by *Scytalidium sp.*[28] The oxygenation pattern of (46) strongly suggested that its formation occurs through a pentaketide such as (47), (although other folding patterns are possible). The naphthoquinone mompain (48) has a close structural resemblance to a co-metabolite of (46), flaviolin (49). However,

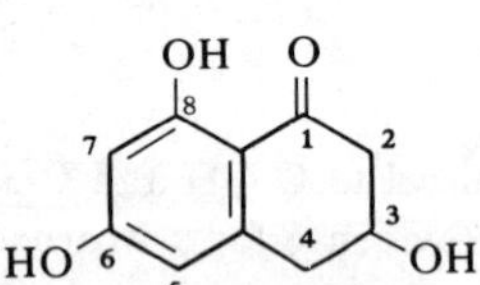

(46)

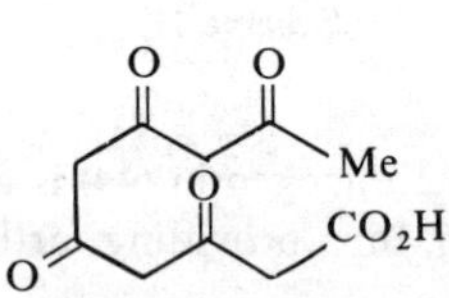

(47)

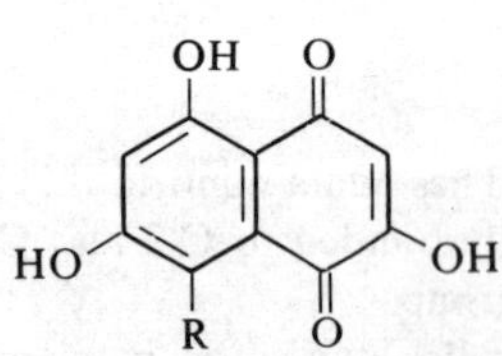

(48) R = OH
(49) R = H

biosynthetic studies of (48) gave results that were inconsistent with its formation by a pentaketide pathway;[29] further biosynthetic work was therefore undertaken with ^{13}C n.m.r. to confirm its polyketide origin.[27]

The ^{13}C n.m.r. spectrum of scytalone shows ten well-resolved signals which were assigned as shown in Table 9. The table reveals reasonable agreement between the observed aromatic chemical shifts and the values calculated using published substituent chemical shift values for substituted benzenes.[9c]

27 D. C. Aldridge, A. B. Davies, M. R. Jackson, and W. B. Turner, *J.C.S. Perkin I*, 1974, 1540.
28 J. A. Findlay and D. Kwan, *Canad. J. Chem.*, 1973, 51, 1617.
29 S. Natori, Y. Inouye, and H. Nishikawa, *Chem. and Pharm. Bull.* (*Japan*), 1967, **15**, 380.

Table 9 ^{13}C *N.m.r. data for scytalone*

Carbon	*δc*	*δ calc.*
1	202.6	—
2	48.7[a]	—
3	66.2	—
4	39.3[a]	—
5	109.6	109.2
6	166.1[b]	160.9
7	101.6	100.2
8	160.9[b]	156.8
4a	146.3	140.3
8a	111.6	118.3

[a, b] These shifts may be interchanged.

The ^{13}C n.m.r. spectrum of scytalone biosynthesized from sodium [1-^{13}C]acetate showed signal intensities about twice natural abundance at C-1, C-3, C-4a, C-6, and C-8, in full accord with a pentaketide route.

Colchicine.—The ^{13}C n.m.r. technique for biosynthetic studies has only been used so far with micro-organisms or with partially purified enzyme systems. The difficulty in applying the technique to higher plants arises not only from the generally lower incorporations of labelled precursors in these systems but also from dilution of the labelled metabolite with a large pool of unlabelled endogenous metabolite, which can often offset satisfactory incorporations. However, the use of 90 atom-% enriched precursors will yield an answer whenever dilutions of 1:90 or less are obtained.

Two successful studies on the incorporation of ^{13}C-labelled precursors into higher plants have recently been achieved with the alkaloids colchicine[30] and camptothecin.[31] The use of relatively complex and advanced ^{13}C-enriched precursors in these studies gave assurance of higher incorporation yields.

^{14}C-Labelling studies have proved that colchicine (50) is biosynthesized in *Colchicum autumnale* from the tetrahydroisoquinoline (*S*)-autumnaline (51) by oxidative phenolic coupling to the dienone and subsequent ring expansion of this dienone,[32] as shown in Scheme 18. For the ^{13}C n.m.r. study, (*RS*)-[1-^{13}C]autumnaline (51) was prepared from the alcohol (52) by the method shown in Scheme 19.[33] This alcohol was converted into the mesylate, and ^{13}C-label was introduced with [^{13}C]cyanide. The nitrile was hydrolysed to the acid (53) which was converted into the acid chloride and then condensed with 4-benzyloxy-3-methoxyphenethylamine to yield the amide (54). Cyclization of the amide afforded the 3,4-dihydroisoquinoline (55), which was *N*-methylated, reduced with sodium borohydride, and catalytically debenzylated to complete the synthesis of (*RS*)-[1-^{13}C]autumnaline hydrochloride (51). This labelled precursor (300 mg) was injected as an aqueous solution into the seed capsules of *C. autumnale* (1 mg per capsule), and after two weeks 1.24 g of colchicine (50) was isolated from them.

The upfield signals in the ^{13}C n.m.r. spectrum of unlabelled colchicine were assigned

[30] A. R. Battersby, P. V. Sheldrake, and J. A. Milner, *Tetrahedron Letters*, 1974, 3315.
[31] C. R. Hutchinson, A. H. Heckendorf, P. E. Daddona, E. Hagaman, and E. Wenkert, *J. Amer. Chem. Soc.*, 1974, **96**, 5609.
[32] A. C. Barker, A. R. Battersby, E. McDonald, R. Ramage, and J. H. Clements, *Chem. Comm.*, 1967, 390.
[33] A. R. Battersby, R. B. Herbert, E. McDonald, R. Ramage, and J. H. Clements, *J.C.S. Perkin I*, 1972, 1741.

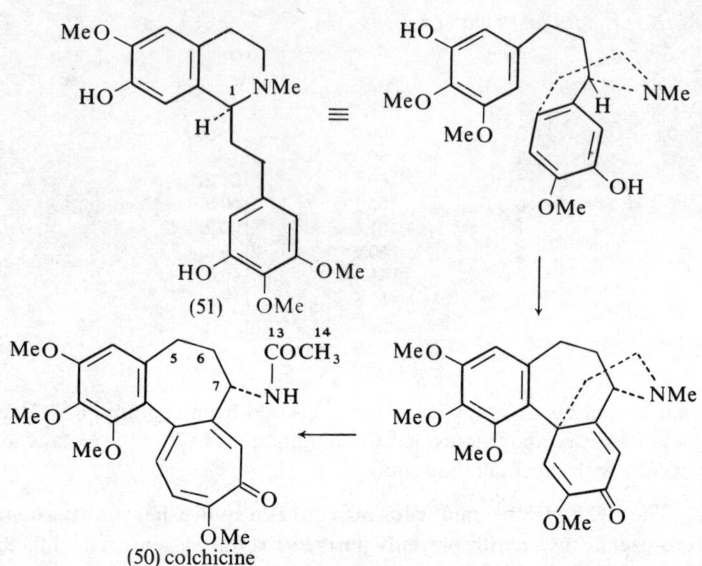

Scheme 18

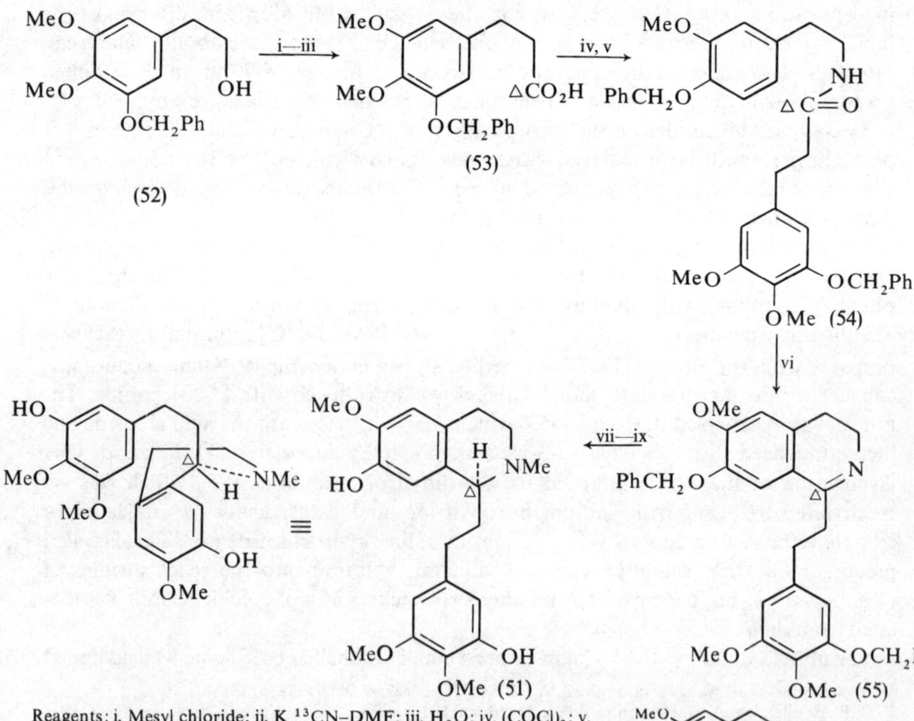

Reagents: i, Mesyl chloride; ii, K ^{13}CN–DMF; iii, H_2O; iv, $(COCl)_2$; v, 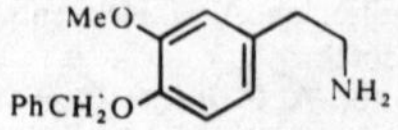
vi, $POCl_3$; vii, MeI; viii, $NaBH_4$; ix, H_2–Pd/C

Scheme 19

as shown in Table 10. Comparison with the enriched colchicine clearly showed that the signal from C-7 is enhanced 2.5-fold for an incorporation of about 6% without randomization of the label.

Table 10 ^{13}C *N.m.r. data for colchicine*

δc	*Position*	*Ratio of Integrals Labelled/Unlabelled*
22.4	C-6	0.93
29.6	C-5	1.07
36.0	C-14	1.00
52.4	C-7	2.54
55.8	MeO	1.03
56.1	MeO	1.18
61.0	MeO	0.87
61.2	MeO	0.90

Camptothecin.—At the same time as the colchicine studies, experiments on the biosynthesis of the alkaloid camptothecin (56) were initiated in *Camptotheca acuminata Decne* (Nyssacene) plants.[31] From the results on the biosynthesis of related indole alkaloids a possible route for the biosynthesis of (56) was proposed as shown in Scheme 20. The condensation of tryptamine with secologanin (57) yields vincoside (58a) and isovincoside (58b), which are cyclized to the epimeric lactams (59a) and (59b) These lactams can be converted by straightforward biosynthetic transformations into desoxycamptothecin (60).

To obtain the necessary corroboration of the site-specific incorporation of (59) into (56), feeding experiments with ^{13}C-labelled (59) were conducted. The ^{13}C-label was introduced into (59) by condensation of [^{13}C]tryptamine with seco-loganin (57), yielding vincoside (58) and, *via* cyclization, the ^{13}C-labelled lactam (59); the lactam contained >84% ^{13}C as shown by mass spectrometry, and ^{13}C n.m.r. showed that it was solely labelled at C-5 (δc 40.9). The ^{13}C n.m.r. shifts of camptothecin are shown in Table 11, and under the operating conditions of the spectrometer, all but the signals from the carbonyl groups were found. The aromatic carbon assignments were made by comparison with the model compounds: quinoline, α-pyridone, and its *N*-methyl derivative. All the non-aromatic carbons of (56) are unique. There is a single methyl group (C-18) and a single quaternary carbon (C-20). Both methylene carbons, C-17 and C-19, are in different environments.

Wick feeding of 3.5 mg of (59) to 206 g of intact plants after 19 days yielded 20 mg of labelled (56). Comparison of the relative signal intensities of the labelled *versus* natural abundance (56), determined under conditions as close to identical as possible to obtain comparable *S*/*N* ratios and peak heights, revealed that only C-5 is labelled by [5-^{13}C]-(59). The signal enhancement corresponding to an about 55% increase over natural abundance represents a specific incorporation of 1.6% that agrees with the specific incorporation of 1.3—1.8% of radioactive (59). The specific labelling of C-5 of (56) is in agreement with the route proposed in Scheme 20.

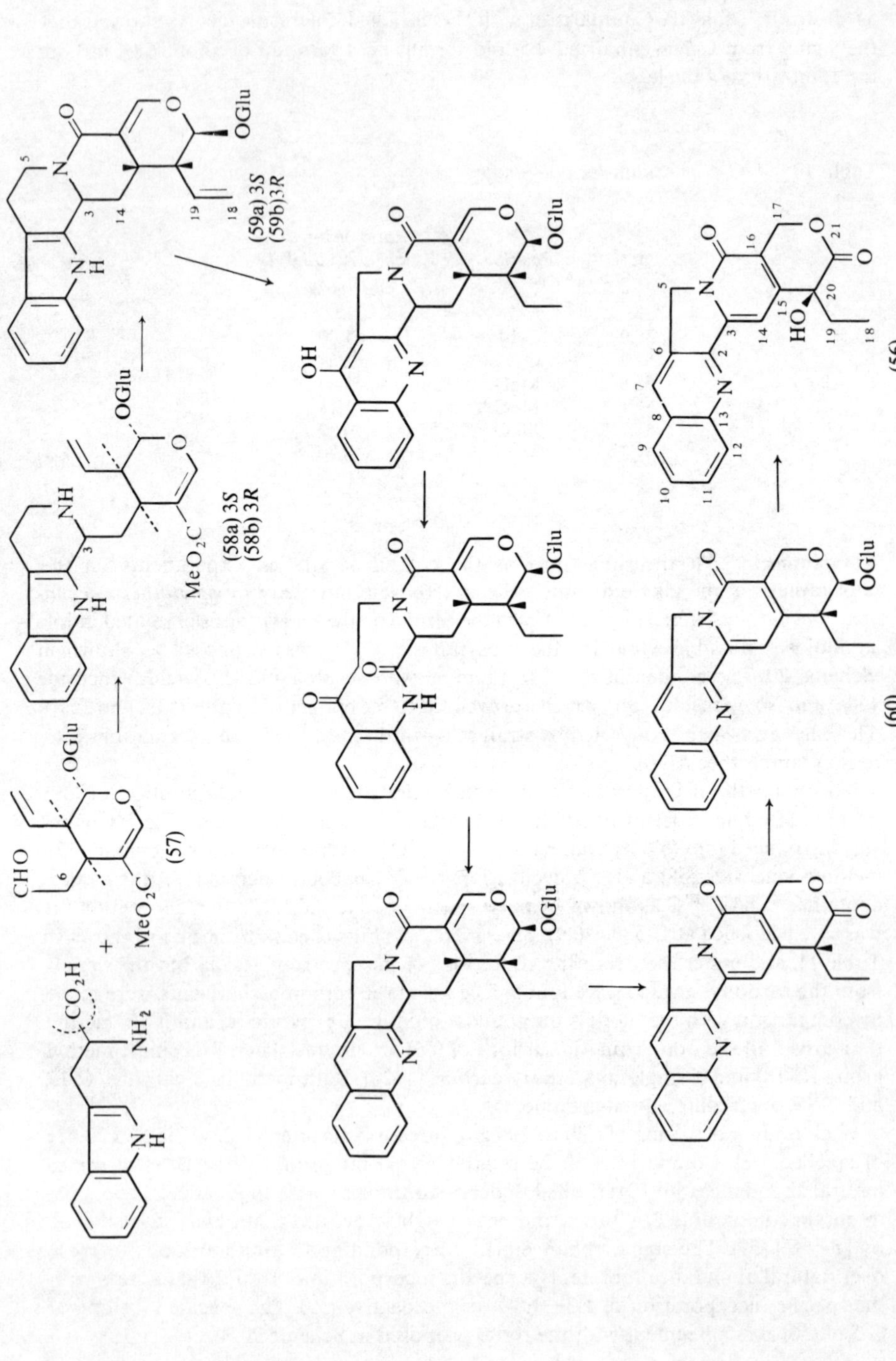

Scheme 20

Table 11 ^{13}C *N.m.r. data for camptothecin*

Position	δc^a	*Relative peak height Natural Abundance*[b]/ ^{13}*C-enriched*[c]
2	156.8	—
3	145.4[d]	—
5	50.2	97/150
6	129.7	—
7	131.4[e]	—
8	127.9	—
9	128.4	—
10	127.5	—
11	129.0	—
12	130.2[e]	—
13	149.9[d]	—
14	96.7	113/90
15	147.9[d]	—
16	119.0	—
17	65.4	100/100
18	7.8	77/63
19	30.6	103/87
20	72.4	52/55

[a] Relative to external TMS: $\delta^{MSO} = \delta^{TMS} + 39.6$ p.p.m.; [b] *ca.* 0.1 mol l^{-1}; [c] *ca.* 0.03 mol l^{-1}; [d, e] Assignments may be interchanged.

Neomycins.—Previous studies on the biosynthesis of this important class of aminoglycoside antibiotics demonstrated that [1-^{14}C]glucose, [6-^{14}C]glucose, and [1-^{14}C]-glucosamine were very well incorporated into neomycin B (61) and C (62).[34] However, the unavailability of crystalline radioactive derivatives, as well as the lack of a satisfactory degradative procedure for the 2-deoxystreptamine moiety, led to ambiguities in the biosynthetic pathway. The biosynthesis of this commercially important group of antibiotics was therefore reinvestigated using [1-^{13}C]glucosamine and [6-^{13}C]-glucose.[35]

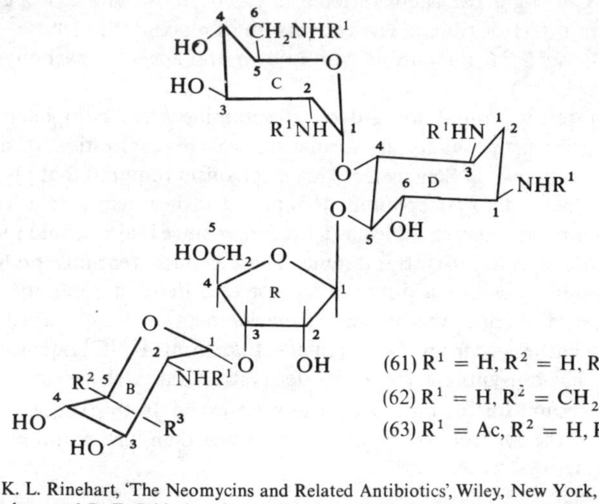

(61) $R^1 = H, R^2 = H, R^3 = CH_2NH_2$
(62) $R^1 = H, R^2 = CH_2NH_2, R^3 = H$
(63) $R^1 = Ac, R^2 = H, R^3 = CH_2NHAc$

[34] (*a*) K. L. Rinehart, 'The Neomycins and Related Antibiotics', Wiley, New York, 1964, pp. 98–114; (*b*) K. L. Rinehart and R. F. Schimbor, in 'Antibiotics', ed. D. Gottlieb and P. D. Shaw, Springer-Verlag, Berlin, 1967, Vol. 2, pp. 359–372.

[35] K. L. Rinehart, J. M. Malik, R. S. Nystrom, R. M. Stroshane, S. T. Truitt, M. Taniguchi, J. P. Rolls, W. J. Haak, and B. A. Ruff, *J. Amer. Chem. Soc.*, 1974, **96**, 2263.

Neomycin B (61) was isolated as the major antibiotic from two separate feeding experiments in *Streptomyces fradiae*. The antibiotic was converted into *N*-hexa-acetylneomycin (63) for n.m.r. examination.

The specific carbon resonances in the natural-abundance p.n.d.-^{13}C n.m.r. spectrum of (63) were assigned by off-resonance decoupling experiments to obtain the multiplicities of each carbon. The resonances were divided into absorptions by standard chemical shift data based on carbon *vs.* nitrogen *vs.* dioxygen-substitution patterns.[36] Specific chemical shift data for other sugars such as ribostamycin (64) were also useful in assigning the chemical shifts shown in Table 12.

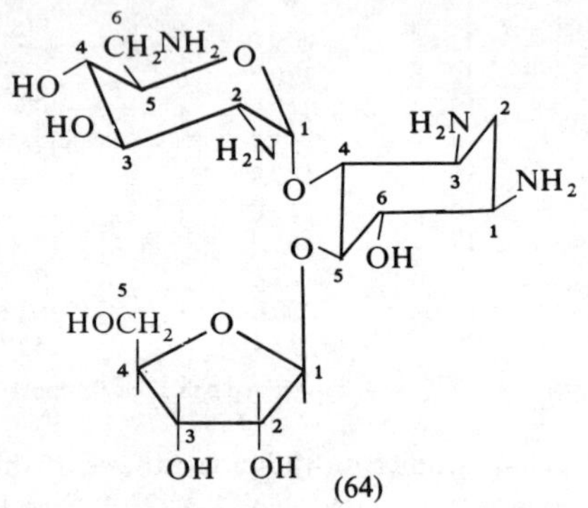

(64)

The ^{13}C n.m.r. spectra of (63) enriched by [1-^{13}C]glucosamine and [6-^{13}C]glucose revealed that C-1 of each of the subunits, B-1, C-1, D-1, and R-1 was labelled by [1-^{13}C]-glucosamine, while carbons B-6, C-6, D-2, and R-5 were enriched by [6-^{13}C]glucose. These labelling results showed that both C-1 and C-6 of glucose and glucosamine were converted into these respective carbons in the neosamines. For the ribose moiety, the results show that C-6 of glucose is converted into C-5 of ribose, and C-1 of glucosamine is converted into C-1 of ribose. The conversion of glucose into ribose[37] and glucosamine into glucose[38] are compatible with known and accepted carbohydrate metabolic pathways.

The incorporation results found for 2-deoxystreptamine were completely unexpected since an earlier proposal for its formation was by cyclization of a 2,6-diamino-5-oxohexose as shown in Scheme 21. This mechanism required that [1-^{13}C]-glucosamine should label 2-deoxystreptamine (65) at D-2 rather than at D-1 as was found. 5-Oxoglucosamine was also excluded as a precursor since it also would require the label to appear at D-2 in 2-deoxystreptamine. These results were interpreted to indicate that glucosamine was not a direct precursor of 2-deoxystreptamine. This conclusion was further verified by the higher ^{13}C-enrichment at D-2 observed in 2-deoxystreptamine biosynthesized from [6-^{13}C]glucose than from [1-^{13}C]glucosamine, as shown in Table 12. The biosynthetic route to 2-deoxystreptamine shown in Scheme 22 was proposed to account for the labelling pattern observed. In this route, glucose was directly cyclized to the cyclitol deoxyinosose which was then further aminated in stepwise fashion to 2-deoxystreptamine.

36 S. Omoto, S. Inoue, M. Kojima, and T. Niida, *J. Antibiotics*, 1973, **26**, 717.
37 B. Magasanik in 'The Bacteria', ed. I. C. Gunsalus and R. Y. Stanier, Academic Press, New York, Vol. 3, p. 297.
38 J. B. Wolfe, B. B. Britton, and H. I. Nakada, *Arch. Biochem. Biophys.*, 1957, **66**, 333.

Table 12 ^{13}C *N.m.r. data for hexa*-N-*acetylneomycin B*

Position[a]	δc/p.p.m.[b]	Enrichment[c] [1-^{13}C]*Glucosamine derived*	Enrichment[c] [6-^{13}C]*Glucose derived*
C-1	97.0	7.07	0.78
C-2	54.2	0.86	1.00
C-3, C-4, C-5	71.4	1.25	1.30
C-6	40.8	0.74	1.94
D-1	50.4	2.17	1.21
D-2	33.2	1.26	3.72
D-3	48.9	1.23	0.96
D-4	76.7	0.98	0.93
D-5	86.0	0.96	0.74
D-6	74.5	0.98	0.92
R-1	109.3	1.60	0.83
R-2	74.5	0.98	0.92
R-3	77.2	0.94	0.87
R-4	82.4	0.99	0.88
R-5	62.2	1.03	2.28
B-1	98.5	6.32	0.76
B-2	51.8	1.03	1.14
B-3	70.3	1.14	1.15
B-4	68.5	1.07	0.86
B-5	74.0	1.13	1.07
B-6	40.8	0.74	1.94

[a] See formula (61) in text for numbering scheme; [b] p.p.m. from TMS, determined using D_2O (solvent) as internal lock, and 1,4-dioxan ($\delta = 67.4$ p.p.m.) as internal standard; [c] Calculated as times natural abundance, by dividing intensities of individual peaks (computer calculated) in the enriched spectrum relative to average methyl carbon atom intensity in that spectrum by intensities of individual peaks in the natural abundance spectrum relative to methyl carbon atom intensity in that spectrum. Values of 1.0 ± 0.3 are regarded as within natural abundance.

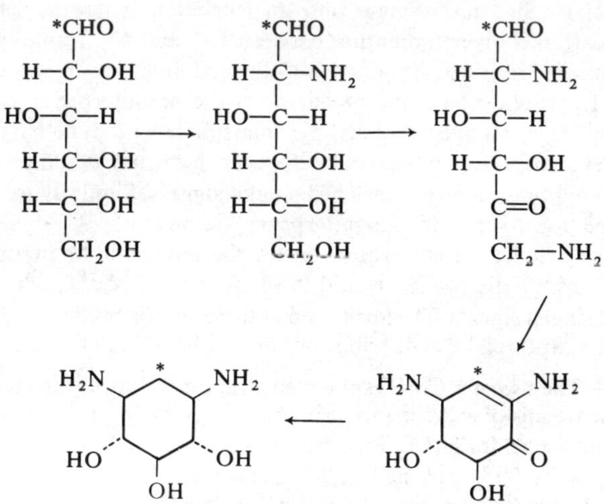

(65) 2-Deoxystreptamine

Scheme 21

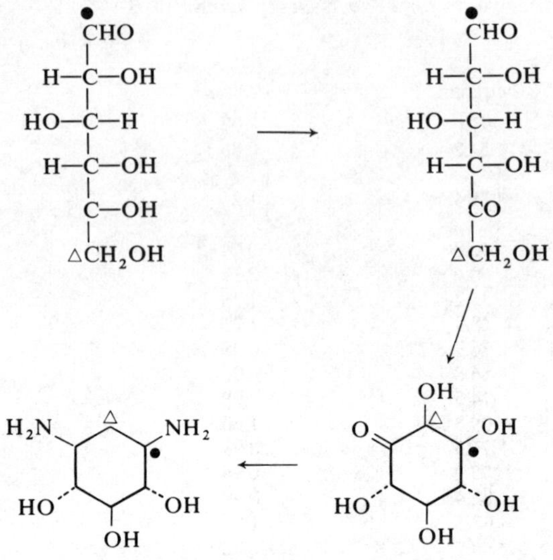

(65)

Scheme 22

Terpenes.—Application of the ^{13}C n.m.r. method to solve terpene biosynthesis problems is expanding. These studies add definition to biosynthesis yielding information about chemically reasonable cyclization pathways from the classic acyclic precursors. Studies carried out with [1,2-$^{13}C_2$]acetate as a precursor can be particularly fruitful in fungal terpene biosynthesis since [1,2-$^{13}C_2$]acetate is biochemically converted into mevalonic acid with a net incorporation of three [1,2-$^{13}C_2$]acetate units. Conversion of labelled mevalonate into an isopentenyl pyrophosphate unit then occurs by decarboxylative elimination between C-1 and C-2 resulting in the loss of one doubly labelled acetic acid unit with C-2 remaining as a single enriched carbon. Thus, geranyl pyrophosphate, the precursor of the monoterpenes, is biochemically formed by condensation of isopentenyl pyrophosphate with dimethyl allyl pyrophosphate. Its label distribution in the p.n.d. ^{13}C n.m.r. spectrum would provide four pairs of ^{13}C—^{13}C couplings and two enriched singlet signals. Similarly for farnesyl pyrophosphate, the precursor of the sesquiterpenes, six pairs of ^{13}C—^{13}C couplings and three enhanced singlet signals would occur. Geranylgeraniol pyrophosphate, the C_{20} precursor of the diterpenes, would have eight pairs of ^{13}C–^{13}C couplings and four enriched singlet signals. The formation of these acyclic precursors from [1,2-^{13}C]-acetate and the expected label distributions are outlined in Scheme 23.

Ascochlorin.—The use of ^{13}C–^{13}C spin–spin coupling relationships to yield information about the details of a cyclization process is illustrated in the biosynthesis of the triprenylphenol ascochlorin (66). Enriched (66) was isolated from cultures of *Nectria coccinea* containing 90% enriched [1,2-$^{13}C_2$]acetate.[39] In the p.n.d. FT-^{13}C n.m.r. spectrum of labelled (66), 18 signals with ^{13}C–^{13}C coupling were detected, indicating

[39] M. Tanabe and K. T. Suzuki, *J.C.S. Chem. Comm.*, 1974, 445.

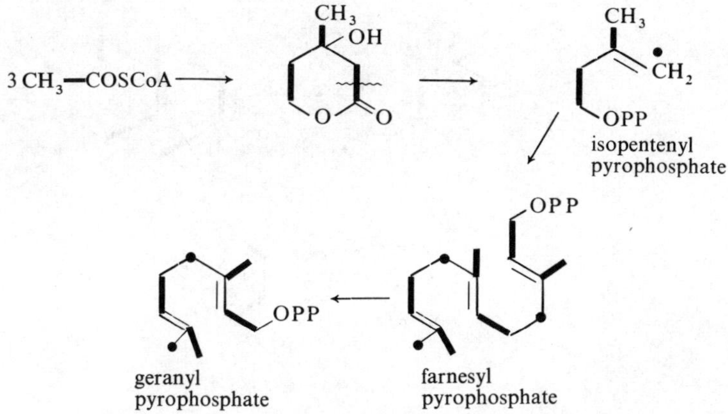

Scheme 23

that nine intact acetate units were incorporated. The chemical shift assignments of the metabolite were made by comparison with published results.[9b] The ^{13}C–^{13}C coupling values assigned to the labelled metabolite were in accord with literature values[9] and are given in Table 13. The labelling results are consistent for a biosynthetic pathway for ascochlorin as shown in Scheme 24.

Table 13 ^{13}C *N.m.r. data*[a] *for ascochlorin*

Carbon chemical shifts[a, b]		^{13}C–^{13}C *Coupling constants*/Hz
C-1 (d) 53.6		
C-2 (s) 212.8	C-3 (t) 41.6	37
C-4 (t) 31.1		
C-5 (d) 40.8	C-13 (q) 16.3[c]	37
C-6 (s) 48.5	C-7 (d) 135.6	45
C-8 (d) 133.2		
C-9 (s) 134.1	C-15 (q) 14.4	46
C-10 (d) 127.6	C-11 (t) 22.2	43
C-12 (q) 8.9[c]		
C-14 (q) 10.3[c]		
C-1′ (q) 12.6	C-2′ (s) 137.8	44
C-5′ (s) 113.8	C-4′ (s) 156.3	70
C-3′ (s) 113.2	C-6′ (s) 162.2	71
C-7′ (s) 111.7	C-8′ (d) 193.2	57

[a] Determination in deuteriochloroform downfield from internal TMS; [b] s = singlet, d = doublet, t = triplet, q = quartet; [c] In the ^1H n.m.r. spectrum, the C-12, C-13, and C-14 methyl protons appear at δ 0.71, 0.82, and 0.86 respectively. These carbons were assigned by selective proton decoupling at δ 0.70, 0.85, and 1.00 p.p.m.

Detection of only five ^{13}C–^{13}C couplings, rather than the six expected for the farnesyl pyrophosphate-derived part of (66), confirmed cleavage of a ^{13}C—^{13}C bond in the biosynthesis. The C-14–C-1 coupling was absent in the labelled metabolite, and both the C-14 and C-1 signals appeared as enriched singlets. This result established

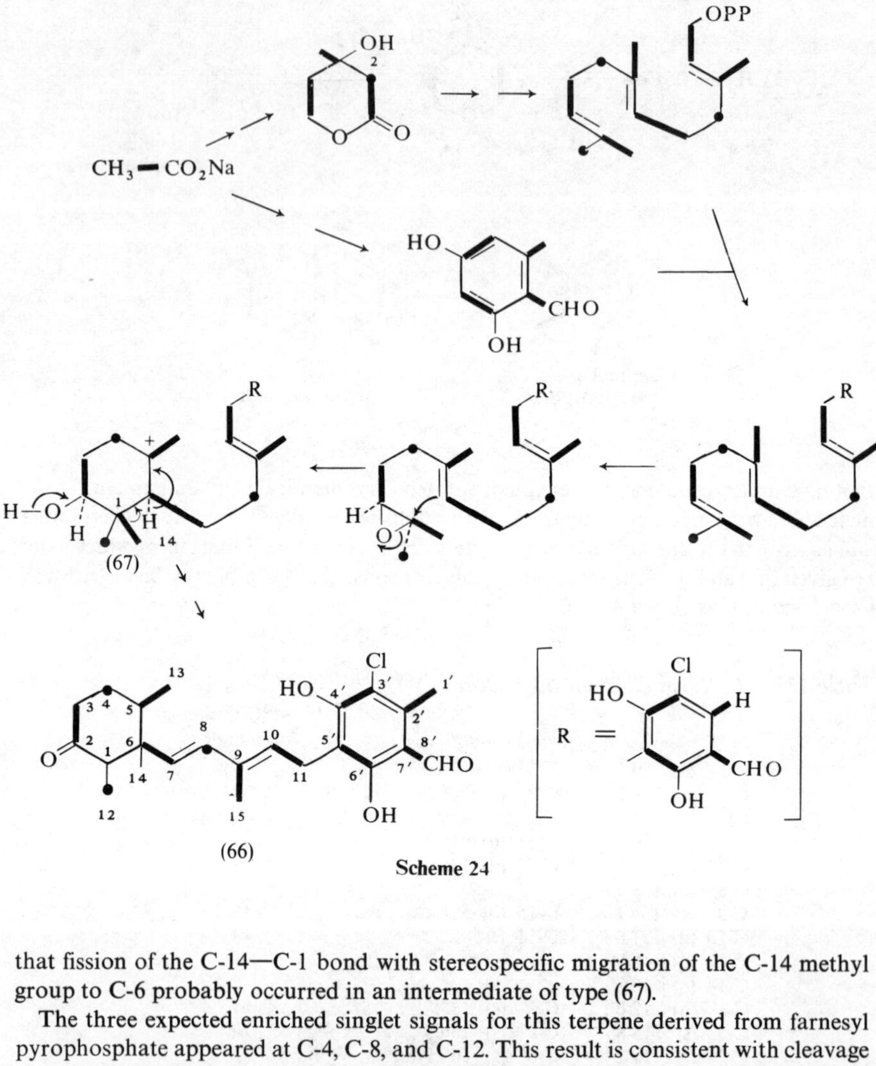

Scheme 24

that fission of the C-14—C-1 bond with stereospecific migration of the C-14 methyl group to C-6 probably occurred in an intermediate of type (67).

The three expected enriched singlet signals for this terpene derived from farnesyl pyrophosphate appeared at C-4, C-8, and C-12. This result is consistent with cleavage of the C-1—C-2 bond in mevalonic acid during conversion into farnesyl pyrophosphate and establishes the biosynthetic origin of these three carbons from C-2 of mevalonate. The four ^{13}C–^{13}C couplings in the orsellinic aldehyde portion (66) confirm that its biosynthetic origin is through a polyketide and is the same as that of orsellinic acid.[40]

Ovalicin.—Elucidation of the biosynthesis of this cytotoxic fungal sesquiterpene is another illustration of the effective use of the [1,2-$^{13}C_2$]acetate method.[41]

Pseudeurotium ovalis was cultured in media fortified with 90% enriched [1,2-$^{13}C_2$]-

[40] W. B. Turner, 'Fungal Metabolites', Academic Press, New York, p. 87.
[41] M. Tanabe and K. T. Suzuki, *Tetrahedron Letters*, 1974, 4417.

acetate that was diluted two-fold with unlabelled acetate. In the FT-^{13}C n.m.r. spectrum of the isolated labelled metabolite, 12 characteristic carbon satellite signals caused by ^{13}C–^{13}C spin–spin coupling were seen, establishing that six molecules of doubly labelled acetic acid had been incorporated. The position of incorporation of these six acetate units shown in Scheme 25 was determined from known carbon

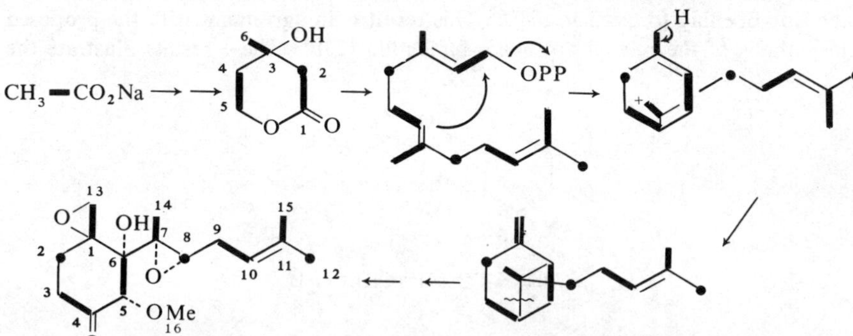

(68) Ovalicin (69) α-bergamotene

Scheme 25

chemical shifts, carbon–carbon coupling constants, and the carbon signal multiplicities in the off-resonance decoupled spectrum of the unlabelled metabolite. The chemical shift and coupling data for ovalicin enriched by [1,2-$^{12}C_2$]acetate are shown in Table 14.

Table 14 ^{13}C *N.m.r. data*[a] *for ovalicin*

Carbon number, multiplicity[b], *and chemical shift*						^{13}C–^{13}C *Coupling constants*/Hz
C_1	(s)	(60.8)	$-C_{13}$	(t)	(51.4)	32
C_2	(t)	(30.4)				
C_3	(t)	(36.8)	$-C_4$	(s)	(207.1)	38
C_5	(d)	(86.4)	$-C_6$	(s)	(78.9)	38
C_7	(s)	(60.5)	$-C_{14}$	(q)	(14.4)	46
C_8	(d)	(59.4)				
C_9	(t)	(27.2)	$-C_{10}$	(d)	(118.5)	43
C_{11}	(s)	(135.8)	$-C_{15}$	(q)	(18.0)	42
C_{12}	(q)	(25.8)				
C_{16}	(q)	(57.0)				

[a] Determined in deuteriochloroform in p.p.m. downfield from internal TMS; [b] s = singlet; d = doublet; t = triplet; q = quartet. Obtained from off-resonance decoupled spectrum of unlabelled ovalicin.

The signals for the C-2, C-8, and C-12 carbons of (68) have enhanced intensities relative to the natural abundance C-16 methoxy-signal and appear as singlet signals without carbon–carbon coupling. The C-16 methoxy-signal is judged to be at natural abundance since it is derived from methionine and not acetate. This result confirms the acetate–mevalonate pathway in ovalicin biosynthesis, since cleavage of the C-1—C-2 bond in mevalonic acid occurs during its conversion to isopentenyl pyrophosphate, and the C-2, C-8, and C-12 carbons of (68) are derived from C-2 of mevalonic acid. The pattern of acetate incorporation is consistent with the acetate→mevalonate→farnesylpyrophosphate route. The retention of six pairs of ^{13}C–^{13}C couplings along with three enhanced singlet signals in (68) enriched by [1,2-$^{13}C_2$]acetate can be accommodated by the biosynthetic route shown in Scheme 25. This route proceeds by cyclization of the farnesol derivative to α-bergamotene (69) followed by cyclobutane ring opening to ovalicin (68).[42] This result is in agreement with the proposed biosynthetis of the related antibiotic fumagillin (70).[43] These results illustrate the

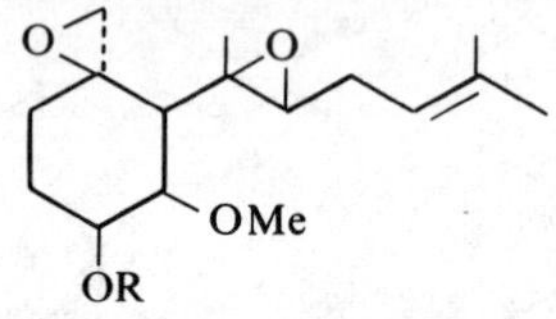

(70) R = CO(CH=CH)$_4$CO$_2$H

economy of the [1,2-$^{13}C_2$]acetate method in the study of terpenoid biosynthesis, since with a single feeding experiment the origin of all 15 carbons of (68) was established.

Coriolins.—This group of sesquiterpene metabolites is composed of three known members (71), (72), and (73).[44] Likely pathways begin with the cyclization of all-*trans*-

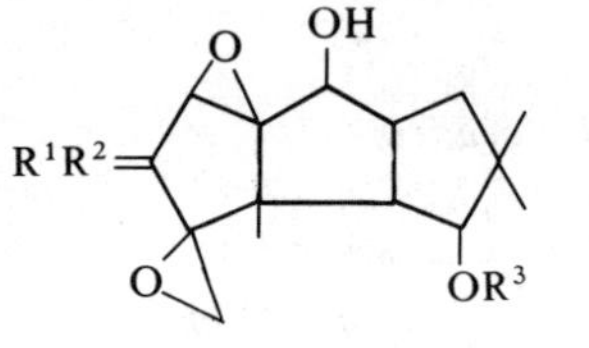

(71) R^1R^2 = O, R^3 = H
(72) R^1 = OH, R^2 = H, R^3 = COC$_7$H$_{15}$
(73) R^1R^2 = O, R^3 = COCHOHC$_6$H$_{13}$

farnesyl pyrophosphate to humulene (74), from which there exist three possible cylization routes, differing only in type and degree of concertedness of the cyclization.[45] Each of the proposed pathways (a, b, and c in Scheme 26) involves the initial protonation of humulene at C-10 followed by formation of the C-2 and C-3 bond. The resulting C-3 cationic species could then react further by one of these pathways to give, after a series of Wagner–Meerwein shifts, the coriolin ring system.

The correct pathway was determined, without recourse to elaborate degradative

[42] Ref. 40, p. 231.

[43] A. J. Birch and S. F. Hussan, *J. Chem. Soc.* (*C*), 1969, 1474.

[44] T. Takeuchi, H. Iinuma, J. Iwanaga, S. Takahashi, T. Takita, and H. Umezawa, *J. Antibiotics*, 1969, **22**, 215; T. Takeuchi, S. Takahaiha, H. Iinuma, and H. Umezawa, *ibid.*, 1971, **24**. 631; S. Takahashi, H. Naganawa, H. Iinuma, T. Takita, K. Maeda, and H. Umezawa, *Tetrahedron Letters*, 1971, 1955.

[45] (*a*) F. W. Comer, F. McCapra, I. H. Quereshi, and A. I. Scott, *Tetrahedron*, 1967, **23**, 4761; (*b*) ref. 40, p. 228; (*c*) W. Parker, J, S. Roberts, and R. Ramage, *Quart. Rev.*, 1967, **21**, 331.

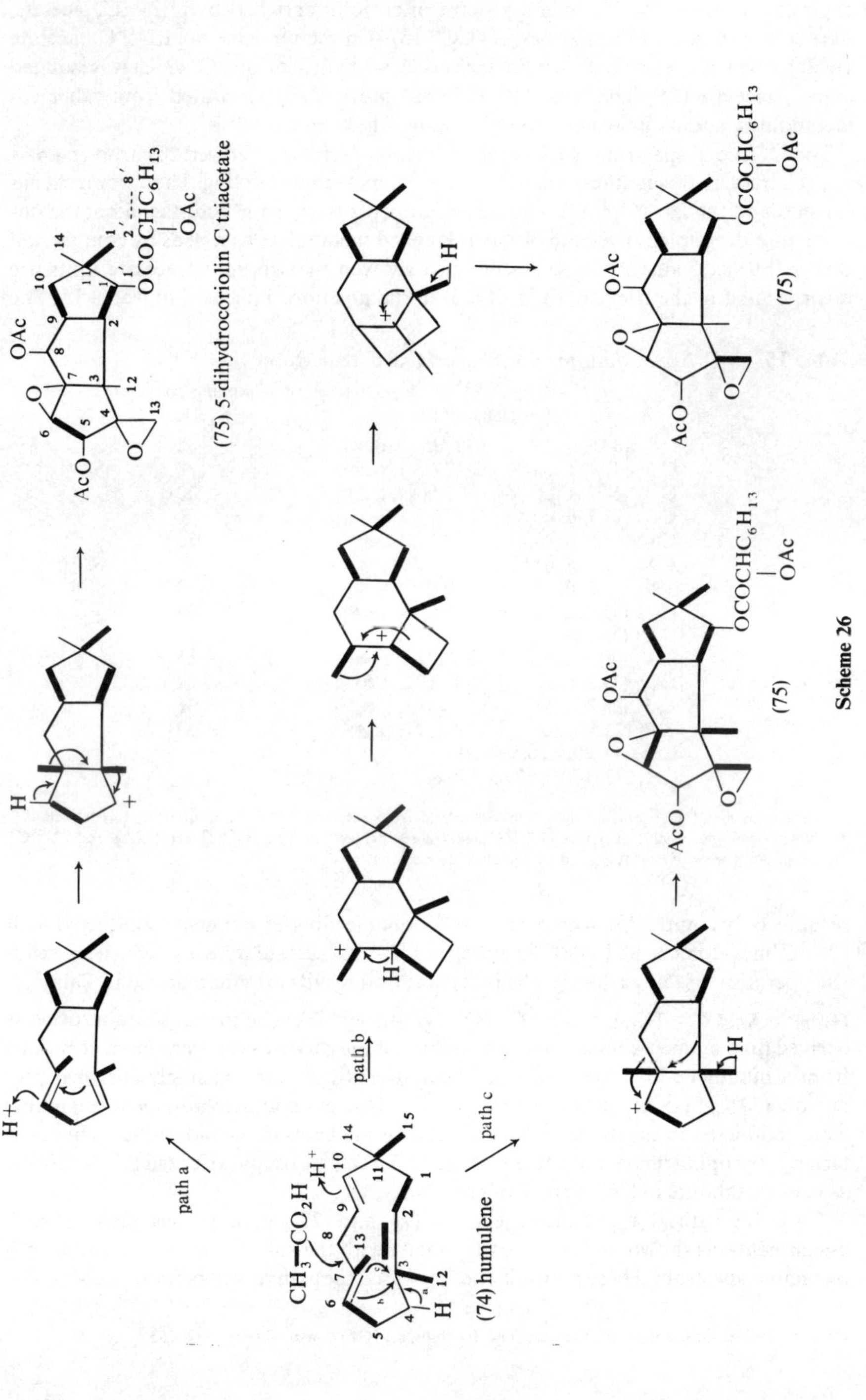
(74) humulene
path a
path b
path c
(75) 5-dihydrocoriolin C triacetate
(75)
(75)

Scheme 26

experiments, from the ^{13}C n.m.r. spectra of coriolin enriched by [1,2-$^{13}C_2$]acetate. Fermentation of *Coriolus consors* (ATCC 11574) in the presence of [1,2-$^{13}C_2$]acetate yielded a new ^{13}C-enriched coriolin derivative 5-dihydrocoriolin C, which was isolated as its triacetate (75). The triacetate (75) was more easily separated from other co-metabolites, and its increased solubility aided the n.m.r. analysis.

The ^{13}C n.m.r. spectrum of the labelled coriolin derivative showed 20 carbon signals with characteristic satellites caused by ^{13}C-^{13}C spin–spin coupling. The carbon chemical shift and the 20 ^{13}C–^{13}C coupling assignments were made with the aid of the off-resonance decoupled spectrum of the unlabelled metabolite, as well as by comparison with published values.[9] These results also showed that six intact acetate units are incorporated in the tricyclic rings of (75) in the positions indicated in Table 15. The

Table 15 ^{13}C *N.m.r. data for dihydrocoriolin C triacetate*

Position, δ_C^{a,b,c}		Coupling constants J/Hz
C-1 (80.8, d, △)^d	– C-2 (51.0, d, ●)	42
C-3 (47.1, s, △)	– C-12 (13.1, q, ●)	38
C-4 (64·5, s, △)	– C-13 (44.9, t, ●)	30
C-5 (71.7, d, ●*)^d		
C-6 (60.9, d, △)	– C-7 (73.5, s, ●)	26
C-8 (72.7, d, ●*)^d		
C-9 (41.3, d, △)	– C-10 (37.6, t, ●)	33
C-11 (43.9, s, △)	– C-15 (21.4, q, ●)	34
C-14 (26.5, ●*)		
C-1′ (170.0, s, △)	– C-2′ (72.4, d, ●)	65
C-3′ (31.3, t, △)	– C-4′ (25.2, t, ●)	34
C-5′ (28.8, t, △)	– C-6′ (31.5, t, ●)	35
C-7′ (22.5, t, △)	– C-8′ (14.0, q, ●)	34
CH_3CO (20.5, 20.5, 20.9)		
CH_3*C*O (170.0, 170.5, 170.8)		

[a] Measured in $CDCl_3$ solution, in p.p.m. downfield from internal TMS; [b] q = quartet; t = triplet; d = doublet; s = singlet; [c] 90% enriched [1,2-^{13}C]acetate, [2-^{13}C]acetate (●), [1-^{13}C]acetate (△), and [2-^{13}C]-mevalonate (*) were used; [d] Assigned by selective decoupling.

result is only compatible with path c. Additional feeding experiments conducted with [2-^{13}C]mevalonic acid, [1-^{13}C]acetate, and [2-^{13}C]acetate were also consistent with the operation of this pathway. The incorporation results are summarized in Table 15.

Hirsutic Acid C.—Hirsutic acid C (76), a prototype tricyclic fungal sesquiterpene, is derived (in a manner similar to the formation of the related sesquiterpenoid coriolins) from a humulene-like intermediate. Labelling studies with the singly enriched precursors (77%) [1-^{13}C]acetate and (90%) [2-^{13}C]acetate in *Stereum complicatum* (Fr) were conducted to clarify the nature of the intermediates in the biosyntheses between farnesyl pyrophosphate and the hirsutane skeleton.[46] Complicatic acid (77) is formed as a co-metabolite of (76) in the fermentation.

From the natural abundance spectra of (76) and (77) unambiguous chemical shift assignments, as shown in Table 16, were made with the aid of multiplicities in the off-resonance spectrum. These multiplicities provided the assignment of the C-4, C-7, C-12,

[46] T. C. Feline, G. Mellows, R. B. Jones, and L. Phillips, *J.C.S. Chem. Comm.*, 1974, 63.

Table 16 ^{13}C *N.m.r. data for hirsutic acid C and complicatic acid*[a]

Carbon	*Hirsutic acid C/δc*	*Complicatic acid/δc*
1	36.6[c]	36.9
2	48.5	49.6[b]
3	48.5	46.4[c]
4	158.3[c]	152.8
5	74.0[b]	197.8
6	63.7[c]	60.9
7	75.4[b]	76.5
8	30.0[b]	29.8
9	39.2[c]	39.3
10	46.3[b]	46.1
11	53.2[c]	53.3
12	183.7[b]	183.8
13	24.2[b]	24.2
14	17.0[b]	17.4
15	111.8[b]	120.5

[a] In $CDCl_3$ solutions; [b] Enriched by [2-^{13}C]acetate; [c] Enriched by [1-^{13}C]acetate.

and C-15 resonances. Comparison between the spectra of (76) and (77) led to the assignment of the C-5 and C-6 signals, the quaternary C-3 and C-11 centres, and the C-13 and C-14 methyl groups. The C-2 signal was assigned downfield of C-9 because of the greater substitution at C-3 than at C-8. Use of $Eu(fod)_3$ with 30—250 mg of (76) produced lanthanide-induced shifts which were approximately linear with the concentration of shift reagent. The carboxylic acid function was considered the major complexation site; this was reflected in the larger induced shifts at C-1 and C-10 than at C-8. The C-1 resonance was assigned at a higher field than C-10 because of steric interactions between C-1 and C-14.

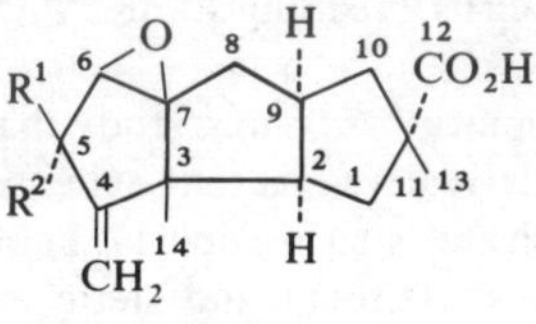

(76) R^1 = OH, R^2 = H
(77) R^1R^2 = O

The p.n.d. ^{13}C n.m.r. spectrum of (76) enriched by [1-^{13}C]acetate showed about 100% peak enhancements for C-1, -4, -6, -9, and -11. Signals due to C-5, -7, -8, -10, -12, -13, -14, and -15 were similarly enhanced in (76) enriched by [2-^{13}C]acetate. The overlapping of the C-2 and C-3 signals of (76) at 48.5 p.p.m. obscured the enrichment data; however, it was demonstrated that C-2 was labelled by [2-^{13}C]acetate by oxidatively converting the allylic alcohol function with manganese dioxide into the ketone (77). The C-2 and C-3 signals then became well separated by about 3 p.p.m. in (77). Similarly, addition of $Eu(fod)_3$ to both enriched samples of (77) separated the overlapping C-3 and C-10 signals and showed that C-3 was enriched by [1-^{13}C]acetate.

This label distribution is consistent with a mechanism involving the formation of a tricyclic cation (78) from farnesyl pyrophosphate which proceeds to the hirsutane skeleton by Wagner–Meerwein shifts of the type shown in Scheme 26. The possible direct conversion of the protoilludane cation (79) into the tricyclic cation (78) by a Wagner–Meerwein shift (Scheme 27) was also considered; such a pathway is analogous to that proposed for the formation of the coriolins.

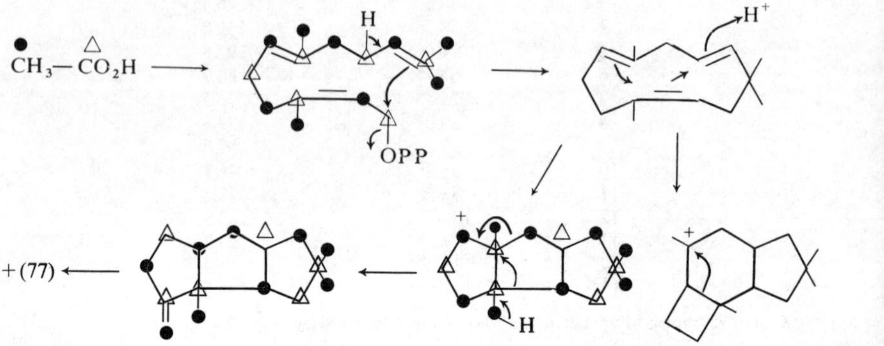

(78) (79)

Scheme 27

Trichothecolone.—An earlier biosynthetic study of the fungal sesquiterpene trichothecin (80) indicated that [2-^{14}C]mevalonic acid labelled the C-10 position.[47] From this it was concluded that the precursor farnesyl pyrophosphate was folded in the manner (81) shown in Scheme 28. More recent work on the trichothecanes established that C-10 retained a [4-*pro-R*-^{3}H]mevalonic acid label.[48] Thus, the implication was that either a hydrogen migration occurs in the biosynthesis, or that farnesyl pyrophosphate is folded as in (82) and not as originally thought. To differentiate between these two possible folding patterns, labelling studies with [2-^{13}C]mevalonic acid were conducted.[49]

A comprehensive and systematic ^{13}C n.m.r. study that examined eight other structurally related trichothecane derivatives led to the chemical shift assignments for all 15 trichothecolone carbons which are listed in Table 17. The relevant trichothecane carbon resonances were assigned by substituent and steric effects, and off-resonance and specific single-frequency proton decoupling experiments. [2-^{13}C]Mevalonic acid was prepared as shown in Scheme 29. Reformatsky reaction of methyl [2-^{13}C]bromoacetate with 4-chlorobutan-2-one produced methyl-5-chloro-3-hydroxy-3-methylpentanoate, which was hydrolysed to [2-^{13}C]mevalonate in 24% overall yield.

Preparation of ^{13}C-labelled trichochecin (80) followed a time *versus* incorporation study of [2-^{13}C]mevalonate into trichothecin by *T. roseum*, which showed a sharp maximum between the fourth and fifth day after incubation and a mass *versus* incorporation study which revealed an optimum concentration to be 100 mg of [2-^{13}C]mevalonate in one litre of culture broth. On the basis of those studies ^{13}C-enriched (80) was isolated after a five-day growth period. The crude metabolite was hydrolysed to trichothecolone (83) which was easier to purify.

[47] E. R. H. Jones and G. Lowe, *J. Chem. Soc.*, 1960, 3959.

[48] (*a*) B. Achilladelis, P. M. Adams, and J. R. Hanson, *Chem. Comm.*, 1970, 511; *J.C.S. Perkin I*,, 1972, 1425; (*b*) S. Nozoe and Y. Machida, *Tetrahedron*, 1972, 28, 5105.

[49] J. R. Hanson, T. Marten, and M. Siverns, *J.C.S. Perkin I*, 1974, 1033.

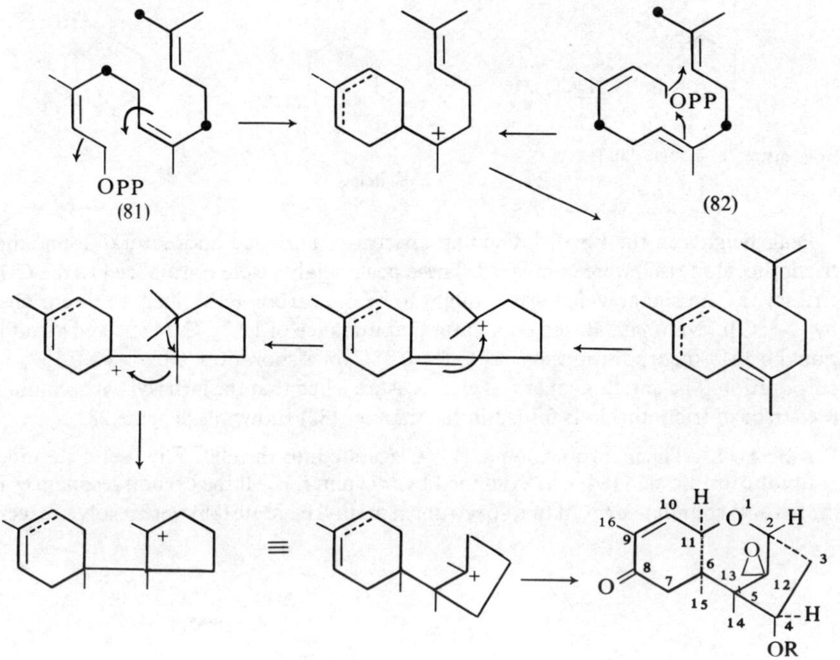

(81)

(82)

(80) R = O–C(=O)–CH=CHMe
(83) R = OH

Scheme 28

Table 17 ^{13}C *N.m.r. data for trichothecolone*

Carbon	δc^{a}
2	70.2 (d)
3	40.2 (t)
4	73.1 (d)[b]
5	49.5 (s)
6	43.6 (s)
7	42.35 (t)
8	199.4 (s)[b]
9	138.5 (s)
10	137.7 (d)
11	79.6 (d)
12	66.0 (s)
13	47.3 (t)
14	6.2 (q)[b]
15	15.6 (q)
16	18.5 (q)

[a] s = singlet, d = doublet, t = triplet, q = quartet; [b] Labelled by [2-^{13}C]mevalonate.

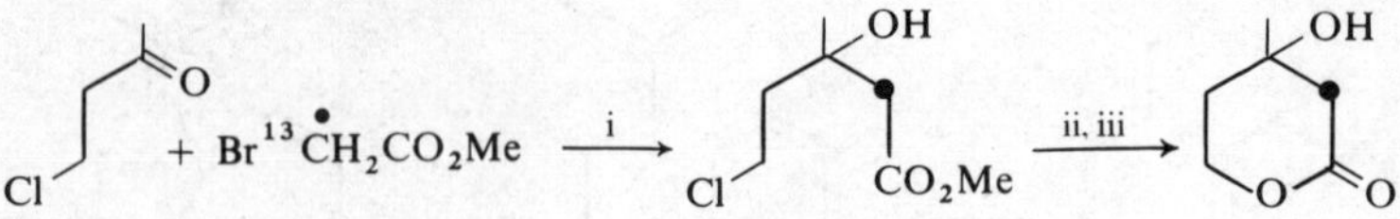

Reagents; i, Zn; ii, OH^-; iii, H^+

Scheme 29

Peak heights in the p.n.d. ^{13}C n.m.r. spectra of enriched and natural-abundance trichothecolone (83) were compared. These peak heights were normalized to the C-12 tertiary carbon signal which was thought to be the carbon least likely to be enriched by [2-^{13}C]mevalonate. Based on a natural abundance of 1.1 %, C-14 showed an additional 0·36 % incorporation of ^{13}C, C-4 a 0.35 % incorporation, and C-8 a 0.35 % incorporation. The enrichment at C-8 clearly established that the farnesyl pyrophophate precursor of trichothecin is folded in the manner (82) shown in Scheme 28.

Fusidic Acid.—The incorporation of [1-^{13}C]acetate into the clinically useful steroidal antibiotic fusidic acid (84) was examined by ^{13}C n.m.r.[50] All the carbon resonances in the natural abundance ^{13}C n.m.r. spectrum of methyl fusidate (85) were resolved except

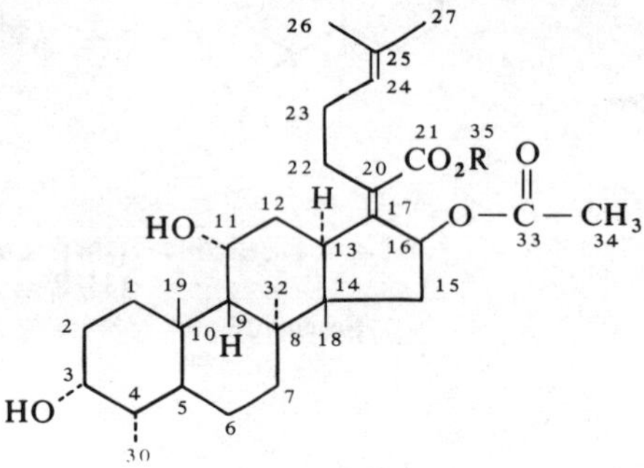

(84) R = H (Fusidic acid)
(85) R = Me (Methyl fusidate)

for the overlap of the C-18 and C-26 signals. Comparison with the ^{13}C-spectra of a series of methyl fusidate derivatives, the 3-keto-, 11-keto-, 3,11-diketo-, 7α-hydroxy-, and 16-desacetyl-compounds made the complete assignment of the individual shifts of (85) possible by use of substituent effects and substituent parameters. Specific proton decoupling of all the protonated carbons as well as the observed multiplicities in the off-resonance decoupled spectrum of (84) and (85) were also used to make the individual chemical shift assignments that are shown in Table 18.

Selective decoupling of long-range ^{13}C–^{1}H couplings is a useful technique for making specific carbon shift assignments between nearly identical ester carbonyl signals. The shift assignment for the C-21 ester carbonyl carbon ($\delta c = 174.22$ p.p.m.) was made by selective decoupling of long range ^{13}C–^{1}H couplings between the methoxyl protons of the methyl ester ($\delta H = 3.65$ p.p.m.) and the C-21 carbonyl group. Specific

[50] T. Riisom, H. J. Jakobsen, N. R. Andersen, and H. Lorck, *Tetrahedron Letters*, 1974, 2247.

Table 18 *^{13}C and ^{1}H-chemical shifts of fusidic acid*

Carbon	δc	Carbon	δc
1	30.21	17	150.51
2	29.91[b]	18	17.83
3	71.59	19	23.08
4	36.51[b]	20	129.85[b]
5	36.02	21	174.22
6	20.96[b]	22	28.82
7	32.12	23	28.49[b]
8	39.57[b,c]	24	123.19
9	49.46	25	132.53[b]
10	36.98[b]	26	17.83
11	68.29[b,c]	27	25.71
12	35.69[b,c]	30	15.92
13	44.31	32	23.89
14	48.79[b,c]	33	170.79[b]
15	39.04	34	20.62
16	74.54[b]		

[a] Solutions are in $CDCl_3$ and chemical shifts are p.p.m. relative to TMS; [b] Labelled by [1-^{13}C]acetate; [c] Carbon–carbon coupled. J = 36 Hz.

proton decoupling of the methyl group (δH = 1.98 p.p.m.) from the C-33 acetoxy-carbonyl carbon (δc = 170.79 p.p.m.) yielded its assignment since their respective proton shifts are well separated in the ^{1}H n.m.r. spectrum of (85).

^{13}C-Enriched fusidic acid was isolated from cultures of the fungus *Fusidium coccineum* Fuckel-K. Tubaki in a growth medium fortified with 0.6% [1-^{13}C]acetate. Peak-height enhancement in the p.n.d. ^{13}C n.m.r. spectrum of the ^{13}C-enriched antibiotic indicated about 1% incorporation of [1-^{13}C]acetate. The spectrum also showed that enrichment had taken place in the positions indicated in Table 18. As expected, this labelling result is fully consistent with the biosynthesis of fusidic acid through a squalene intermediate.

The Reporter wishes to thank K. Baker, L. Cheney, M. I. Dawson, G. Detre, V. Fung, R. H. Peters, and D. M. Yasuda for their help in the preparation of this Report.

8

Biosynthesis of Penicillins and Cephalosporins

BY PATRICIA A. FAWCETT AND E. P. ABRAHAM

1 Introduction

The biosynthesis of the penicillins has now been studied for nearly thirty years and that of the related cephalosporins for more than ten. Although the biogenetic precursors of these molecules were established with relative ease, attempts to elucidate the mechanisms by which the precursors are incorporated into the fused ring systems have met with considerable difficulties. Recent progress has been made possible by use of broken cell systems in place of intact mycelium, and of ^{1}H-n.m.r., ^{13}C-n.m.r., and mass spectrometry, and of mutants of antibiotic-producing strains of *Penicillium chrysogenum* and *Cephalosporium acremonium* which are blocked at certain stages of the biosynthetic pathways.

The present chapter, unlike others in this volume, covers a restricted topic rather than a limited period, but attention has been concentrated upon the recent developments in the field.

2 Production and Structures of Penicillins and Cephalosporins

The penicillins (1) contain a fused β-lactam–thiazolidine ring system, and are *N*-acyl derivatives of 6-aminopenicillanic acid (6-APA)(2).

The naturally-occurring cephalosporins (3) contain a fused β-lactam–dihydrothiazine ring system and an *N*-acyl side-chain consisting of δ-linked D-α-aminoadipic acid. The nucleus of cephalosporin C, 7-aminocephalosporanic acid (7-ACA), has the structure (4).

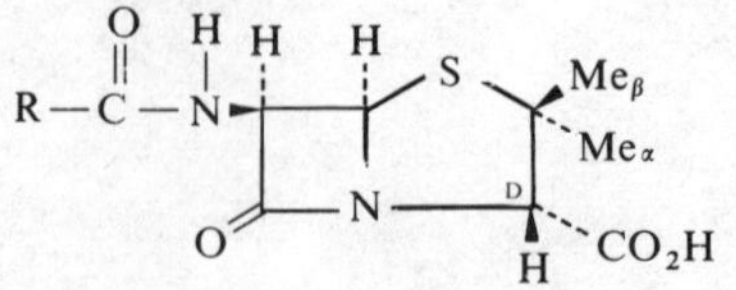

(1)

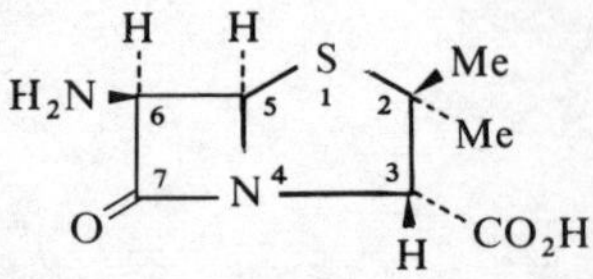

(2)

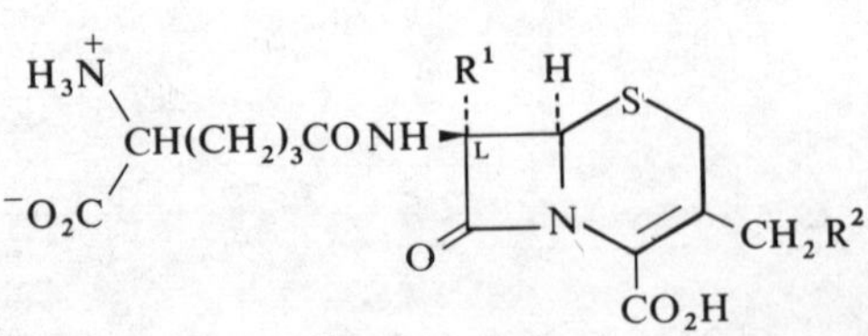

(3)

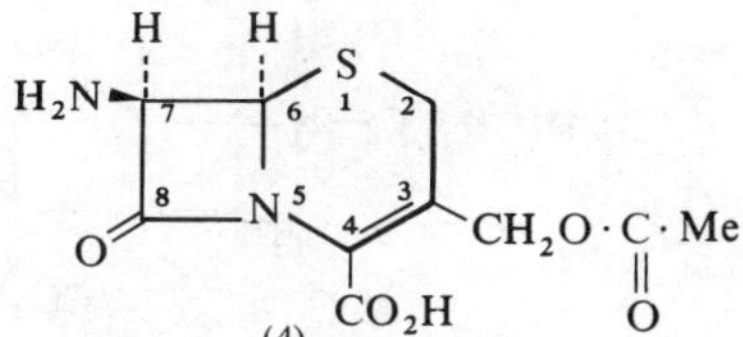

(4)

Penicillium chrysogenum produces a number of different penicillins with various non-polar side-chains [*i.e.* various RCO in (1)] the nature of which depends on the particular side-chain precursor added to the fermentation medium.[1,2] Monosubstituted acetic acids, or some of their derivatives, are the most effective precursors. When the side-chain precursor is phenylacetic acid, for example, benzylpenicillin [penicillin G, (1) R = Ph = $PhCH_2$] is produced. This type of fermentation, characterized by the potential production of an extensive range of penicillin antibiotics, can be classified as the *Penicillium* type.[3] Two other products accumulate in this type of fermentation, especially when the availability of side-chain precursors is limited. These are 6-APA (2)[4—6] and isopenicillin N[7,8] [(1), R = $^{-}O_2C\cdot CH(\overset{+}{N}H_3)(CH_2)_3CO$], which has a δ-linked L-α-aminoadipyl side-chain.

Cephalosporium acremonium typifies the organisms which show a second pattern of fermentation which can be called the *Cephalosporium* type. The *Cephalosporium* type of fermentation yields penicillin N, which has a δ-linked D-α-aminoadipyl side-chain, as the sole penicillin.[9—12] It is this type of fermentation which also yields a number of cephalosporin compounds. *C. acremonium* produces cephalosporin C [(3) R^1 = H, R^2 = OCOMe][13,14]. It also produces deacetylcephalosporin C [(3) R^1 = H, R^2 = OH][15,16] and deacetoxycephalosporin C [(3) R^1 = R^2 = H].[17] In addition, a thiomethyl derivative of deacetoxycephalosporin C [(3) R^1 = H, R^2 = SMe] has recently been isolated from a cephalosporin C-producing strain.[18]

Other cephalosporins have been obtained from certain species of *Streptomyces*. 7-Methoxycephalosporin C (5) is produced by *Streptomyces lipmanii*. Two related

[1] O. K. Behrens, J. Corse, J. P. Edwards, L. Garrison, R. G. Jones, Q. F. Soper, F. R. van Abeele, and C. W. Whitehead, *J. Biol. Chem.*, 1948, **175**, 793.

[2] J. A. Thorn and M. J. Johnson, *J. Amer. Chem. Soc.*, 1950, **72**, 2052.

[3] P. A. Lemke and D. R. Brannon, in 'Cephalosporins and Penicillins—Chemistry and Biology', ed. E. H. Flynn, Academic Press, New York and London, 1972, Chap. 9.

[4] M. Kato, *J. Antibiotics*, 1953, **6**, 130, 184.

[5] F. R. Batchelor, F. P. Doyle, J. H. C. Naylor, and G. N. Rolinson, *Nature*, 1959, **183**, 257.

[6] F. R. Batchelor, E. B. Chain, and G. N. Rolinson, *Proc. Roy. Soc.*, 1961, **B154**, 478.

[7] E. H. Flynn, M. H. McCormick, M. C. Stamper, H. De Valeria and C. W. Godzeski, *J. Amer. Chem. Soc.*, 1962, **84**, 4594.

[8] M. Cole and F. R. Batchelor, *Nature*, 1963, **198**, 383.

[9] K. Crawford, N. G. Heatley, P. F. Boyd, C. W. Hale, B. K. Kelley, G. A. Miller and N. Smith, *J. Gen. Microbiol.*, 1952, **6**, 47.

[10] E. P. Abraham, G. G. F. Newton, K. Crawford, H. S. Burton, and C. W. Hale, *Nature*, 1953, **171**, 343.

[11] E. P. Abraham, G. G. F. Newton, and C. W. Hale, *Biochem. J.*, 1954, **58**, 94.

[12] G. G. F. Newton and E. P. Abraham, *Biochem. J.*, 1954, **58**, 103.

[13] G. G. F. Newton and E. P. Abraham, *Biochem. J.*, 1956, **62**, 651.

[14] E. P. Abraham and G. G. F. Newton, *Biochem. J.*, 1961, **79**, 377.

[15] J. D'A. Jeffery, E. P. Abraham, and G. G. F. Newton, *Biochem. J.*, 1961, **81**, 591.

[16] Y. Fujisawa, H. Shirafuji, M. Kida, K. Nara, M. Yoneda, and T. Kanzaki, *Nature New Biol.*, 1973, **246**, 154.

[17] C. E. Higgens, R. L. Hamill, T. H. Sands, M. M. Hoehn, N. E. Davis, R. Nagarajan, and L. D. Boeck, *J. Antibiotics*, 1974, **27**, 298.

[18] T. Kanzaki, T. Fukita, H. Shirafuji, T. Fujisawa, and K. Kitano, *J. Antibiotics*, 1974, **27**, 361.

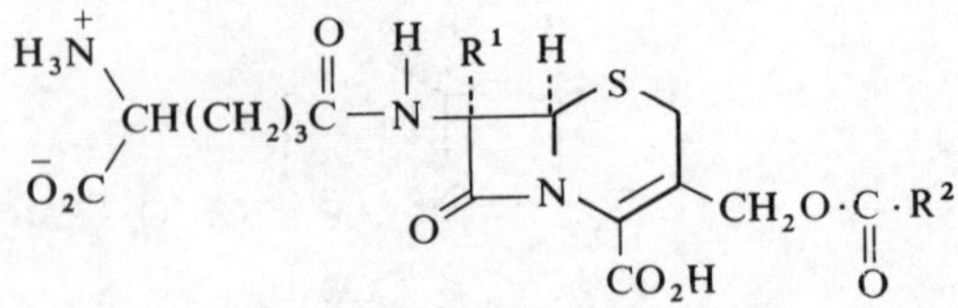

(5) R^1 = OMe, R^2 = Me
(6) R^2 = H, R^2 = NH_2
(7) R^1 = OMe, R^2 = NH_2
(8) R^1 = OMe, R^2 = 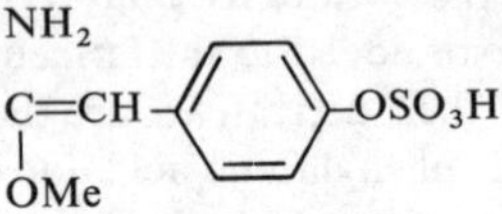
(9) R^1 = OMe, R^2 = 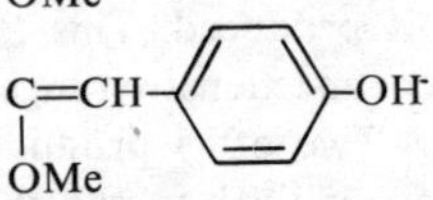

antibiotics (6) and (7) are produced by *S. clavuligerus*, both containing a carbamoyl function and the latter also a 7-methoxy-group.[19–21]

Other 7-methoxycephalosporins, produced by one or more of eight different species of *Streptomyces*, have been named cephamycins A (8), B (9), and C (7).[22–24] Cephamycin C, from *S. lactamdurans*, is identical to the antibiotic obtained from *S. clavuligerus.*[19] A mixture of cephamycins A and B is produced by the other seven species, the best characterized of which is a strain of *S. griseus*.

Deacetoxycephalosporin C is also present in the fermentation broths of a number of *Streptomyces* species[17] and of micro-organisms belonging to a number of different genera.[17, 25] *Streptomyces* species which produce cephalosporins also produce penicillin N.[19, 26]

The *Cephalosporium* type of fermentation so far encountered differs from the *Penicillium* type in several characteristics. It is insensitive to the addition of side-chain precursors to the fermentation medium. The cephalosporins and the penicillin which are produced invariably have the D-α-aminoadipyl side-chain. Moreover, the addition of methionine, especially the D-isomer, has a marked stimulatory effect on the production of penicillin N and cephalosporin C, whereas it does not stimulate penicillin production by *P. chrysogenum.* The amount of 6-APA or 7-APA which is formed in a *Cephalosporium* fermentation appears to be minimal or nil. Whereas all organisms

19 R. Nagarajan, L. D. Boeck, M. Gorman, R. L. Hamill, C. E. Higgens, M. M. Hoehn, W. M. Stark, and J. G. Whitney, *J. Amer. Chem. Soc.*, 1971, **93**, 2308.
20 M. Gorman, M. M. Hoehn, R. Nagarajan, L. D. Boeck, E. A. Presti, J. G. Whitney, and R. L. Hamill. Abstracts 11th Interscientific Conference on Antimicrobial Agents and Chemotherapy, American Society for Microbiology, 1971, p. 7.
21 R. L. Hamill, C. B. Carrell, M. M. Hoehn, W. M. Stark, L. D. Boeck, and M. Gorman, in ref. 20, p 7.
22 E. O. Stapley, M. Jackson, S. Hernandez, S. B. Zimmerman, S. A. Currie, S. Mochales, J. M. Mata, H. B. Woodruff, and D. Hendlin, *Antimicrob. Agents Chemother.*, 1972, **2**, 122.
23 G. Albers-Schonberg, B. H. Arison, and J. L. Smith, *Tetrahedron Letters* 1972, **29**, 2911.
24 T. W. Miller, R. T. Goegelman, R. G. Weston, I. Putter, and F. J. Wolf, *Antimcrob. Agents Chemother.*, 1972, **2**, 132.
25 K. Kitano, K. Kintaka, S. Suzuki, K. Katamoto, K. Nara, and Y. Nakao, *Agric. and Biol. Chem.* (*Japan*), 1974, **38**, 1761.
26 I. M. Miller, E. O. Stapley, and L. Chaiet, *Bacteriol. Proc.*, 1962, 32.

which are known to have a *Penicillium*-type fermentation are eukaryotic, the *Cephalosporium* type includes both eukaryotic and prokaryotic organisms.

3 Amino-acid and Side-chain Precursors

The penicillin structure (1) may be divided (as shown below) into the residues of a mono-substituted acetic acid, L-cysteine, and D-valine; similarly the cephalosporin

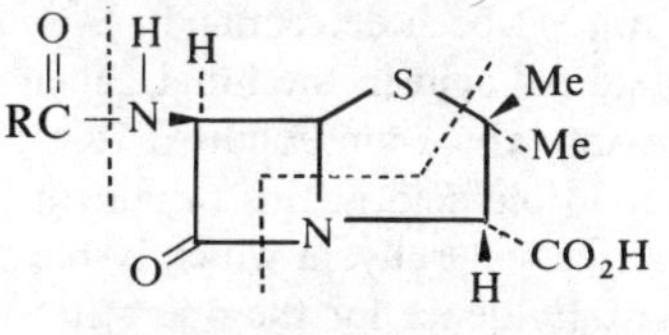

(1)

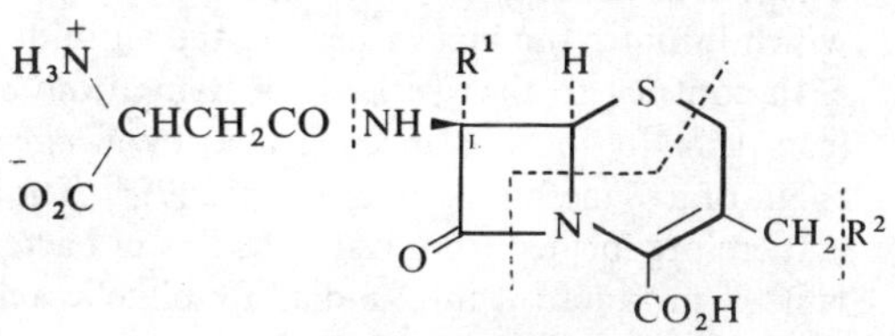

(3)

structure (3) may be divided into residues of D-α-aminoadipic acid, L-cysteine, αβ-dehydrovaline and a side-chain which is acetoxy in cephalosporin C and contains a carbamoyl or cinnamoyl function in the cephalosporins produced by *Streptomyces* spp. The addition of isotopically labelled compounds to fermentations or mycelial suspensions of the antibiotic-producing organisms has shown that these segments correspond to the biogenetic units from which the penicillins and cephalosporins are formed.

Side-chain Precursors. Phenylacetic acid is the direct precursor of the benzylpenicillin side-chains.[27] The D-α-aminoadipyl side-chain of penicillin N and cephalosporin C was shown to be derived from α-aminoadipic acid, and the L-isomer was used more efficiently than the D-isomer.[28, 29] The high dilution of radioactivity on the incorporation of D-α-amino[^{14}C]adipic acid into the antibiotics suggested that the free D-isomer was not involved directly in biosynthesis.[29] The α-aminoadipic acid synthesized by washed suspensions of the *Cephalosporium* sp. was predominantly, if not exclusively, the L-isomer,[29] as is the α-aminoadipic acid which is present in the mycelium of *P. chrysogenum.*[30] Therefore the L-α-aminoadipyl side-chain of isopenicillin N produced by the latter organism also is presumably derived from L-α-aminoadipic acid.

The carbon skeleton of the α-aminoadipyl side-chain in the eukaryotes arises from condensation of α-oxoglutarate and acetyl-CoA, since [1-^{14}C]acetate led to labelling of C-1 and C-6 of the D-α-aminoadipic acid residue as well as the acetoxy-group of cephalosporin C;[31] C-6 was also labelled by [5-^{14}C]-α-oxoglutarate.[32] These results have been confirmed and extended by the use of [1-^{13}C]- and [2-^{13}C]-acetate as precursors and the determination of the position of ^{13}C in cephalosporin C by n.m.r. spectroscopy.[33] The isotope was incorporated from [1-^{13}C]acetate into the carbonyl carbon of the acetoxy-group and into C-1 and C-6 of the side-chain. With [2-^{13}C]-

[27] W. J. Halliday and H. R. V. Arnstein, *Biochem. J.*, 1956, **64**, 380.
[28] P. W. Trown, B. Smith, and E. P. Abraham, *Biochem. J.*, 1963, **86**, 284.
[29] S. C. Warren, G. G. F. Newton, and E. P. Abraham, *Biochem. J.*, 1967, **103**, 891.
[30] H. R. V. Arnstein and D. Morris, *Biochem. J.*, 1960, **76**, 357.
[31] P. W. Trown, E. P. Abraham, G. G. F. Newton, C. W. Hale, and G. A. Miller, *Biochem. J.*, 1962, **84**, 157.
[32] P. W. Trown, M. Sharp, and E. P. Abraham, *Biochem. J.*, 1963, **86**, 280.
[33] N. Neuss, C. H. Nash, P. A. Lemke, and J. B. Grutzner, *J. Amer. Chem. Soc.*, 1971, **93**, 2337, 5314 (correction).

acetate ^{13}C was incorporated into the methyl carbon of the acetoxy-group and into C-2, C-3, C-4, and C-5 of the side-chain. These findings can be explained by the circulation of acetate through the citric acid cycle.

In both *Penicillin* and *Cephalosporium* spp. α-aminoadipic acid functions as an intermediate in lysine biosynthesis, as it does in other fungi, including yeast.[34,35] α-Aminoadipic acid δ-adenylate is an intermediate in the conversion of α-aminoadipic acid to α-aminoadipic δ-semialdehyde,[36] but the activated form of α-aminoadipic acid which initiates the biosynthesis of the antibiotics has not yet been determined.

In contrast to the synthesis of lysine from α-aminoadipic acid in the fungi, label from [^{14}C]lysine was incorporated by *S. clavuligerus* into the α-aminoadipyl side-chain of a 7-methoxycephalosporin (7).[37] The α-aminoadipic acid in this organism is therefore formed from lysine, as it is in bacterial systems generally, in which lysine is itself produced from 2-6-diaminopimelic acid (DAP). Evidence for the operation of the DAP pathway in *S. lipmanii*, which produces penicillin N and a 7-methoxy-cephalosporin (5), has come from the finding that a mutant which was defective in diaminopimelate decarboxylase required lysine in the medium for growth.[38]

Precursors of the Ring Systems.—*Cysteinyl Moiety*. Evidence for the involvement of L-cysteine in the formation of the 6-APA ring system was provided by the finding that L-cystine suppressed the incorporation of ^{35}S from $Na_2{}^{35}SO_4$ into benzylpenicillin, whereas D-cystine and penicillamine did not.[39] Cysteine is incorporated intact into the molecule, since benzylpenicillin formed by *P. chrysogenum* in the presence of [3-^{14}C,^{15}N,^{35}S]cystine contained the three labels in proportions similar to those in which they were present in the precursor.[40,41]

Experiments with *C. acremonium* have indicated that [^{14}C]cysteine is incorporated directly into cephalosporin C.[28] On the addition of DL-[3,3′-^{14}C]cystine and L-[3,3′-^{3}H]cystine to a fermentation of *S. clavuligerus*, 9% ^{14}C and 7% ^{3}H were incorporated into the 7-methoxycephalosporin (7).[37] The methyl carbon of the methoxy-group was labelled from [*methyl*-^{14}C]methionine.

In the case of *P. chrysogenum* the sulphur in cysteine may be derived from sulphate *via* a reduction pathway, but it may also be derived *via* reverse trans-sulphuration from methionine.[42] In the *Cephalosporium* sp. all of the sulphur for cephalosporin C synthesis may be derived from methionine.[43] The incorporation of label from [^{14}C]serine and [^{35}S]methionine into the 7-methoxycephalosporin (7) produced by *S. clavuligerus* suggests that cysteine is produced by a reverse trans-sulphuration route in this organism also.

The stimulation of penicillin N and cephalosporin C biosynthesis by methionine has not been adequately explained. The addition of methionine appears to affect the

34 M. Strassman and L. N. Ceci, *Biochem. Biophys. Res. Comm.*, 1964, **14**, 262.
35 M. Strassman, L. N. Ceci, and B. E. Silverman, *Biochem. Biophys. Res. Comm.*, 1964, **14**, 268.
36 A. K. Sinha and J. K. Bhattacharjee, *Biochem. Biophys. Res. Comm.*, 1970, **39**, 1205.
37 J. G. Whitney, D. R. Brannon, J. A. Mabe, and K. J. Wicker, *Antimicrob. Agents Chemother.*, 1972, **1**, 247.
38 J. R. Kirkpatrick, L. E. Doolin, and O. W. Godfrey, *Antimicrob. Agents Chemother.*, 1973, **4**, 542.
39 C. M. Stevens, P. Vohra, E. Inamine, and O. A. Roholt, *J. Biol. Chem.*, 1953, **205**, 1001.
40 H. R. V. Arnstein and P. T. Grant, *Biochem. J.*, 1954, **57**, 353.
41 H. R. V. Arnstein and P. T. Grant, *Biochem. J.*, 1954, **57**, 360.
42 C. M. Stevens, P. Vohra, J. E. Moore, and C. W. De Long, *J. Biol. Chem.*, 1954, **210**, 713.
43 J. Nüesch, H. J. Treichler, and M. Liersch, 'Genetics of Industrial Microorganisms: Actinomycetes and Fungi', ed. Z. Vanek, Z. Hostalěk, and K. Cudlin, Academia, Prague, 1973, p. 309; P. G. Caltrider and H. F. Niss, *Appl. Microbiol.*, 1966, **14**, 746.

metabolism of the *Cephalosporium* sp. in a number of ways during the phases of both growth and antibiotic production. Although methionine may provide the source of sulphur for cephalosporin C synthesis, this does not appear to be the reason for its stimulatory effect. DL-Norleucine can replace methionine as a stimulant in defined media in which it was taken up by the mycelium.[44] No support has been obtained for an earlier suggestion that methionine inhibits or represses a cysteine-degrading enzyme, thus increasing the amount of cysteine available for antibiotic synthesis. Further information, however, has been provided by genetic studies. In a mutant of *C. acremonium* which was blocked early in the pathway of cystine and methionine formation and also in the pathway from cystine to methionine, cephalosporin C production was dependent upon the addition of excess methionine or norleucine. A double mutant, blocked in both the sulphate reduction and trans-sulphuration pathways to cysteine, required added cysteine for growth. The cysteine inhibited methionine transport, but methionine peptides were taken up by the mycelium and stimulated cephalosporin C production.[45] It was concluded that the role of methionine is that of an inducer or activator of an enzyme involved in antibiotic biosynthesis.

Valinyl moiety. L-Valine is the precursor of the D-penicillamine fragment of benzylpenicillin[46,47] and of penicillin N.[48] D-Valine was found to be a poor precursor, even after correction was made for its slow uptake into the mycelial cells,[49] and was shown, in fact, to inhibit the production of benzylpenicillin[50] and penicillin N.[48]

Label from DL-[1-^{14}C]valine was incorporated into the $\alpha\beta$-dehydrovaline fragment of cephalosporin C[28] and again the L-isomer was found to be utilized more efficiently than the D-isomer.[40] However, in contrast to its effect on penicillin N synthesis, the addition of D-valine to a suspension of *C. acremonium* did not affect the synthesis of cephalosporin C. The yield of cephalosporin C, but not of penicillin N, was decreased on the addition of γ-hydroxyvaline.[48] It therefore appeared that γ-hydroxyvaline is not a precursor of cephalosporin C and that the oxidation of the valine methyl group occurs after the valine has been incorporated into an intermediate.

4 Possible Intracellular Intermediates

The first compound to be isolated from mycelium with a structure which suggested that it was an intermediate in the biosynthesis of penicillins was the tripeptide, δ-(α-aminoadipyl)cysteinylvaline (10).[51,52] This tripeptide was present in the mycelium of *P. chrysogenum* in very small amounts and the optical configurations of its amino-acid residues were not determined.[30] The isolation of this compound, following the elucidation of the structure of penicillin N,[12] gave rise to the tripeptide theory of biosynthesis. It was suggested that δ-(α-aminoadipyl)cysteinylvaline was a precursor

[44] S. W. Drew and A. L. Demain, *Biotechnol. and Bioeng.*, 1973, **15**, 743.
[45] S. W. Drew and A. L. Demain, Proceedings 1st Intersectional Congress of the International Association of Microbiological Societies, Tokyo, September, 1974, Science Council of Japan, 1975, Vol. 5, p. 539.
[46] C. M. Stevens, E. Inamine, and C. W. De Long, *J. Biol. Chem.*, 1956, **219**, 405.
[47] H. R. V. Arnstein and M. E. Clubb, *Biochem. J.*, 1957, **65**, 618.
[48] S. C. Warren, G. G. F. Newton, and E. P. Abraham, *Biochem. J.*, 1967, **103**, 902.
[49] H. R. V. Arnstein and H. Margreiter, *Biochem. J.*, 1958, **68**, 339.
[50] A. L. Demain, in 'Biosynthesis of Antibiotics', ed. J. F. Snell, Academic Press, New York and London, 1966, Vol. 1, p. 29.
[51] H. R. V. Arnstein, D. Morris, and E. J. Toms, *Biochim. Biophys. Acta*, 1959, **35**, 561.
[52] H. R. V. Arnstein, M. Artman, D. Morris, and E. J. Toms, *Biochem. J.*, 1960, **76**, 353.

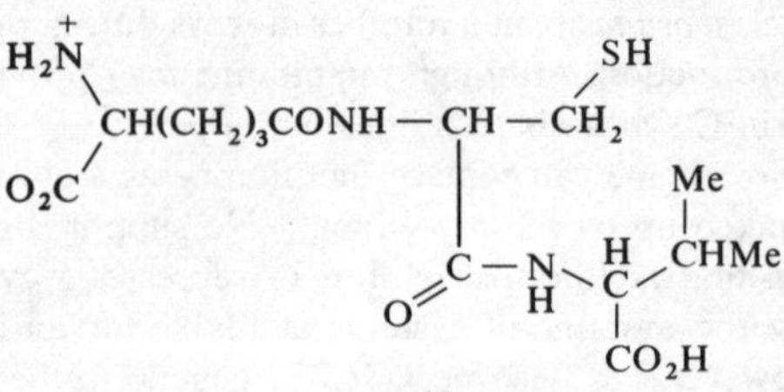

(10)

of penicillin N and that other penicillins might be obtained from the latter by the action of a side-chain transferase.[30, 53] When it was discovered later that isopenicillin N, with a δ-(L-α-aminoadipyl) side-chain, was produced by *P. chrysogenum*[7, 8] it seemed possible that isopenicillin N, rather than penicillin N, was the precursor of the penicillins with non-polar side-chains.

Material, which after oxidation with performic acid possessed an electrophoretic mobility similar to that of the sulphonic acid of δ-(α-aminoadipyl)cysteinylvaline was subsequently detected in the mycelium of *C. acremonium*.[54] This material was resolved by chromatography into three peptides having different R_F values; the peptides were named P1, P2, and P3.[55] These thiol-containing peptides were isolated from the mycelium, together with glutathione, by a procedure developed for the isolation of glutathione from animal tissues.[56] After oxidation and separation peptides P1—3 were found to yield the following amino-acids on acid hydrolysis:

P3; α-aminoadipic acid, cysteic acid, valine

P2; α-aminoadipic acid, cysteic acid, valine, glycine

P1; α-aminoadipic acid, cysteic acid, β-hydroxyvaline, glycine.

M.s. analysis of peptide P3 and determination of the c.d. spectra of its constituent amino-acids showed that it was δ-(L-α-aminoadipyl)-L-cysteinyl-D-valine.[55] The structures of peptides P2 and P1 have not been finally determined, but comparison with synthetic tetrapeptides by paper chromatography and electrophoresis indicates that they may be δ-(α-aminoadipyl)cysteinylvalylglycine and δ-(α-aminoadipyl)-cysteinyl-β-hydroxyvalylglycine respectively. The mixture of peptides P1, P2, and P3 will subsequently be referred to as the ACV complex.

Recent work has shown that the LLD and all-L forms of δ-(α-aminoadipyl)cysteicyl-valine can be separated by paper electrophoresis at pH 1.8. This has made it possible to conclude that the α-aminoadipyl tripeptide from *P. chrysogenum*, like that from the *Cephalosporium* sp, is in the LLD form.[57] The same, or a closely related peptide has been detected in the mycelium of *S. clavuligerus*.[57] In other experiments it has been found that high-yielding strains of *C. acremonium* yield much larger amounts of the tripeptide, relative to glutathione, than do low-yielding strains.[58]

5 Biosynthesis of Intracellular Peptides

The formation of δ-(α-aminoadipyl)cysteinylvaline was reported to occur in a cell-

[53] E. C. Wolff and H. R. V. Arnstein, *Biochem. J.*, 1960, **76**, 375.
[54] E. P. Abraham and G. G. F. Newton, *Adv. Chemotherapy* 1965, **2**, 23.
[55] P. B. Loder and E. P. Abraham, *Biochem. J.*, 1971, **123**, 471.
[56] E. Waelsch and D. Rittenberg, *J. Biol. Chem.*, 1941, **139**, 761.
[57] P. A. Fawcett, J. J. Usher, and E. P. Abraham, unpublished experiments, 1975.
[58] P. A. Fawcett, S. Queener and E. P. Abraham, unpublished experiments, 1975.

free extract of *P. chrysogenum.*[59] However, neither the sequence of the incorporation of the constituent amino-acids into the tripeptide nor the optical configurations of the amino-acids in the product were determined.

It was suggested that the initial reaction in penicillin biosynthesis is the condensation of α-aminoadipic acid and cysteine to form a dipeptide.[30] An attempt was made with intact mycelium of *C. acremonium* to study the involvement of the L- and D-isomers of δ-(α-aminoadipyl)-L-cysteine in the biosynthesis of the tripeptide. However, the dipeptides were hydrolysed at the surface of the cell and only the free amino-acids were taken up.[60] A system was therefore required which lacked the permeability barriers of the intact mycelium to the dipeptide. In an extract prepared by ultrasonic treatment the synthesis of peptides P2 and P3 could be demonstrated, but only when δ-(L-α-aminoadipyl)-L-cysteine was added.[61] No synthesis of tripeptide could be detected in the presence of either δ-(D-α-aminoadipyl)-L-cysteine and [^{14}C]valine or of DL-α-amino-[^{14}C]adipic acid and L-cysteinyl-L-valine or L-cysteinyl-D-valine. It appears, therefore, that the biosynthesis of P3 begins at the *N*-terminal amino-acid, as it does with glutathione.[62] However, whereas the synthesis of glutathione occurs in the supernatant fraction of the extract, peptides P2 and P3 were synthesized in the particulate fraction.

The dipeptide δ-(L-α-aminoadipyl)-L-cysteine has not yet been detected in the mycelium and it may be that this dipeptide is enzyme-bound. An enzyme-bound dipeptide intermediate, γ-glutamyl-α-aminobutyrylphosphate, is involved in the synthesis of γ-glutamyl-α-aminobutyrylglycine (ophthalmic acid).[62, 63] It has also been shown that enzyme-bound γ-glutamylphosphate is an intermediate in dipeptide synthesis.[64, 65]

6 The Role of the Tripeptide in Biosynthesis

A number of findings have given indirect support to the suggestion that an α-amino-adipyl peptide is an intermediate in the biosynthesis of β-lactam antibiotics.

The involvement of α-aminoadipic acid in the biosynthesis of benzylpenicillin by *P. chrysogenum* became apparent from the observation that the synthesis was inhibited by lysine and the inhibition was reversed by α-aminoadipic acid.[66, 67] Lysine was shown to exert end-product control of the formation of α-aminoadipic acid, thereby decreasing the amount of the latter available for penicillin biosynthesis.[68] This effect of lysine was reported to be caused by feedback inhibition of homocitrate synthase, the first enzyme of the α-aminoadipic acid-lysine biosynthetic pathway.[69, 70]

A second effect of lysine is that of stimulation of penicillin production by suspensions of washed mycelium of *P. chrysogenum* after prior growth in lysine.[71] This has

[59] K. Bauer, *Z. Naturforsch*, 1970, **25b**, 1125.
[60] P. B. Loder, E. P. Abraham, and G. G. F. Newton, *Biochem. J.*, 1969, **112**, 389.
[61] P. B. Loder and E. P. Abraham, *Biochem. J.*, 1971, **123**, 477.
[62] E. D. Mooz and A. Meister, *Biochemistry*, 1967, **6**, 1722.
[63] J. S. Nishimura, E. A. Dodd, and A. Meister, *J. Biol. Chem.*, 1964, **239**, 2553.
[64] M. Orlowski and A. Meister, *J. Biol. Chem.*, 1971, **246**, 7095.
[65] J. S. Davis, J. B. Balinsky, J. S. Harington, and J. B. Shepherd, *Biochem. J.*, 1973, **133**, 667.
[66] A. L. Demain, *Arch. Biochem. Biophys.*, 1957, **67**, 244.
[67] N. L. Somerson, A. L. Demain, and T. D. Nunheimer, *Arch. Biochem.*, 1961, **93**, 238.
[68] S. A. Goulden and F. W. Chattaway, *Biochem. J.*, 1968, **110**, 55P.
[69] A. L. Demain and P. S. Masurekar, *J. Gen. Microbiol.*, 1974, **82**, 143.
[70] P. S. Masurekar and A. L. Demain, *Appl. Microbiol.* 1974, **28**, 265.
[71] P. S. Masurekar and A. L. Demain, *Canad. J. Microbiol.*, 1972, **18**, 1045.

been attributed to repression of enzyme(s) converting α-aminoadipic acid into lysine. In yeast, lysine in high concentration inhibits homocitrate synthase, whereas in low concentration it inhibits lysine biosynthesis at a stage following α-aminoadipic acid formation.[72] In a *Streptomyces* sp. D-lysine stimulated the production of a 7-methoxy-cephalosporin but in this case the mechanism of action is uncertain.[72a]

In *C. acremonium* the ACV complex began to be labelled from [^{14}C]valine at about the same time in the fermentation as that at which penicillin N and cephalosporin C began to be synthesized at a maximum rate.[48] Genetic studies also provided evidence for a biosynthetic role for the tripeptide in *C. acremonium.*[73] A mutant which was auxotrophic for lysine did not synthesize ACV or β-lactam antibiotics in a medium supplemented with lysine alone. However, small amounts of antibiotic, as well as ACV, were synthesized when α-aminoadipic acid was also present. Further mutants of *C. acremonium* which were blocked for synthesis of β-lactam antibiotics could be divided into two classes. One class did not incorporate label from valine or α-amino-adipic acid into cysteine-containing peptides while the other did so. These findings could be accounted for if mutants in the first class were blocked in the biosynthetic pathway at a stage before the synthesis of the peptides, and those in the second class were blocked in the conversion of peptides into antibiotics. The two classes were complementary: heterokaryons formed between certain mutant pairs were capable of antibiotic synthesis.[74] In other antibiotic-negative strains of *C. acremonium* the —S—S— dimer of δ-(L-α-aminoadipyl)-L-cysteinyl-D-valine and also the disulphide of this tripeptide and methanethiol accumulated in the fermentation medium.[75]

The role of δ-(α-aminoadipyl)cysteinylvaline in the biosynthesis of β-lactam antibiotics could not be tested directly with intact mycelium because the tripeptide failed to be taken up by the mycelial cell. Extracts of *C. acremonium* obtained by use of ultrasonication or abrasives failed to show detectable antibiotic synthesis. However, a cell-free system obtained from protoplasts of *C. acremonium* has proved to have somewhat higher biosynthetic capabilities. Metabolically active protoplasts were prepared from mycelium harvested at an early stage of antibiotic production[76] and disrupted by osmotic lysis. To the crude extract was added an energy-generating system consisting of ATP, phosphoenolpyruvate, and pyruvate kinase. This system could synthesize the α-aminoadipyl tripeptide from α-amino[^{14}C]adipic acid, cysteine, and valine, and did not require the addition of preformed δ-(L-α-amino-adipyl)-L-cysteine. It was also able to incorporate traces of [^{14}C]valine into penicillin N.[77]

For use in this system the LLD isomer of δ-(α-aminoadipyl)cysteinylvaline was synthesized with a tritium label in the methyl groups of its valine residue and also with tritium only in the α-position of its valine residue. The all-L isomer of the tripeptide

[72] A. F. Tucci and L. N. Ceci, *Arch. Biochem. Biophys.* 1972, **153**, 751.
[72a] E. Inamine, unpublished experimental results.
[73] P. A. Lemke and C. H. Nash, *Canad. J. Microbiol.*, 1972, **18**, 255.
[74] C. H. Nash, N. De La Higuera, N. Neuss, and P. A. Lemke, in 'Developments in Industrial Microbiology', (Proceedings 13th General Meeting Soc. Industrial Microbiology, Illinois, August, 1973) ed. E. D. Murray and A. W. Bourquin, American Institute of Biological Sciences, 1974, Vol. 15, p. 114.
[75] T. Kanzaki, H. Shirafuji, Y. Fujisawa, T. Fukita, K. Nara, K. Kitano, and M. Kida, Abstracts 2nd. International Symposium on the Genetics of Industrial Microorganisms, August, 1974, Sheffield, England.
[76] P. A. Fawcett, P. B. Loder, M. J. Duncan, T. J. Beesley, and E. P. Abraham, *J. Gen. Microbiol.*, 1973, , 293.
[77] P. A. Fawcett, J. J. Usher, and E. P. Abraham, in 'Proceedings of the Second International Symposium on the Genetics of Industrial Microorganisms, Sheffield, August 1974', ed. K. D. Macdonald, Academic Press, London and New York, 1975, p. 131.

containing methyl-labelled valine was also synthesized.[57] After incubation of each of these labelled peptides with the cell-free extract the products were separated by electrophoresis and chromatography on paper, before and after oxidation with performic acid, and the radioactivity was determined in spots corresponding to penicillin N and penicillaminic acid respectively. Determinations were also made of the loss of radioactivity in the position of penicillin N and the gain in that of the penicilloic acid of penicillin N, after treatment of samples of the product with penicillinase and paper electrophoresis at pH 7.0.[77]

The results of these experiments showed that tritium was incorporated into a penicillin N-like compound from the LLD tripeptide labelled in either the methyl groups or the α-position of its valine residue, but that it was not incorporated from the methyl-labelled all-L tripeptide. No significant amount of hydrolysis of the LLD tripeptide occurred and the results could not be accounted for by the incorporation of free labelled valine. Although further experiments are desirable, with doubly labelled LLD tripeptide and with the DLD isomer, it appears that δ-(L-α-aminoadipyl)-L-cysteinyl-D-valine is a precursor of penicillin N in the cell-free system. Isopenicillin N would not have been distinguished from penicillin N by the methods used, but there is no evidence for the production by the *Cephalosporium* of an antibiotic with an L-α-aminoadipyl side-chain.

The D-valine residue of the tripeptide may be derived from enzyme-bound D-valine which is itself formed from L-valine. The labelled tripeptide produced by a suspension of *C. acremonium* in the presence of L-[2,3-^{3}H]valine has been isolated as its sulphonic acid and hydrolysed to its constituent amino-acids. The radioactivities of the resulting D-valine and of the α-oxoisovaleric acid obtained from it on treatment with D-amino-acid oxidase showed that tritium had been retained at C-3 but not at C-2.[78] Hence, it could be concluded that the epimerization of the α-centre of valine during the formation of the LLD tripeptide does not involve an αβ-dehydrovalinyl intermediate or a direct internal hydride shift of the type which occurs with certain hydroxy-acid racemases.[79]

7 The Final Stage in Penicillin Biosynthesis by *P. chrysogenum*

6-Aminopenicillanic Acid and Isopenicillin N.—The finding of 6-aminopenicillanic acid and isopenicillin N in fermentations of *P. chrysogenum* in which the availability of the side-chain precursor was limited raised the question whether one or both of these compounds are precursors of the penicillins with non-polar side-chains produced by this organism.

A number of enzymic reactions have been shown to occur in extracts from certain bacteria and fungi in which 6-APA is formed by hydrolysis of penicillins, or penicillins are synthesized from 6-APA (for examples see reference 3). An extract of a strain of *P. chrysogenum* deacylated a number of penicillins with non-polar side-chains, but failed to hydrolyse penicillin N or isopenicillin N, although hydrolysis of the latter penicillins was observed when the carboxy-group of the side chain was esterified.[80] These findings gave no support to the hypothesis that 6-APA is formed during fermentation by hydrolysis of free isopenicillin N.

[78] F-C Huang, J. A. Chan, C. J. Sih, P. Fawcett, and E. P. Abraham, *J. Amer. Chem. Soc.*, 1975, **97**, 3858.
[79] S. S. Shapiro and D. Dennis, *Biochemistry*, 1965, **4**, 2283.
[80] H. Vanderhaeghe, M. Claesen, A. Vlietinck, and G. Parmentier, *Appl. Microbiol.*, 1968, **16**, 1557.

Penicillin acyltransferase activity which catalysed exchange of ^{35}S between certain penicillins and 6-APA, and also between benzylpenicillin and phenoxymethylpenicillin, was observed in extracts of *P. chrysogenum.*[81] No exchange between the side-chain of penicillin N and two other penicillins was observed, and isopenicillin N was not tested. Homogenates and crude extracts of *P. chrysogenum* were reported to contain an enzyme which transferred the acyl group from phenylacetyl- or phenoxyacetyl-CoA to 6-APA at pH 7.0.[82] Acylation of 6-APA by the CoA derivatives of phenylacetic acid, phenoxyacetic acid, *p*-methoxyphenylacetic acid, and octanoylacetic acid was shown to occur in the presence of a partially purified enzyme from *P. chrysogenum*, whose activity was at a maximum at pH 8.5 and zero at pH 7.0.[83]

It was also reported briefly that an acyltransferase from *P. chrysogenum* showed a number of activities whose ratios appeared to remain constant during the purification. These were penicillin–penicillin acyltransferase, 6-APA acyltransferase, penicillin acylase, and phenylacetyl-CoA hydrolase activity.[84]

An indication of the possible involvement of isopenicillin N in an acyltransferase reaction was obtained from later experiments.[85] When isopenicillin N was added to a crude extract of *P. chrysogenum* prepared by grinding the mycelium with sand, ^{14}C from phenylacetyl-CoA was incorporated into benzylpenicillin. In the absence of isopenicillin N, or in the presence of penicillin N, no incorporation was detected. The incorporation also occurred in the presence of 6-aminopenicillanic acid, but it appeared that 6-APA and isopenicillin N were substrates for enzymes which occupied different locations in the cell. When isocephalosporin C was added to this system there was no detectable formation of 7-phenylacetamidocephalosporanic acid.

With a view to obtaining more direct evidence for the ability of isopenicillin N to act as a substrate for an acyltransferase in an extract of *P. chrysogenum*, isopenicillin N, penicillin N, and 6-APA were synthesized with tritium in their 2β-methyl groups.[86] The method used for the introduction of tritium into the ring system was based on an earlier finding that one deuterium atom was incorporated into the 2β-methyl group of phenoxymethylpenicillin sulphoxide methyl ester when the ester was refluxed in benzene and deuterium oxide.[87]

In a crude extract of mycelium grown in the presence of potassium phenylacetate and *N*-methylphenylacetamide, which presumably contained available side-chain precursor, tritium from labelled isopenicillin N was incorporated into solvent-extractable penicillin.[57]

In a protein-containing fraction of the extract, from which low molecular weight material had been removed by filtration through Sephadex G-25, similar incorporation was observed in the presence of phenylacetyl CoA, but not in its absence. In a further experiment in which phenylacetyl CoA was used, unlabelled benzylpenicillin was added to the product and radioactive benzylpenicillin isolated in crystalline form.

In contrast to the findings with isopenicillin N, no detectable radioactivity was

[81] D. L. Pruess and M. J. Johnson, *J. Bacteriol.*, 1967, **94**, 1502.
[82] R. Brunner, M. Röhr, and M. Zinner, *Z. physiol. Chem.*, 1968, **349**, 95.
[83] S. Gatenbeck and U. Brunsberg, *Acta Chem. Scand.*, 1968, **22**, 1059.
[84] B. Spencer and C. Maung, *Biochem. J.*, 1970, **118**, 29P.
[85] P. B. Loder, *Postepy Hig. Med. Doswiadczalnej*, 1972, **26**, 493.
[86] J. J. Usher, P. B. Loder, and E. P. Abraham, unpublished experiments, 1974.
[87] R. D. G. Cooper, *J. Amer. Chem. Soc.*, 1970, **94**, 1018.

incorporated into a solvent-soluble penicillin from labelled penicillin N.[57] In agreement with earlier findings, however, incorporation was observed from labelled 6-APA.

These results are consistent with the suggestion that isopenicillin N is a precursor of the penicillins with non-polar side-chains produced by *P. chrysogenum*, and with the evidence that α-aminoadipic acid has a role in the biosynthesis of benzylpenicillin. Whether there is a transient formation of free 6-APA in a pathway from isopenicillin N to benzylpenicillin, and whether isopenicillin N and 6-APA are substrates for two different enzymes, remain to be determined. The origin of the 6-APA which may accumulate in fermentations of *P. chrysogenum* is also uncertain. One possibility is that some of it, at least, is produced by the action of an acylase on the so-called natural penicillins with aliphatic side-chains which are produced when no side-chain precursors are added.[88]

8 Late Stages in the Biosynthesis of Cephalosporins

Deacetylcephalosporin C.—During the fermentation of *C. acremonium* deacetylcephalosporin C [(3), $R^1 = H$, $R^2 = OH$] is found in both the extracellular fluid and the mycelium. When [^{14}C]valine was added to a re-suspension of the *Cephalosporium* sp. the label in deacetylcephalosporin C appeared slightly later than that in cephalosporin C, suggesting that it was mainly a product rather than a precursor of cephalosporin C.[48] Under these conditions deacetylcephalosporin C may have been formed from cephalosporin C by non-enzymic hydrolysis and by the action of an extracellular esterase.[89,90]

Evidence has now been obtained for the view that deacetylcephalosporin C is the precursor of cephalosporin C. A number of *Cephalosporium* mutants have been isolated which produce deacetylcephalosporin C but no cephalosporin C.[16,91,75] The accumulation of the former has been shown not to be due to intracellular hydrolysis of the latter. Cell-free extracts of a cephalosporin C-producing strain of the *Cephalosporium* sp. contain a deacetylcephalosporin C acetyl-transferase which incorporates ^{14}C from [1-^{14}C]acetyl-CoA into cephalosporin C in the presence of deacetylcephalosporin C, but this transferase was not found in strains which accumulated deacetylcephalosporin C. These results appear to indicate that intracellular deacetylcephalosporin C is a precursor of cephalosporin C, while deacetylcephalosporin C in the culture filtrate is produced by hydrolysis of cephalosporin C.

Extracellular cephalosporin C acetylesterase was found in the culture filtrates of *S. clavuligerus*, which produces cephalosporins (6) and (7) containing a carbamoyl function.[92] However, the presence of an intracellular enzyme could not be demonstrated. Moreover, no evidence was obtained for either intra- or extra-cellular conversion of cephalosporin C into the carbamoyl-containing antibiotics.

Deacetoxycephalosporin C.—Although it was originally considered to be absent from the fermentation broth of the *Cephalosporium* sp., deacetoxycephalosporin C [(3), $R^1 = R^2 = H$] has now been detected in the fermentation of a number of fungi

88 E. P. Abraham, 'Biosynthesis and Enzymic Hydrolysis of Penicillins and Cephalosporins', University of Tokyo Press, Tokyo, 1974.
89 F. M. Huber, R. H. Baltz, and P. G. Caltrider, *Appl. Microbiol.*, 1968, **16**, 1011.
90 J. Nüesch, H. Hinnen, M. Liersch, and H. J. Treichler, 2nd in ref. 77, p. 451.
91 M. Liersch, J. Nüesch, and H. J. Treichler, in ref. 77, p. 179.
92 D. R. Brannon, D. S. Fukuda, J. A. Mabe, F. M. Huber, and J. G. Whitney, *Antimicrob. Agents Chemother.*, 1972, **1**, 237.

including *Cephalosporium* spp. and also in the cephalosporin-producing *Streptomyces* spp.[17]

Circumstantial evidence for the involvement of deacetoxycephalosporin C in the biosynthesis of cephalosporin C has come from the finding of mutants of *C. acremonium* which are blocked for cephalosporin C production but which accumulate deacetoxycephalosporin C either alone or together with an amount of deacetylcephalosporin C which is less than that from the parent strain.[93] Further evidence has been provided by the discovery of a labile intracellular deacetoxycephalosporin C hydroxylase in an extract of *C. acremonium.*[91] In the presence of this enzyme and an acetyl transferase deacetoxycephalosporin C is converted into cephalosporin C. Direct support for the participation of a hydroxylase in the formation of deacetylcephalosporin C has been obtained from an experiment in which cephalosporin C was produced by a suspension of *C. acremonium* shaken in oxygen enriched in ^{18}O. M.s. analysis of the methyl ester of the *N*-acetyl derivative of the product revealed the incorporation of one atom of ^{18}O and indicated that the isotope was localized at C-17 of cephalosporin C.[94]

It thus appears likely that deacetoxycephalosporin C is the first cephalosporin on the biosynthetic pathways, and that it is the precursor of deacetylcephalosporin C which is converted into cephalosporin C itself. If this is so, inversion of the L-α-aminoadipic acid residue of an intermediate occurs at a stage prior to, or during, the formation of deacetoxycephalosporin C, and not at the final stage of the biosynthesis of cephalosporin C itself. The small amount of intracellular cephalosporin C in *C. acremonium* contains a D-α-aminoadipyl side-chain.[88]

9 Mechanisms of Ring Closure

The β-Lactam Ring.—It has been postulated that the β-lactam is the first ring to be formed in the biosynthesis of the *β*-lactam antibiotics, but no experimental evidence for this hypothesis has been obtained. The mechanism of ring closure is uncertain, but it does not involve a free intermediate containing a dehydrocysteine residue. The incorporation of tritium from DL-[2,2′-^{3}H]cystine and DL-[3,3′-^{3}H]cystine into benzylpenicillin by *P. chrysogenum* was reported in 1957 to be accompanied by retention of one tritium atom at the C-6 as well as the C-5 position.[95]

This finding has been confirmed and extended by more recent experiments. When L-[*U*-^{14}C, 2-^{3}H]cystine hydrochloride was incubated with the mycelium of *P. chrysogenum* the ^{14}C and ^{3}H were incorporated into benzylpenicillin with only a small reduction in the ^{3}H/^{14}C ratio. Hydrolysis of benzylpenicillin labelled from L-[2-^{3}H]cystine showed that the label was located in the resulting 6-aminopenicillanic acid. Conversion of the latter into the corresponding 6-chloro-compound *via* a diazo-intermediate resulted in the loss of more than 98% of the label and the loss was only from the 6-position.[96] Similarly, deuterium from L-[2-^{2}H]cystine was shown by ^{2}H n.m.r. spectroscopy to be incorporated into the 6-position of benzylpenicillin.

93 S. W. Queener, J. J. Capone, A. B. Radue, and R. Nagarajan, *Antimicrob. Agents Chemother.*, 1974, **6**, 334.
94 C. M. Stevens, E. P. Abraham, F. -C. Huang, and C. J. Sih, Abstracts, Annual Meeting of the Federation of American Societies for Experimental Biology and Medicine, April, 1975.
95 H. R. V. Arnstein and J. C. Crawhall, *Biochem. J.*, 1957, **67**, 180.
96 B. W. Bycroft, C. M. Wels, K. Corbett, and D. A. Lowe, *J.C.S. Chem. Comm.*, 1975, 123.

In another experiment, DL-[2,2′-3,3′-$^2H_6$]cystine was incubated in the presence of dithiothreitol with a suspension of mycelium of *C. acremonium* and the cephalosporin C obtained was converted into the dimethyl ester of its *N*-acetyl derivative. M.s. analysis of this derivative showed that it yielded a significant quantity of a fragment containing two deuterium atoms and indicated that cysteine was incorporated without loss of deuterium from C-2 and with loss of only one deuterium atom from C-3.[97] The relative incorporation of ^{14}C and ^3H into the 7-methoxycephalosporin (7) on addition of a mixture of DL-[3, 3′-^{14}C] cystine and L-[3, 3′-^3H]cystine to a fermentation of *S. clavuligerus* indicated that no dehydrocysteine residue was involved in the introduction of the 7-methoxy-group.[37]

The synthesis of chirally labelled cysteines has enabled it to be shown that the incorporation of C-3 of cysteine into benzylpenicillin by *P. chrysogenum* probably occurs with overall retention of configuration.[98] Poor retention of tritium (relative to ^{14}C) was observed from (2*R*, 3*S*)-[3-^3H]cysteine and good retention from (2*R*,3*R*)-[2,3-$^3H_2$]cysteine[(11), X = ^3H] when these compounds were mixed with L-[U-^{14}C]-cysteine and the mixtures fed separately to cultures of the organism. Degradation of

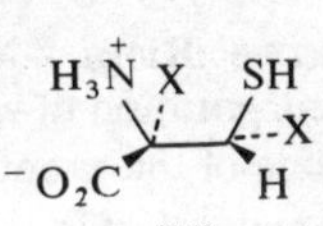

(11)

the penicillin labelled from (11) to its penilloaldehyde and oxidation of the latter to the corresponding acid indicated that incorporation of tritium at C-5 was high but that some loss of tritium had occurred from C-6, possibly as a consequence of loss of the α-proton of the cysteine by transamination.

Soon after the first indication that no free intermediate containing a dehydrocysteine residue was involved in penicillin biosynthesis two mechanisms were proposed for the formation of the β-lactam ring. The first[95] involved oxidation to a thioaldehyde (12), and the second[99] assumed that the thiol group induced the adjacent carbon to become a nucleophilic centre (13). A more recent proposal[100] has been based on the

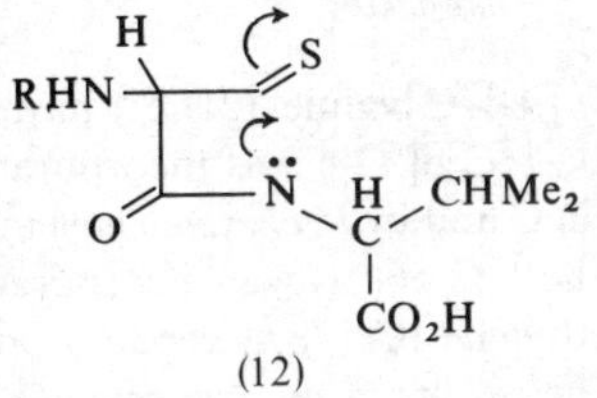

(12)

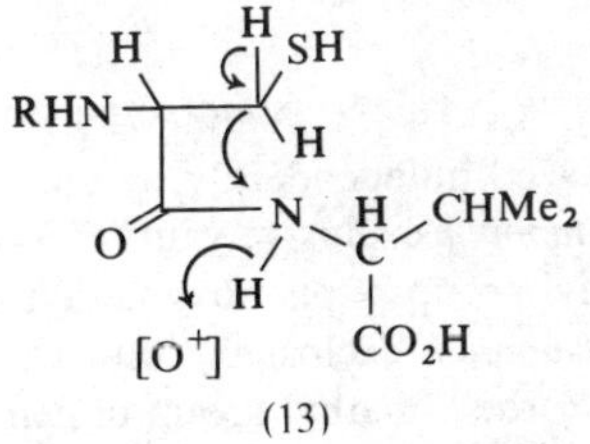

(13)

finding that cysteinylvaline and cysteinyldehydrovaline derivatives can readily be oxidized to the isothiazolidinones (14) and (15) respectively. It was suggested that such

[97] F-C Huang, F. C. Sih, P. Fawcett, and E. P. Abraham, unpublished experiments, 1975.

[98] D. J. Morecombe and D. W. Young, *J.C.S. Chem. Comm.*, 1975, 198.

[99] A. J. Birch and H. Smith, in 'Amino Acids and Peptides with Antimetabolite Activity', Ciba Foundation Symposium, Churchill, London, 1958, p. 247.

[100] J. E. Baldwin, S. B. Haber, and J. Kitchin, *J.C.S. Chem. Comm.*, 1973, 790.

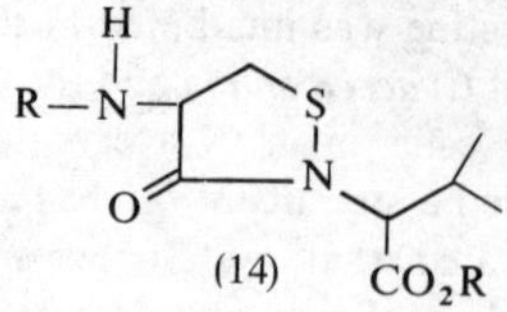

(14)

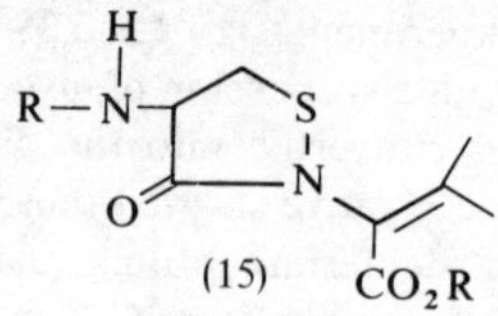

(15)

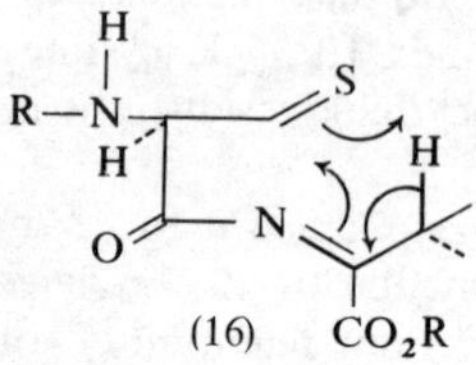

(16)

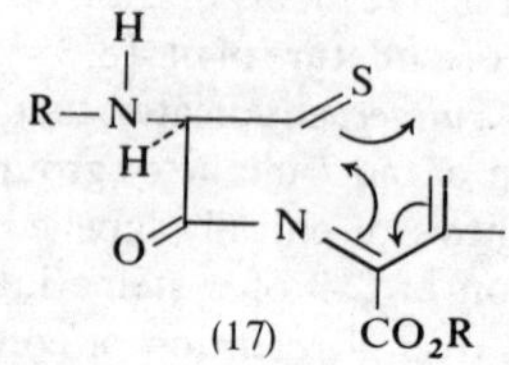

(17)

isothiazolidinones are transformed to the thioaldehydes (16) and (17) and thence to the penicillin and cephalosporin ring systems. However, an attempt to demonstrate the cyclization of synthetic thioaldehydes of type (12) was unsuccessful.[101]

The Thiazolidine and Dihydrothiazine Rings.—More detailed information has recently been obtained about the incorporation of valine into the sulphur-containing rings of the β-lactam antibiotics. Some of the progress in this area has depended on the interpretation of the ^{1}H n.m.r. and ^{13}C n.m.r. spectra of penicillins (1) and of cephalosporin C (3)[102] and the finding that fragmentation of derivatives of these antibiotics in the mass spectrometer yielded ions of relatively high abundance such as the well-stabilized thiazolidine ion (18)[103] with *m/e* 174 and the dihydrothiazine ion (19) with *m/e* 230.[102]

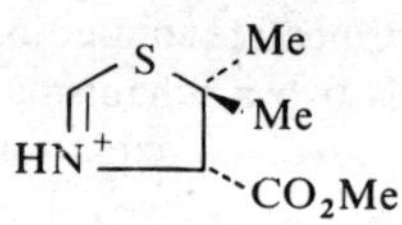

(18)

m/e 174

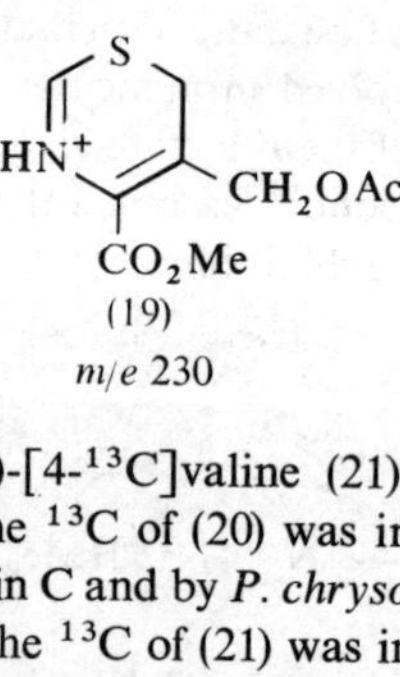

(19)

m/e 230

(2*RS*, 3*R*)-[4-^{13}C]Valine (20)[104] and (2*S*, 3*S*)-[4-^{13}C]valine (21)[105] have been synthesized independently by different routes. The ^{13}C of (20) was incorporated by *C. acremonium* exclusively into C-2 of cephalosporin C and by *P. chrysogenum* into the β-methyl group of phenoxymethylpenicillin.[106] The ^{13}C of (21) was incorporated by *C. acremonium* exclusively into the exocyclic methylene (C-17) of cephalosporin C and into the α-methyl group of penicillin N.[105] The results of the two sets of experi-

[101] J. Cheney, C. J. Moores, J. A. Raleigh, A. I. Scott, and D. W. Young, *J. C. S. Perkin I*, 1974, 986.
[102] P. V. Demarco and R. Nagarajan, in 'Cephalosporins and Penicillins—Chemistry and Biology', ed. E. H. Flynn, Academic Press, New York, 1972, Chap. 8.
[103] W. Richter and K. Biemann, *Monatsh.*, 1964, **95**, 766.
[104] J. E. Baldwin, J. Löliger, W. Rastetter, N. Neuss, L. L. Huckstep, and N. de La Higuera, *J. Amer. Chem. Soc.*, 1973, **95**, 3796, *ibid.*, p. 6511 (correction).
[105] H. Kluender, C. H. Bradley, C. J. Sih, P. Fawcett, and E. P. Abraham, *J. Amer. Chem. Soc.*, 1973, **95**, 6149.
[106] N. Neuss, C. H. Nash, J. E. Baldwin, P. A. Lemke, and J. B. Grutzner, *J. Amer. Chem. Soc.*, 1973, **95**, 3797, *ibid.*, p. 6511 (correction).

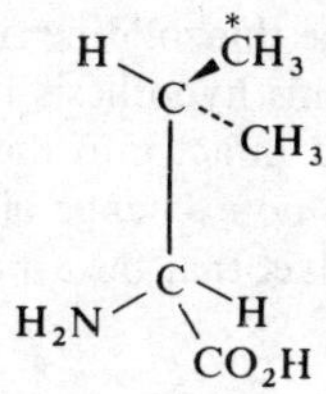

(20)

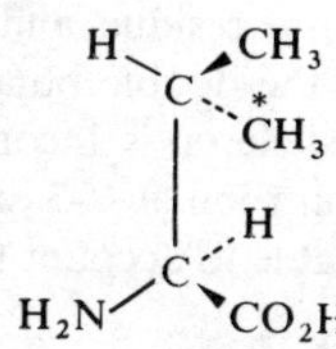

(21)

ments were therefore complementary. They demonstrated that the incorporation of the chiral isopropyl group of valine into these β-lactam antibiotics was stereospecific and that the incorporation into the penicillin ring system occurred with retention of configuration at C-3.

Chirally labelled CD_3-valines were synthesized in order to examine the fate of the isopropyl hydrogen atoms during the biosynthesis of β-lactam antibiotics.[107] (2*S*,3*R*)-[4,4,4-$^2H_3$]Valine (22) and (2*S*,3*S*)-[4,4,4′-$^2H_3$]valine were incubated separately with suspensions of washed mycelium of *C. acremonium*. The labelled penicillin N and cephalosporin C were converted into their *N*-benzoyl methyl ester and *N*-acetyl methyl ester derivatives respectively and the latter compounds were analysed by mass spectroscopy. In addition, labelling from (2*RS*)-[$^2H_6$]valine in the C-2 methylene of cephalosporin C was studied by ^1H n.m.r. spectroscopy of the *N*-benzoyl derivative of the latter.

The results of these experiments showed that all three deuterium atoms of both chirally labelled valines were incorporated into penicillin N [(23), R = $^-O_2C{\cdot}CH\overset{+}{N}H_3{\cdot}(CH_2)_3CO$]. Two deuterium atoms from (2*S*, 3*S*)-[4,4,4-$^2H_3$]valine were retained in the exocyclic C-17 position of cephalosporin C and two from (2*S*, 3*R*)-[4,4,4-$^2H_3$]valine (22) in the endocyclic methylene at C-2 [(24), R = $^-O_2C{\cdot}CH{\cdot}\overset{+}{N}H_3(CH_2)_3CO$]. It is thus clear that ring closure in the formation of cephalosporin C does not involve the oxidation of the (*R*)-methyl group of valine to the state of an aldehyde and unlikely that the formation of the Δ^3-cephalosporin ring system proceeds through a Δ^2-cephalosporin intermediate.[107]

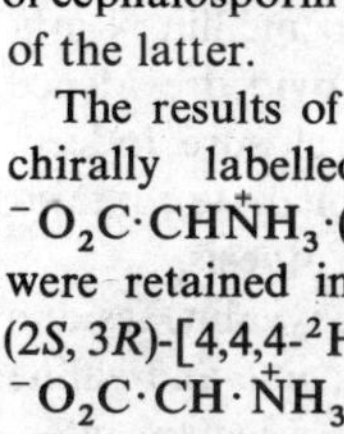

(22)

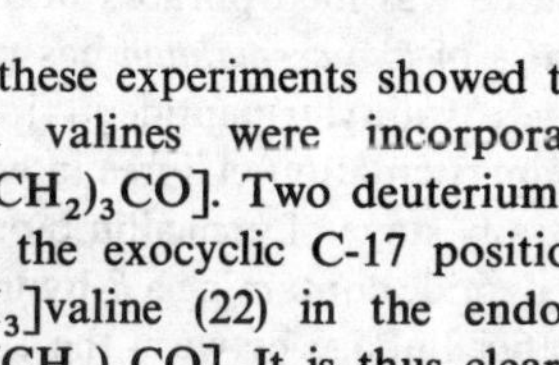

(23)

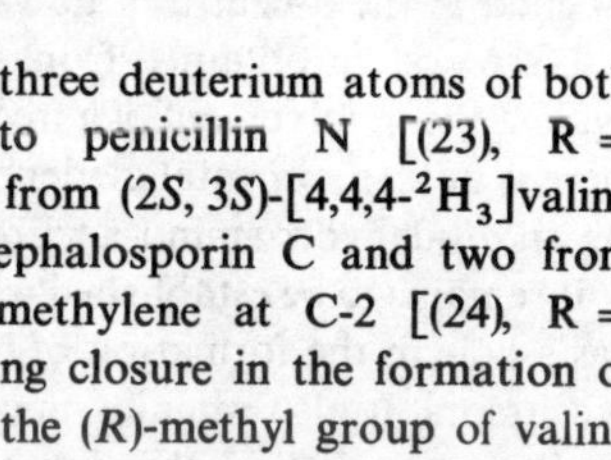

(24)

A similar experiment with 2*S*-[^{15}N]-3*S*-[$^2H_3$]valine has demonstrated that the entire valine skeleton is incorporated intact into penicillin N and cephalosporin C.[78]

10 Current Problems

A chemically plausible mechanism for the final stage in the biosynthesis of the penicillin ring system assumes that a monocyclic intermediate (25) is converted into (26),

[107] H. Kluender, F.-C. Huang, A. Fritzberg, H. Schnoes, C. J. Sih, P. Fawcett, and E. P. Abraham, *J. Amer. Chem. Soc.*, 1974, **96**, 4054.

containing a dehydrovaline residue, and that the thiazolidine ring is then formed by addition of the thiol to the double bond.[108] This hypothesis is consistent with the finding that the valine skeleton is incorporated intact into the penicillin molecule, with retention of configuration at C-3 and with no exchange of its methyl hydrogen atoms. It has also been able to account for the fact that the α-position of an L-valine

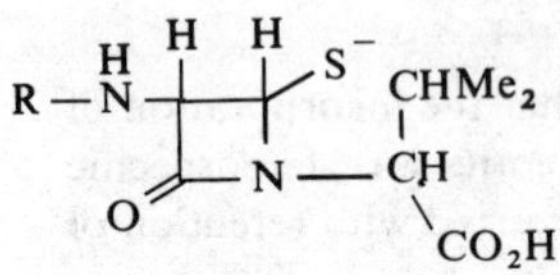

(25)

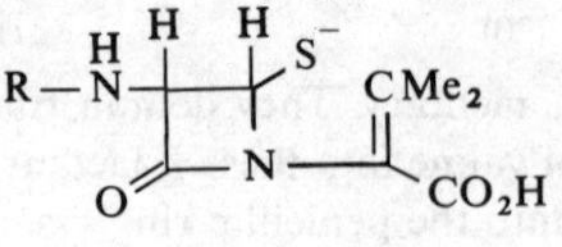

(26)

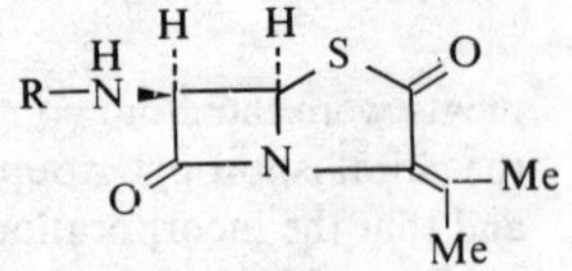

(27)

precursor becomes a D-centre (C-3) in the penicillins.[99] Thus phenoxymethylanhydropenicillin (27) was reported to rearrange spontaneously in aqueous solution to give trace amounts of phenoxymethylpenicillin itself.[109] The formation of the D-centre would require either a *cis*-dehydrogenation of valine followed by *trans*-addition of SH or *vice versa*.[110]

However, the finding that tritium from δ-(L-α-aminoadipyl)-L-cysteinyl-D-valine labelled at the α-centre of its valine residue was incorporated into penicillin N in a cell-free system obtained from protoplasts of *C. acremonium* has provided evidence against the involvement of a free αβ-dehydrovalinyl tripeptide in ring closure. Indeed, there is no experimental evidence for the participation of a free intermediate containing an αβ-dehydroamino-acid residue at any stage of penicillin biosynthesis.

It remains to be established whether a peptide containing a β-hydroxyvaline residue has a role in the formation of the thioether linkage between the cysteine and valine precursors. It also remains uncertain whether there is a monocyclic β-lactam intermediate, or whether the first cyclic compound on the biosynthetic pathway is an

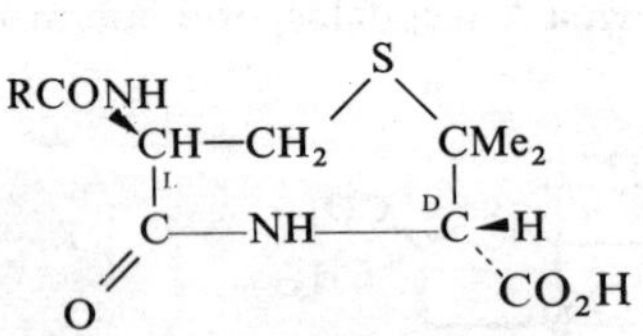

(28)

N-acyl derivative of 'cyclic cysteinylvaline' (6-amino-2,2-dimethyl-5-oxo-hexahydro-1,4-thiazepine-3-carboxylic acid) [(28) where RCO could be δ-(α-aminoadipyl)]. The negative results so far obtained with compounds of this type and intact mycelium of *P. chrysogenum*[111] are inconclusive.

Direct evidence for the involvement of a common tripeptide intermediate in the formation of penicillin N and cephalosporin C is lacking, but the hypothesis that this

108 H. R. V. Arnstein, *Ann. Reports*, 1957, **54**, 339.
109 R. Thomas, *J.C.S. Chem. Comm.*, 1972, 478.
110 D. J. Aberhart and L. J. Lin, *J. C. S. Perkin I*, 1974, 2320.
111 B. Sjöberg, H. Thelin, L. Nathorst-Westfelt, E. E. Van Tamelen, and E. R. Wagner, *Tetrahedron Letters*, 1965, 281.

is so is consistent with the finding that the isotopic ratios of the valinyl fragments of both antibiotics labelled from 2*S*-[^{15}N]-3*S*[$^2H_3$]valine were very similar although the valine underwent considerable transamination.[78] Also consistent with this hypothesis is the observation that an increase in oxygen tension can lead to higher production of cephalosporin C by *C. acremonium* and lower production of penicillin

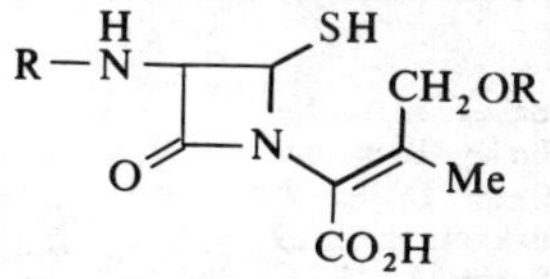

(29)

N.[94] Early suggestions for the immediate precursor of cephalosporin C included monocyclic β-lactams in which one or both of the methyl groups of the valinyl moiety had been converted into oxymethylenes. Evidence for the view that cephalosporin C is formed from deacetoxycephalosporin C has now led to the proposal that (29) is a precursor of the latter.[93]

A further possibility is that deacetoxycephalosporin C is derived from penicillin N *via* an oxidative ring expansion analogous to that which has been achieved chemically.[112, 113] This would be consistent with the fact that mutants of *C. acremonium* which are blocked for cephalosporin production may or may not produce penicillin N, but that no mutants have been reported to have been isolated to date which produce cephalosporins but not penicillin N. However, the apparent ability of D-valine to inhibit the production of penicillin N but not that of cephalosporin C was more readily accounted for on the assumption that the pathways to the two antibiotics diverge before penicillin N itself has been formed.

An attempt to determine directly whether penicillin N is a precursor of cephalosporin C, by addition of ^{14}C-labelled penicillin N to a suspension of *C. acremonium*, was unsuccessful because the penicillin was not taken up by the mycelial cells.[114] It is possible that this question will be answered by further experiments with cell-free preparations.

[112] R. B. Morin, B. G. Jackson, R. A. Mueller, E. R. Lavagnino, W. B. Scanlon, and S. L. Andrews, *J. Amer. Chem. Soc.*, 1963, **85**, 1896.
[113] J. A. Webber, E. M. Van Heyningen, and R. T. Vasileff, *J. Amer. Chem. Soc.*, 1969, **91**, 5674.
[114] B. Smith, S. C. Warren, G. G. F. Newton, and E. P. Abraham, *Biochem. J.*, 1967, **103**, 877.

Author Index

SCIENTIFIC INFORMATION DEPT.

JAN 16 1986

BRISTOL-MYERS/EVANSVILLE

PROPERTY OF BRISTOL-MYERS SQUIBB CO.
RESEARCH LIBRARY
WALLINGFORD, CT